GEOLOGIC REFERENCE SOURCES

A subject and regional bibliography
of publications and maps in the geological sciences

by

Dederick C. Ward
University of Colorado Libraries

and

Marjorie W. Wheeler
Lamar University Library

With a section on geologic maps
by

Mark W. Pangborn, Jr.
U.S. Geological Survey Library

The Scarecrow Press, Inc.
Metuchen, N.J. 1972

Library of Congress Cataloging in Publication Data

Ward, Dederick C
Geologic reference sources.

1. Geology--Bibliography. I. Wheeler, Marjorie W., joint author. II. Pangborn, Mark White. III. Title.
Z6031.W35 1972 016.55 71-190223
ISBN 0-8108-0477-8

Preface

The purpose of this guide is to introduce the reader to the literature of a geologic discipline or to the publications and maps which deal with the general geology of a country or region. The publications selected for the guide range from introductory to highly technical, and serve to suggest, in addition to the more formal reference works (bibliographies, dictionaries, etc.), some representative texts, treatises and serials.

Most of the books contain good, often excellent, bibliographies. Annotations are minimal, since inclusion of a title indicates that the work was judged useful to summarize a topic, to indicate the direction of current research or to introduce fundamentals. The guide is worldwide in scope, but as can be expected in a work by American authors, English language publications predominate. Current reference sources are stressed, although classical works were selectively retained to give the guide bibliographical completeness. Most of the books listed have been published since 1950, many in 1970.

This edition of Geologic reference sources is more authoritative than the first* by virtue of joint authorship and assistance from more persons engaged in science information work. It substantially up-dates the earlier work, as evidenced by the appearance of publications in areas such as new global tectonics, environmental geology, remote sensing, mathematical geology, and lunar geology. A valuable addition to the new work is a section on geologic maps.

The guide is divided into four sections: general, subject, regional, and maps. The table of contents serves as a classified subject index. Alphabetized subject and geographic indexes appear at the end.

---

*Ward, Dederick C. Geologic reference sources. Boulder, Colo.: University of Colorado Press, 1967. 114 p. (Univ. Colo. Studies. Series in the earth sciences no. 5.)

We are indebted to colleagues who have given of their time to read, criticize and supplement portions of the guide. We wish to extend our appreciation to Robert R. Wheeler, who served as consultant and informal editor; to Harriet W. Smith and Robert McAfee, who performed extended literature searches for us; and to Edwin E. Larson, who contributed to the introduction.

A bibliography of this type necessarily draws upon the knowledge and comment of others. It is a pleasure to acknowledge the individual assistance of Claude Albritton, Albert V. Carozzi and George W. White for the history and philosophy of geology; Martha L. Andrews and Kathleen A. Blackie for Arctic and glacial geology; Judith Licea Ayala for Mexico; Joanne Bannister for Guyana; D. Bright and Patricia Sheeher for New Zealand; Joseph J. Kohut for paleontology; Harriet K. Long for the United States; Grace Matthaei for Australia and the Pacific; and Doreen M. Sutherland for Canada. Many individuals answered questions about specific titles or editions, others permitted us to draw freely from articles they had written or bibliographies they had compiled. To these colleagues collectively, we extend our sincere gratitude.

Special thanks are also due members of the geology and library faculties of Lamar University and the University of Colorado, who freely offered advise when needed. Ward is particularly indebted to Sylvia McCracken for her assistance in the library at Colorado.

We wish to extend cordial thanks to Ruth M. Ward, who prepared the manuscript, and to the staff of our publisher for their editorial advice and guidance.

Most of all, we wish to extend our warmest appreciation to Mark Pangborn for providing the map section, for without his kind offer, this outstanding contribution to the guide would not have been compiled.

D. C. W.
M. W. W.

Contents

Geology and Geological Information in the 1970's

This book is principally a bibliographical guide to the literature of geology. In preparing the book, however, the authors deemed it appropriate to preface the main work with a statement regarding the changing face of geology and geological information, and to append a list of selected articles on communication in the geological sciences.

THE GEOLOGICAL SCIENCES*

A century ago, natural philosophy included all of the fields that we generally associate with science: physics, chemistry, mathematics, astronomy, geology, geography, biology, etc. There were few scientists and accordingly not a great deal of data, so it was possible for an individual to concern himself with large areas of natural philosophy.

As more and more individuals concerned themselves with natural history, the variety of scientific inquiry increased at an accelerating rate. Geology, like the other major subdivisions, was split into a number of categories of smaller size: paleontology, petrology, structural geology, mineralogy, stratigraphy and sedimentation. Each of these units was likewise subdivided, and geologists concentrated on very narrow portions of the total geologic spectrum. Geologists still do specialize (and some of our most meaningful results have come from these scientists), but it has become increasingly clear that if the earth is to be really understood we cannot permit overly stringent compartmentalization of our studies and our thinking. This is not to say that we have all become generalists--with the information explosion now in progress, this becomes an all-but-impossible realization. However, it is now apparent that many scientists are involving themselves to varying degrees in realms peripheral to their principal concentration.

*Written by Edwin E. Larson, Associate Professor of Geological Sciences, University of Colorado, Boulder, Colorado.

As many individual geologists now try to include more peripheral areas in their studies, we see that the whole field of geology has expanded, and it now includes essentially all of the fields previously included in natural philosophy and many new ones which have recently developed. Geology can now be more properly termed earth science or geoscience: that is, any aspect of science concerning the earth is within its scope.

Earth science is primarily based on physics, chemistry, and mathematics. Most earth processes are physical or chemical. In addition, specific application of physics and chemistry to earth problems has resulted in the formulation of two subfields--geophysics and geochemistry. Included within their domains of study are such subjects as the physical properties and behavior of earth materials, exploration of the earth's deep interior, fields of force, mineral resources, experimental petrology, solution geochemistry, and soil formation.

In the 1960's the study of the oceans continued to grow in importance as we realized that the oceans occupy a key role in understanding earth history (particularly the possibility of continental drift) and perhaps represent one factor in the future survival of mankind. Ocean studies include all aspects of marine geophysics, geochemistry, biology, and ocean-floor geology.

Geophysics is largely a remote sensing technique inasmuch as the object of interest is not directly observable but can be "seen" only after suitable interpretation of geophysical data. Recently, the field of remote sensing in general has burgeoned. New techniques in aerial photography, radar imagery, and electromagnetic sensing have made this one of the fastest growing of the new fields in earth science.

Similarly, the field of space exploration has recently added a new dimension to earth science. Although the data returned to earth from rocket fly-bys to other planets, from orbiting astronomic stations and satellites, and from Apollo moon landings normally concern investigation of nearby planetary bodies and the sun, they have enriched our total knowledge about our solar system and its origin and, thereby, our understanding of the earth. This field will continue to expand.

In earth science we deal with data, at times in prodigious amounts. The data are, of course, best treated statistically and mathematically, and normally by computer. The earth scientist today must keep abreast of statistical and computational methods and be familiar with computers and available computer programs, even be able to write or modify programs to his needs. No field of earth science seems beyond the scope of computer applications.

Even paleontology, long considered to be merely the description, evolution and classification of extinct faunal groups, has become modified. Today it also incorporates the modern aspects of genetics, biophysics, anthropology, and ecology.

The latest field to be included in earth science is that of environmental geology. Actually, environmental geology is nearly synonomous with earth science. It encompasses all those fields needed to provide a basic understanding of earth processes so that man can learn to live in accord with nature. Within environmental geology, however, lies the concept of the physical and biologic entities of the earth, including man, as part of an integrated whole. This concept states that man has a responsibility to utilize the limited resources of the earth so as to insure that this whole will last. As man continues to increase in numbers and demands, and the store of natural resources continues to dwindle, we can be sure that this field will expand accordingly.

Geology, then, has changed in response to shifts in scientific emphasis and the expansion of the entire scientific domain. Presently the field is moving toward a broadened scope, and the modern geologist finds himself crossing traditional subject boundaries in order to satisfy his professional needs.

GEOLOGICAL INFORMATION*

Traditional requirements and methods of disseminating scientific information are changing as the volume of publication accelerates. Historically and today, it has been the continuous exchange of scholarly communication by which science, technology and public interest have grown and bene-

*Written by the authors.

fited. A century ago a scientist could keep abreast of major discoveries and evolving concepts in the general realm of natural history. In the following decades, he could keep pace with most research reports in the field of geoscience and even marvel at new discoveries in related sciences. Today he can, with much effort, keep current in a specialized field in geoscience. Tomorrow he will be even more dependent on new methods of gathering, storing and disseminating information.

Increasingly, the question of quality of geoscience publication and the problems of retrieval are at issue. Without purporting to judge these issues, the authors note the following trends in reporting: (1) an increasing duplication of published information; (2) a diminishing scope and relevance of research; (3) piecemeal reporting of progress rather than fully interpreted and integrated final results; (4) inadequate standards of selection, editing and indexing; (5) a growing proliferation of conferences, symposia and transactions conducted in publicity and concluded in obscurity in so far as existing methods of bibliographic search are concerned. We hope that, with more critical analysis on the part of editors to assure continuity, and more care for the problems of referencing, our scientific record, now in danger of neglect, will be restored to the dignity it once enjoyed.

Modern Technology and Innovation. The traditional information pattern of geologists is changing. The modern geologist requires that the total amount of science information be sifted and analyzed before he uses it. This sifting and analysis of information, or re-packaging of it, is the basic idea behind the new innovative concepts which are now being adopted by scientists, librarians and other information specialists to help facilitate information transfer in science and technology. These information systems must be capable of recalling information from a variety of sources and from a large store of data and then culling out only the highly relevant data. These systems must also present the resultant data in a format which is most useful to the scientists so that he will elect to use the formal information systems rather than depend on limited individual collections. The computer has provided the best possible opportunity for scientists to control the information explosion.

Two examples, employing a computer to analyze a

large store of information for an individual scientist, will illustrate new ways that are being adopted to handle information. SDI (Selective Dissemination of Information) is an information service by which the scientist is provided with relevant information on a narrow subject field. The computer searches a backfile of information on magnetic tape against the scientist's "profile" or interest, and provides him with a retrospective bibliography of all information pertinent to his "profile." The computer can then keep him up-to-date (current awareness) by supplying him with new information as it is made available to the master file. Today, SDI is being performed for scientists by special information centers, chiefly in government and industry, but SDI services have not reached public and academic libraries to any appreciable extent. The American Geological Institute has recently offered geologists the Geologic Reference File (GEO-REF) designed to provide, at small cost, a high-velocity retrospective search of AGI's tape file, which now (1970) covers 80, 000 citations with approximately 3, 500 being added each month.

SDI selects certain data from a file for the scientist. However, the user himself can query the tape files directly by a method designed to pull data from a file or data bank. This pull system is called an on-line or time-sharing approach. The advantage over SDI is that the user should be able to obtain more relevant data by direct communication with the computer. On-line systems have not been developed as extensively as SDI, and for today's geologist, they are remote, but potentially important.

The information systems just described are only two examples of components which may in the future link a vast geoscience network of libraries and special information centers, from which a geologist, regardless of location, can retrieve the information he needs. Individual components of such a network exist today, both in this country and in others--customized indexing and abstracting services; geoscience library collections in universities, industry and government; special data centers, as exist for petroleum and oceanography; a center that catalogs on-going research (Science Information Exchange); and a center that refers scientists to colleagues doing similar research (National Referral Center)--but the components have not been tied into a network. Network cooperation is not easy to come by, and critical problems involving the transmittal of data must be solved--e.g., compatible computer language, compatible scientific terms, methods of data display, and problems of expense-sharing and owner compensation.

In 1967, the American Geological Institute and its member societies began a program to develop a network concept for the geological sciences in the United States. Implicit in the program was coordination with other scientific and technical information efforts, as geology cannot progress alone in its efforts to solve its information problems. AGI is studying the following broad areas: bibliographic services and publications; translation publications and services; vocabularies, definitions and nomenclature; meetings, symposia and personal information exchange; library sources and services; specialized information and data services; primary and informal publications; and services providing information on current research and development. It is not surprising that one of the newest societies to become a member of AGI is the Geoscience Information Society, whose membership of librarians, documentalists and geoscientists are devoted to the exchange of information in the geological sciences.

The Classical Services. In contrast to the innovative channels of information transfer, which include the computer-related specialized services and networks just described, the classical or traditional channels of information exchange are thought of as libraries, indexing and abstracting services, formal and informal publication, and meetings. Labeling the established channels of communication "classical" or "traditional" does not imply that they are archaic or ineffective. On the contrary, they have come a long way in their ability to preserve the scientific record reasonably intact. As the "information explosion" progressed, however, the traditional information channels were meeting with considerably less success in delivering the most relevant information to the right scientist on time. Consequently, scientists and information specialists began to seek new ways, through the new technology, to update traditional channels and make them more responsive to modern needs.

Meetings. Direct oral communication has always constituted a significant part of geological information exchange. The sponsorship of various types of meetings--conferences, colloquia, symposia, etc.--has chiefly been a major function of the scientific societies. In the U. S. alone, there are well over 100 major meetings each year in the geosciences.

Effective announcement of and the identity of the proceedings from these sessions has always been a problem, particularly in relation to the smaller conferences. Inadequate communication between societies and institutions sponsoring scientific meetings has led to program overlap and duplication. For these reasons, geologists are looking for new ways to make meetings more effective and to improve documentation of the scientific information presented at them. It is obvious that the speed of the computer can be a great asset in the rapid announcement of meetings and in relating a scientist's interest to appropriate meeting papers.

Formal and Informal Publication. Journals, monographs, bulletins and proceedings have traditionally provided geologists with their most popular and most effective scientific record, and of these, the primary journal remains "king." For the library, acquisition of major science journals is a relatively easy task. For the scientist, however, acquisition of an increasing number of pertinent scientific journals is no easy task. His narrow specialty requires only single articles from a wide variety of journals, and his pocketbook prohibits purchase of all these journals. Therefore, he wonders if there is a better way to supply journal information.

The American Mathematical Society is currently experimenting with a plan whereby an individual member scientist selects only the papers relevant to his work from a wide variety of mathematical serials. He creates, in effect, his own journal. He may select articles from a monthly list of titles in his chosen field, or he may let the computer select his articles automatically by matching his "profile" to the data file. He pays for what he chooses. Does publication of separate papers, in lieu of entire journals, offer an attractive alternative to the geologist, who is facing rising costs, limited office space, and narrowing interests?

An increasing amount of new scientific information is being documented and disseminated through informal publications as memoranda, occasional papers, laboratory bulletins, theses, and field and research reports, supported chiefly by federal and state government agencies. Since the bulk of this material does not enter normal distribution channels, obtaining these scientific results becomes a source of frustration to scientist and librarian alike. Not only is the identification, processing and dissemination of this informal information a problem, but also there is the question of its proper role in the whole area of geoscience documentation. Some informal publications are trivia, others are important. New ways of identifying and handling these publications will be necessary if we are to continue to channel this information to the appropriate scientists.

Indexing and Abstracting Services. Indexing and abstracting publications, which are secondary publications, have been most affected by the introduction of the computer. The monthly Bibliography and Index of Geology published by the Geological Society of America and the American Geological Institute, and the U. S. Geological Survey's Abstracts of North American Geology are both produced by a computer. A companion bibliography, the Bibliography of Theses in Geology, has been prepared in the same manner. The information for these services is being placed on magnetic tape to form a permanent geologic data base, from which future innovative services, e.g., retrospective searches, can be prepared and offered to the individual geologist (AGI's GEO-REF, mentioned earlier, is such a service). Further, the bibliographies themselves are composed by a photon process directed by a computer. Since geologists in this country have become accustomed to the hierarchal type of indexing as examplified in the Abstracts of North American Geology and the Bibliography and Index of Geology, automation of the bibliographies was so handled as to keep this form of indexing that geologists prefer.

Another computer-prepared index is KWIC (Key Word In Context). This index is composed of key words from titles of journal articles arranged alphabetically as an index, and surrounded by the remaining words in the title to provide context. The advantage of the KWIC index is that it may be easily prepared by a computer equipped with an optical scanner, but the disadvantage is that key words describing the article may not appear in its title. The KeyAlpha index was developed to overcome this disadvantage by using keywords which may or may not be title words, but which best describe the article. Some knowledge of the subject is required on the input end. The American Association of Petroleum Geologists uses this method to index their bulletin.

Perhaps the most exciting indexing innovation is the Science Citation Index published by the Institute for Scientific Information in Philadelphia. SCI indexes the citations in the bibliographies of every article in a large number of science journals. By using SCI, a scientist can trace a scientific concept throughout its history, both forwards and backwards in time, and across traditional subject boundaries. Subject reference, in effect, is performed by a search of the works of authors writing in their respective fields. SCI has incorporated geology and geophysics into its subject coverage, but not, as yet, paleontology.

Modern indexing and abstracting serials try to be as current as possible. In the geosciences, the principal indexing services are now issued monthly, instead of annually as they were a few years ago. The Geological Society of America has switched from full abstracts to a series of descriptors for each citation in its Bibliography and Index of Geology, thus reducing the lag time between the appearance of an article in a journal and its appearance in a secondary indexing medium. A new current awareness serial entitled Geotitles Weekly has further reduced the publication-to-announcement lag time, and its coverage includes new book announcements, past, recent, and future conference papers, technical reports and papers in process.

Libraries. On the surface, libraries seem to have been least influenced by the application of the new technology. However, it is making in-roads: automated circulation of library materials; machine-readable tapes from the Library of Congress to produce catalog cards; computer-produced serial records and book catalogs; photocopy, microform, reader stations with tape and TV; and to a limited extent, local and regional information networks. In the future, libraries will be expected to become major contributors to national and regional science information networks. As this partnership role grows, and as the traditional information patterns of scientists become more burdensome to them, the pressure to try innovative and special information services should increase.

SELECTED ARTICLES ON GEOSCIENCE COMMUNICATION

Craig, Gordon Y. "Communication in geology." In: Scottish journal of geology. v. 5, pt. 4, 1969. p. 305-321.

Fenner, Peter and Joel Lloyd. "Serials of importance to geologists." In: *Geotimes*. v. 15, no. 6, 1970. p. 14-17.

Geoscience documentation; journal for geoscience information. London: Geosystems, 1969- . Bi-monthly. News, articles and bibliography of all phases of geoscience information.

Geoscience Information Society. *Handling geoscience data and information.* Washington, 1969. 58 p. (Proceedings, v. 1)

Geotimes. v. 12, no. 9, 1967.
A special issue devoted to information in the earth sciences.

Hawkes, Herbert E. "Geology." In: *Library trends.* v. 15, no. 4, 1967.
Results of a study on the primary geological literature.

Hruska, J. *Annotated bibliography on storage, retrieval and coding systems of information in the geo-science, preliminary review.* Prague: Geofond, 1968. 29 p. Up-dated in *Geocom bulletin*. v. 2, no. 1, 1969. p. 25-27.

National Academy of Sciences. National Academy of Engineering. Committee on Scientific and Technical Communication. *Scientific and technical communication; a pressing national problem and recommendations for its solution.* Washington: National Academy of Science, 1969. 322 p. (NAS Publication no. 1707)

National Advisory Committee on Research in the Geological Sciences. Ad Hoc Committee on Storage and Retrieval of Geological data in Canada. *A national system for storage and retrieval of geological data in Canada.* Ottawa: Geological Survey of Canada, 1967. 175 p.

Shirley, Howard B. and Dederick C. Ward. "The Geoscience Information Society - a forum for specialized and scientific communication." In: *American Society for Information Science, Proceedings.* v. 5, 1968. p. 31-35.
A recent commentary on GIS is in *Nature*. v. 224, Nov. 15, 1969. p. 639.

Smith, Foster D., Jr. and others. Developing a coordinated information program for geological scientists in the United States. Springfield, Virginia: National Technical Information Service, 1967. 51 p. (PB 177 290)
"Top journals in earth science." In: Earth science reviews. v. 6, no. 6, 1970. Atlas section, p. A241-243.

Woodford, A. O. "Serial literature used by American geologists, 1967." In: Journal of geological education. v. 17, no. 3, 1969. p. 87-90.

General Section

GEOLOGIC INFORMATION

Geoscience documentation; journal for geoscience information. London: Geosystems, 1969- . Bi-monthly. News, articles and bibliography of all phases of geoscience information. The bibliography, entitled "Geodoc Index," includes citations of reference works (subject and area bibliographies, indexing and abstracting services, dictionaries and encyclopedias, directories, etc.), documentation studies, and geoscience policy.

Hruska, J. and C. F. Burk, Jr. Computer-based storage and retrieval of geoscience information: an annotated bibliography. Ottawa: Geological Survey of Canada, in press.
Up-dates an earlier bibliography by J. Hruska entitled " Machine processable geoscience." In: Geocom bulletin. v. 2, no. 1, 1968. p. 25-27.

INFORMATION SOURCES

Earth Science Curriculum Project. Sources of earth science information. William H. Matthews, III. Englewood Cliffs, N. J.: Prentice-Hall, 1964. 48 p. (Reference Series, RS-1) (available from the American Geological Institute, Washington, D. C.)

Kaplan, Stuart R., ed. Mining, minerals and geosciences. New York: Interscience, 1965. 599 p. (Guides to Information Sources in Science and Technology, v. 2) Pt. 1. Organizations. Over 1,000 organizations in 142 countries, giving their address, purpose, organizational structure and publications; pt. 2. Literature. Over 600 publications, principally journals, in the earth and mineral sciences.

U. S. Library of Congress. National Referral Center for Science and Technology. A directory of information resources in the United States: Physical sciences, biological sciences, engineering. Washington, 1965. 352 p. Being revised as "physical sciences and engineering"; the revision of "biological sciences" to appear as a separate directory.

----------. A directory of information resources in the United States: Water. Washington, 1966. 248 p.

----------. A directory of information resources in the United States: Federal government. Washington, 1967. 411 p.
Being revised.

Vetter, Richard C. Oceanography information sources/70. Washington: National Academy of Sciences, 1970. 51 p.

DEPARTMENTS, SOCIETIES, ORGANIZATIONS

American Geological Institute. Directory of geoscience departments, United States and Canada. Washington: American Geological Institute, 1949- .
Biennial.

Lea, Graham. "The organization of international geoscience." In: Geoscience documentation. v 2, no. 5, 1970. p. 110-113.
Lists and describes the various international geoscience organizations. Articles in future issues of Geoscience documentation will cover policies and publication activities of the organizations.

Scientific and technical societies of Canada. Ottawa: National Science Library - National Research Council of Canada, 1969. 69 p. (NRC Publication no. 10625).

Scientific and technical societies of the United States. 8th ed. Washington: National Academy of Sciences, 1968. 221 p. (Publication no. 1499)
Useful for geological societies; gives their history, purpose, membership, meetings, publications, and activities.

Wöhlbier, Herbert and others. Worldwide directory of mineral industries, education and research. Houston: Gulf Publishing Co., 1968. 451 p.
An international directory of geoscience departments in universities and institutes, with an indication of current research.

PERSONS

American men of science: physical and biological sciences. 11th ed. New York: Bowker, 1965-1967. 6 v.
Kept up-to-date with supplements, the latest (6th) published in 1970. The 12th edition is being prepared, publication planned for mid-1971.

Font-Altaba, M. World directory of mineralogists. 2nd ed. Published for the International Mineralogical Association. Barcelona: Editorial Jover, 1970.

International Union of Crystallography. World directory of crystallographers and of other scientists employing crystallographic methods. 3rd ed. Utrecht: Oosthoeks Uitgevers Mij N.V., 1965. 187 p.
Fourth edition in preparation.

Vetter, Richard C. A directory of oceanographers in the United States. Washington: National Academy of Sciences, 1969. 72 p.

"Volcanology, a directory of names and addresses." In: Tectonophysics. v. 5, no. 2, 1968. p. 161-176.
International in scope.

Westermann, G. E. G. Directory of paleontologists of the world (excl. Soviet Union and continental China). 2nd ed. Published for the International Palaeontological Union. Hamilton, Ontario: McMaster University, 1968. 250 p.

FIELD TRIPS AND MEETINGS

"Calendar," a regular monthly feature in Geotimes, news magazine of the earth sciences, published by the American Geological Institute.
Announcements of meetings and field trips. Worldwide.

"Check list of 19__ field trips in United States, Canada and selected foreign areas," a regular feature in the March and August issues of American Association of Petroleum Geologists. Bulletin.

GUIDES TO GEOLOGIC LITERATURE

Harvey, Anthony P. "Recent developments in geological documentation and bibliography." In: Geoscience Information Society. Proceedings. v. 1, 1969. p. 27-53. Bibliographical essay principally covering years 1963-1967. A new edition is being prepared by the author.

Jenkins, F. B. Science reference sources. 5th ed. Cambridge, Mass.: M.I.T. Press, 1969. 231 p.
Earth sciences, p. 79-99.

Smith, Harriet W. "Guide to geological literature." In: Journal of geological education. v. 18, no. 1, 1970. p. 13-25.
Covers organization of libraries and indexes. Lists important reference works in geology.

Watkins, Jessie B. Information sources in the earth sciences; a guide to selected materials in the (Syracuse University) Natural Sciences Library. New York: Syracuse University Libraries, 1967. 46 p.

GEOLOGIC SERIALS

An inventory of the world's geoscience serials is being prepared by the American Geological Institute's Committee on Geoscience Information. It is estimated that over 10, 000 serials contain papers in the geological sciences.

Results of a recent study on the most-cited geology journals are given in Woodford, A. O. "Serial literature used by American geologists, 1967." In: Journal of geological education. v. 17, no. 3, 1969. p. 87-90.

Serials most cited in the Bibliography and index of geology for 1969 are listed in Fenner, Peter and Joel Lloyd. "Serials of importance to geologists." In: Geotimes. v. 15, no. 6, 1970. p. 14-17.

Two published lists of geologic serials are:

Geoserials 1969; a world list of current geoscience serial publications. London: Geosystems, 1969. 127 p. (Geoscience documentation, v. 1, no. 1.)
Approximately 2, 000 titles presented in 2 lists: current serials and titles lapsed since 1960. Supplements to the main listing are published in each issue of Geoscience documentation, with a new edition of Geoserials planned for 1971. Supplements include revised and lapsed titles.

Lomský, Josef. Soupis periodik geologických věd; periodica geologica, palaeontologica et mineralogica. Prague: Nakladetství Ceskoslovenské Akademie Věd, 1959. 501 p. A world list of geologic serials. 3, 582 citations.

GUIDEBOOKS

Geoscience Information Society. Guidebook and Ephemeral Materials Committee. Geologic field trip guidebooks of North America; a union list incorporating monographic titles. 2nd ed. Houston: Phil Wilson, 1971. 152 p. The guidebook holdings of 57 important libraries, making this bibliography an excellent source to verify elusive guidebook information.

HISTORY AND PHILOSOPHY OF THE GEOLOGICAL SCIENCES

REFERENCE WORKS

Albritton, Claude C., Jr. "Philosophy of geology: a selected bibliography and index." In: Albritton, Claude C., Jr., ed. The fabric of geology. Stanford, California: Freeman, Cooper and Co., 1963 (2nd printing, 1964, with corrections). p. 262-363.

----------. "Second bibliography and index for the philosophy of geology." In: Journal of the Graduate Research Center, Southern Methodist University. v. 33, 1964. p. 73-114.

----------. "Third bibliography and index for the philosophy of geology." In: Journal of the Graduate Research Center, Southern Methodist University. v. 35, 1966. p. 55-87.

Hasse, Adelaide R. Reports of explorations printed in the documents of the United States Government. Washington: Government Printing Office, 1899. 90 p.

LaRocque, Aurèle. Contributions to the history of geology. Columbus: Ohio State University, Department of Geology, 1964. 3 v.
V. 1. Biographies of geologists (out of print); v. 2. Bibliography of the history of geology (out of print; under revision); v. 3. Biographic index (available from the author).

Meisel, Max. A bibliography of American natural history, the pioneer century, 1769-1895. Brooklyn: Premier Publishing Co., 1924-1929. 3 v.
V. 1. Annotated bibliography of the publications relating to the history, biography and bibliography of American natural history and its institutions during colonial times and the pioneer century, which have been published up to 1924; v. 2. Institutions (scientific societies, scientific journals, natural history museums and botanic gar-

dens, state geological and natural history surveys, federal exploring expeditions) which have contributed to the rise and progress of American natural history which were founded or organized between 1769 and 1844; v. 3. Institutions founded or organized between 1845 and 1865.

Schmeckebier, L. M. Catalogue and index of the publications of the Hayden, King, Powell and Wheeler Surveys. Washington: Government Printing Office, 1904. 208 p. (U.S. Geological Survey. Bulletin no. 222.)
Catalogs and indexes the following surveys: Geological and Geographical Survey of the Territories; Geological Exploration of the 40th Parallel; Geographical and Geological Surveys of the Rocky Mountain Region; and Geographical Surveys West of the 100th Meridian.

TEXTS, TREATISES AND COMPILATIONS

Adams, Frank Dawson. The birth and development of the geological sciences. New York: Dover, 1954, c1938. 506 p.
Emphasis on middle ages.

Albritton, Claude C., Jr., ed. The fabric of geology. Stanford, California: Freeman, Cooper and Co., 1963. (2nd printing, 1964, with corrections). 374 p.

----------. Uniformity and simplicity; a symposium on the principle of the uniformity of nature. Boulder, Colorado: Geological Society of America, 1967. 99 p. (Special Paper no. 89.)

American Association for the Advancement of Science. "Symposium on the history of American geology." In: Washington Academy of Science. Journal. v. 49, no. 7, 1959. p. 197-260.

Beringer, Carl C. Geschichte der Geologie und des geologischen Weltbildes. Stuttgart: Enke, 1954. 158 p.

Biswat, Asit K. History of hydrology. New York: Elsevier, 1970. 336 p.

Bubnoff, Serge von. Fundamentals of geology. Translated and edited by W. T. Harry. Edinburgh: Oliver and Boyd, 1963. 287 p.

Translation of the 3rd edition of Grundprobleme der Geologie. Berlin: Akademie-Verlag, 1954. 234 p.

Challinor, John. "The early progress of British geology." In: Annals of Science, London.
Pt. 1. From Leland to Woodward, 1538-1728. v. 9, no. 2, 1953. p. 124-153; pt. 2. From Strachey to Mitchell, 1719-1788. v. 10, no. 1, 1954. p. 1-19; pt. 3. From Hutton to Playfair, 1788-1802. v. 10, no. 2, 1954. p. 107-148.

Chorley, Richard J. and others. The history of the study of landforms, or the development of geomorphology. London: Methuen, 1964- .
V. 1. Geomorphology before Davis. 1964. 678 p.

Cloud, Preston, ed. Adventures in earth history. San Francisco: Freeman, 1970. 922 p.
Pt. 1. Principles, processes, origins; pt. 2. Geologic record. "A volume of significant writings from original sources on cosmology, geology, climatology, oceanography, organic evolution, and related topics of interest to students of earth history from Steno to present."

Davies, Gordon L. The earth in decay; a history of British geomorphology, 1578-1878. New York: Elsevier, 1969. 390 p.

Davison, Charles. The founders of seismology. Cambridge: Cambridge University Press, 1927. 240 p.

Fenton, Carroll L. and Mildred A. Fenton. Giants of geology. New York: Doubleday, 1952. 333 p.

Furon, Raymond. La paléontologie. La science des fossiles, son histoire, ses enseignements, ses curiosités. Nouvelle ed. Paris: Payot, 1951. 291 p.

Geike, Archibald. The founders of geology. 2nd ed. New York: Macmillan, 1905. 486 p. (reprinted by Dover, 1962).

Geological Society of America. Geology, 1888-1938. New York, 1941. 578 p.

Gillespie, C. C. Genesis and geology. Cambridge: Harvard University Press, 1951. 315 p.

Geological thought in Britain from 1790-1850.

Hölder, Helmut. Geologie und palaontologie in Texten und ihrer Geschichte. Freiburg: Karl Alber, 1960. 564 p.

Hooykaas, R. Natural law and divine miracle; an historical-critical study of the principle of uniformity in geology, biology and theology. Leiden: Brill, 1959. 237 p.

Kobell, Franz von. Geschichte der mineralogie von 1650-1860. Munich: Cotta, 1864. 703 p.

Mather, Kirtley. Source book in geology, 1900-1950. Cambridge: Harvard University Press, 1967. 435 p.

---------- and Shirley L. Mason. A source book in geology. New York: McGraw-Hill, 1939. 702 p. (reprinted by Hafner, 1964).
Covers geological literature written prior to 1900.

Merrill, George P. Contributions to a history of American state geological and natural history surveys. Washington: Government Printing Office, 1920. 549 p. (U.S. National Museum. Bulletin no. 109.)

----------. The first 100 years of American geology. New Haven, Conn.: Yale University Press, 1924. 773 p. (reprinted by Hafner, 1964).

Moore, Ruth. The earth we live on; the story of geological discovery. New York: Knopf, 1963. 416 p.

New Hampshire Inter-Disciplinary Conference on the History of Geology. Rye Beach, New Hampshire, 1967. Proceedings. Toward a history of geology. Edited by Cecil J. Schneer. Cambridge: M.I.T. Press, 1969. 469 p.
26 papers on the history of geology before Darwin.

Pestana, Harold R. and Bruce D. Bontana. "Congressional geology." In: Geological Society of America. Bulletin. v. 81, 1970. p. 899-904.
A short discussion of typical geological documents found in the early U.S. Congressional Set, 1818-1907. A bibliography, covering congressional geology in these years, is being prepared by the authors.

Schuchert, Charles and others. "A century of geology." In: A century (1818-1918) of science in America. E. S. Dana and others. New Haven, Conn.: Yale University Press, 1918. p. 60-287.
Covers historical geology in North America, progress in the interpretation of landforms, earth structure, government geological surveys, the development of vertebrate paleontology, the rise of petrology, the growth of mineralogy, and the Geophysical Laboratory of the Carnegie Institution.

Sweet, George Elliott. The history of geophysical prospecting. Los Angeles: Science Press, 1966-1969. 2 v. in 1.

Tikhomirov, V. V. "Development of geological sciences during the first half of the nineteenth century." In: International Geological Congress. 23rd. Prague, 1968. Report. v. 13, 1969. p. 319-329.

Wells, John W. and George W. White. "Biographies of geologists." In: Ohio Journal of Science. v. 58, no. 5, 1958. p. 285-298.

Woodward, Horace B. History of geology. London: Watts, 1911. 154 p.

Zittel, Karl Alfred von. History of geology and paleontology to the end of the nineteenth century. Translated by Maria M. Ogilvie-Gordon. Weinheim: J. Cramer, 1901. 562 p. (reprinted by Hafner, 1962).

SPECIFIC ORGANIZATIONS

Bailey, Edward B. Geological Survey of Great Britain. London: Murby, 1952. 278 p.

Beebe, B. Warren. "Quest for unity in the geological sciences." In: Geotimes, 1967. v. 12, no. 5, p. 18-22; v. 12, no. 6, p. 21-22; v. 12, no. 7, p. 18-20; v. 12, no. 8, p. 19-22.
History of the American Geological Institute. Recent article, "The American Geological Institute: where it stands, what it's doing, where it's going," is published in Geotimes. v. 17, no. 8, 1970. p. 17-21.

Fairchild, Herman L. The Geological Society of America, 1888-1930. New York: Geological Society of America, 1932. 232 p.

Leckwijck, W. P. van. "International Union of Geological Sciences." In: Geotimes. v. 10, no. 8, 1966. p. 16-17.

Manning, Thomas G. Government in science; the U. S. Geological Survey, 1867-1894. Lexington: University of Kentucky Press, 1967. 257 p.
An introduction to the modern survey is: Rabbit, J. C. and M. C. Rabbit. "The U.S. Geologic Survey: 75 years of service to the nation, 1879-1954." In: Science. v. 119, May 28, 1954. p. 741-758.

Morley, Harold T. "A history of the American Association of Petroleum Geologists: first fifty years." A special issue of American Association of Petroleum Geologists. Bulletin. v. 50, no. 4, 1966. p. 668-820.

THE INTERNATIONAL GEOLOGICAL CONGRESS

Sessions of the IGC are held every 4 years, and papers presented at them afford a resume of work taking place in all geological subdisciplines and in every corner of the world. The host country is responsible for publication of the proceedings and a set of authoritative guidebooks about the geology of the region. Post World War II sessions of the congress are: 18th, London, 1948; 19th, Algiers, 1952; 20th, Mexico, 1956; 21st, Copenhagen, 1960; 22nd, New Delhi, 1964; 23rd, Prague, 1968; 24th, Montreal, 1972. Two articles about the IGC are:

"History of the International Geological Congress." In: Geotimes. v. 5, no. 1, 1960. p. 32.

Thurston, William. "The first International Geological Congress." In: Geotimes. v. 13, no. 2, 1968. p. 16-17.

CLASSICAL WORKS

Agassiz, Louis. Studies on glaciers. Translated and edited by Albert V. Carozzi. New York: Hafner, 1967.
Original work published in 1840, entitled Études sur les glaciers.

Agricola, Georgius (1494-1555). De natura fossilium. Translated from the first Latin edition of 1546 by Mark C. Bandy and Jean A. Bandy. New York: Geological Society of America, 1955. 240 p. (Special paper no. 63.)
First systematic classification of minerals by the "Father of Mineralogy."

----------. De re metallica. Translated from the first Latin edition of 1556 by Herbert C. Hoover and Lou H. Hoover. London: Mining Magazine, 1912. (Reprinted by Dover, 1950.)
Prefaced by a discussion of the history of early geology and mineral industry.

Contributions to the history of geology. Edited by George W. White. New York: Hafner, 1957- . Irregular.
A series of facimile reprints of classical works in geology with emphasis on American classics. Each reprint includes bio-bibliographic introduction and commentary. Volumes published: 1. American mineralogical journal, v. 1, 1810-1814; 2. C. F. Volney. View of the soil and climate of the United States of America. 1804; 3. Monthly American journal of geology and natural history. v. 1, 1831-1832; 4. Nicolaus Steno (1638-1686). The prodromus of Nicolaus Steno's dissertation concerning a solid body enclosed by process of nature within a solid. Volumes in preparation: Samuel Akerly. Geology of the Hudson River and the adjacent regions. 1820; Robert Boyle. The origin and virtue of gems. 1672; Amos Eaton. Geological textbook. 1832; James Hutton. Abstract of the theory of the earth. 1785; and Playfair's life of Hutton. 1805; William M. Mather. Elements of geology. 1835; N. F. Moore. Ancient mineralogy. 1834; David Thomas. Travels through the western country. 1819; G. H. Toulmin. Antiquity of the world. 1780; Jeremiah Van Rensselaer. Lectures on geology. 1825.

Cuvier, Georges. A discourse on the revolutions of the surface of the globe, and the changes thereby produced in the animal kingdom. Translated from the French with illustrations and a glossary. London: Whittaker, Treacher, and Arnot, 1829. 261 p.

----------. Essay on the theory of the earth. 5th ed. Translated by Prof. Robert Jameson. Edinburgh: Blackwood, 1827. 550 p.

Hutton, James. Theory of the earth, with proofs and illustrations. Edinburgh: Printed for Cadell and Davies, London, and W. Creech, Edinburgh, 1795-1899. 3 v. (reprinted by Hafner, 1960).
A biography of Hutton is Bailey, Edward B. James Hutton--the founder of modern geology. New York: Elsevier, 1967. 161 p.

Lyell, Sir Charles. Principles of geology. 12th ed. London: Murray, 1875. 2 v.
First published, 1830-1833. A biography of Lyell is Bailey, Edward B. Charles Lyell. New York: Doubleday, 1963. 214 p.

Maclure, William. Observations on the geology of the United States of America. Philadelphia: A. Small, 1817. (Reprinted by Hafner, 1962.)

Maury, Matthew F. The physical geography of the sea and its meteorology. 8th ed., 1861. Edited by John Leighly. Cambridge: Harvard University Press, 1963. 432 p.
First major textbook on oceanography. A biography of Maury is: Williams, Frances L. Matthew Fontaine Maury, scientist of the sea. New Brunswick, N.J.: Rutgers University Press, 1963. 720 p.

Playfair, John. Illustrations of the Huttonian theory of the earth. Edinburgh: Printed for Cadell and Davies, London, and W. Creech, Edinburgh, 1802. 528 p. (reprinted by University of Illinois Press, 1956).

Smithsonian Institution. Exploration of the Colorado River of the West and its tributaries. Washington: Government Printing Office, 1875. 291 p.
Pt. 1 includes John Wesley Powell's account of exploration in 1869-70. This account also published as: Powell, J. W. The Exploration of the Colorado River. Chicago: University of Chicago Press, 1957. 138 p. Two biographies of Powell are: Darrah, William C. Powell of the Colorado. Princeton: Princeton University Press, 1951. 426 p.; and Stegner, Wallace E. Beyond the hundredth meridian: John Wesley Powell and the second opening of the West. Boston: Houghton Mifflin, 1954. 438 p.

Suess, Eduard. The face of the earth (Das Antlitz der Erde). Translated by Hertha Sollas. Oxford: Clarendon Press,

1904-1924. 5 v.

Wegener, Alfred. The origin of continents and oceans. Translated from the 4th revised German edition (1929) by John Biram. New York: Dover, 1966. 246 p.

Werner, Abraham G. On the external characters of minerals. Translated by Albert V. Carozzi. Urbana: University of Illinois Press, 1962. 118 p.
Translation of the 1774 edition of Von den äusserlichen Kennzeichen der Fossilien, considered as the first modern textbook of mineralogy.

CURRENT AWARENESS IN THE GEOLOGICAL SCIENCES

The pace of modern science requires the geologist to keep abreast of current developments in his field. The following media are offered as examples of some of the ways in which geologists keep current. They are similar in that they are oriented toward the rapid dissemination of news and results of geological research.

NEWS

1. Geotimes, published monthly (except June and August) by the American Geological Institute, is the current events magazine for the geological sciences. Includes semi-technical articles on topics of current interest, preliminary results of scientific work, meeting reports, news of societies, government organizations, people, and new publications.

2. News journals, such as EOS, the Transactions of the American Geophysical Union, Oil and gas journal, Engineering and mining journal, etc. are published for specific scientific and technical audiences. They contain a substantial amount of news and timely articles which vary from a semi-technical to technical level. Many of the standard scholarly geological journals include a short section devoted to news events.

3. Newsletters are the news organs of societies or other scientific organizations. Some societies (e.g., Glaciological Society, Geological Society of America) and commercial publishers (Elsevier) issue newsletters (Ice, The geologist, Atlas supplement) in addition to their scholarly journals (Journal of glaciology, GSA Bulletin, Earth science reviews). Other organizations rely solely on a newsletter to communicate, e.g., International Union of Geological Sciences. Geological newsletter; Earth Science Curriculum Project. Newsletter; Geoscience Information Society. Newsletter.

4. Some news is published in card form. The Center for Short-lived Phenomena of the Smithsonian Institution

distributes Event Notification Cards which announce, among other phenomena, earthquakes, volcanic eruptions, mudflows and other geologic events which occur with marked frequency around the world.

RESEARCH AND PUBLICATION

1. In addition to the scientific journal, which heretofore provided geologists with their prime source of current awareness, today's symposia, colloquia and conference proceedings appear in a variety of forms--as special issues of journals, as books, as government reports. Never really controlled to an optimum degree by the secondary bibliographic services in geology or science, they are, nevertheless, the focus of:

Directory of published proceedings; Science, Engineering, Medicine, Technology. White Plains, N. Y.: InterDok Corp., 1965- . Monthly, with annual cumulations.

World meetings (U.S.A. and Canada; Outside U.S.A. and Canada). Newton Centre, Mass.: T.M.I.S., 1963- . Quarterly.

2. Geotitles weekly; current awareness service for geoscience. London: Geosystems, 1969- . Weekly. In addition to conference announcements, Geotitles classified new geoscience publications. Coverage includes journals, trade publications, books, theses, patents and standards, and press releases, which are scanned weekly by regional editors.

3. Two important current awareness magazines, which cover news and announcements, reports, articles, book reviews, and commentary, are:

Nature; international journal of science. London: Macmillan Ltd., Weekly.

Science. Washington: American Association for the Advancement of Science, Weekly.

4. U.S. Department of Commerce. National Technical Information Service. Announcements in science and technology; earth sciences and oceanography. Springfield, Va., 1970- . Semi-monthly. Lists recently published technical reports, many of which are not covered in the

regular bibliographies for geology.

5. Canada. Geological Survey. Abstracts of publications in scientific journals by officers of the Geological Survey of Canada (for the year----). Ottawa, 1966- . Annual. (Issued in the Paper series.)

6. Canada. Geological Survey. Report of activities. Ottawa, 1962- . Annual. (Issued as the first numbered Paper series each year, ex. Paper no. 70-1.) From 1962-1966 issued under the title, Summary of research and activities.

7. National Advisory Committee on Research in the Geological Sciences. Current research in the geological sciences in Canada (for the years ------). Ottawa: Geological Survey of Canada, 1946- . Annual. (Issued in the Paper series.) Lists primarily non-industrial research, including university dissertations. The report for 1968-1969 is C.G.S. Paper no. 69-5.

8. Tulsa University. Information Services Department. Earth sciences research catalog. Tulsa, 1971. In press. Covers 2,200 research-in-progress projects funded during 1969-70, extracted from the files of the Science Information Exchange of the Smithsonian Institution in Washington, D.C.

9. U. S. Geological Survey. Geological Survey research (for the year ----). Washington: Government Printing Office, 1960- . Annual. (Issued in the Professional Paper series.) Annual reviews of the economic and scientific work of the U.S. Geological Survey, accompanied by short papers in the fields of geology, hydrology and related sciences, and by a list of publications released during the year. The report for 1969 is Professional Paper no. 650. The report for 1970 will be Professional Paper no. 700.

REVIEWS

1. Review articles. To keep pace with current research in related fields, geologists refer to review articles, written by specialists considered competent to judge the significance of current research trends. Such articles will not normally include the results of original research or the author's speculations. Unless gathered in a specific volume so labeled (e.g., Advances in geophysics, Oceanography and

marine biology annual review, etc.), review articles are not easy to spot in the primary literature, and they are rarely identified as "reviews" in the secondary literature (bibliographies, abstracts). A paper by Herbert E. Hawkes, "Recent review articles in geology" (*Earth science reviews*, v. 3, 1967. p. 135-155), describes the review article and lists those published between 1964 and 1966.

2. Council on Education in the Geological Sciences. "CEGS short reviews." In: *Journal of geological education.* Irregular. A continuing series of articles, each by a specialist, which cover selected geologic topics. An example is:

Morgan, James P. "Deltas--a resume." In: *Journal of geological education.* v. 18, no. 3, 1970. p. 107-119.

3. State-of-the-art reviews are published irregularly in *Geotimes.* The January 1971 issue, for example, is devoted to "1970 in review," summarizing activity in the various geological fields.

4. Review and outlook type reviews are common to the industry-oriented magazines. *Oil and gas journal* (in January) and *Engineering and mining journal* (in March) devote special issues to review and forecast. The June and August issues of the *A.A.P.G. Bulletin* are devoted to North American and foreign petroleum exploration activity.

5. *AGI short course lecture notes.* In past years the American Geological Institute's Committee on Education in the Geological Sciences has sponsored a series of short courses for the purpose of reviewing, for geologists, current research activity in rapidly developing areas of geoscience. Course notes are published following each session. One example is:

New concepts of continental margin sedimentation. Washington: American Geological Institute, 1969. 1 v.

Commercial publishers have also recognized the need for current, but authoritative, reviews. Springer-Verlag, for example, is issuing a series of monographs entitled *Minerals, rocks and inorganic materials*, which, according to the editor, will "provide detailed reviews on the synthesis and stability of selected mineral groups, supplementing rather than repeating material in mineralogy texts."

BIBLIOGRAPHY AND ABSTRACTING SERVICES

RETROSPECTIVE BIBLIOGRAPHIES

Annotated bibliography of economic geology. Mt. Pleasant, Michigan: Economic Geology Publishing Co., 1929-1971(?). This annotated bibliography of economic and applied geology will cease with v. 39, as the bibliographies by the U. S. Geological Survey and the Geological Society of America now cover the same topics. Cumulative index was published from 1928-1954.

Geological Society of London. Library. Geological literature added to the Geological Society's Library. London, 1895-1935.
This work is the predecessor to the Bibliography and index of geology, covering much of the world's geologic literature during the years 1894-1934.

International catalogue of scientific literature. London: Royal Society, 1903-1920. 14 v. (Reprinted by Mini-Print Corp., Metuchen, New Jersey, 1969.) Covers years 1900-1914. Continues the Royal Society of London's Catalogue of scientific papers (1867-1925, 19 v.), which covered years 1800-1900. The International catalogue is issued in sections: Section G. Mineralogy; Section H. Geology; Section K. Paleontology.

Neues Jahrbuch für Mineralogie, Geologie und Paläontologie. Stuttgart: E. Schweizerbart'sche, 1830-1949.
Important bibliographical serial with long and complicated history. Neues Jahrbuch included both original articles and bibliographies, until, in 1943, the "Referate" or abstracting component of this journal became the Zentralblatt für Geologie und Paläontologie (1943-1949). In 1950, the Zentralblatt split into the Zentralblatt für Geologie und Palaontologie and the Zentralblatt für Mineralogie.

Société géologique de France. Bibliographie des sciences géologique. Paris, 1923-1955. Annual.

During 1948-1955, published as Extrait du Bulletin analytique du Centre national de la recherche scientifique (CNRS). Superceded in 1956 by France. Centre national de la recherche scientifique. Bulletin signalétique.

U. S. Geological Survey. Catalogue and index of Contributions to North American geology, 1723-1891. Nelson H. Darton. Washington: Government Printing Office, 1896. 1045 p. (Bulletin no. 127.)

----------. Geologic literature of North America, 1785-1918. John M. Nickles. Washington: Government Printing Office, 1923-1924. 2 v. (Bulletin nos. 746, 747.)
The Nickles volume is the predecessor to the U. S. Geological Survey's Bibliography of North American geology.

COMPILATIONS OF BIBLIOGRAPHIES

Guelpa, Paulette-Marie. "Sciences de la terre." In: Malclès, L-N. Les sources du travail bibliographique, v. 3. Paris: Minard, 1958. p. 236-296.
Partially annotated. Especially good for European literature.

Kummel, Bernhard. "Compilation of bibliographies of use to paleontologists and stratigraphers." In: Kummel, Bernhard and David M. Raup, eds. Handbook of paleontological techniques. San Francisco: Freeman, 1965. p. 767-832.
Similar to the Guelpa bibliography.

Long, Harriet K. "A bibliography of bibliographies of geology of the states of the United States." A special supplement to Geoscience abstracts. v. 7, no. 7, 1965. p. 111-125.
Up-date in progress.

Margarie, Emmanuel de. Catalogue des bibliographies géologique. Paris: Gauthier-Villars, 1896. 733 p.
Covers 1726-1895. Useful for historical inquiry in geology; includes biography and bibliography for individual geologists. Continued by Mathews.

Mathews, Edward B. Catalogue of published bibliographies in geology, 1896-1920. Washington: National Research Council, 1923. 228 p. (Bulletin v. 6, pt. 5, no. 36.)

Continues, in a more simplified format, the bibliography by Margarie.

CURRENT BIBLIOGRAPHICAL SERIALS

Note: because all bibliographical works cover best the publishing country, some of these international serials are repeated under the country of origin in the regional section. Three monthly bibliographical services--those of East Germany, Romania and Yugoslavia--are not used extensively outside these countries and are listed only in the regional section.

Also, the following specialized bibliographies are repeated in this, the general, section, because they are important and frequently consulted: Geomorphological abstracts, Mineralogical abstracts, Geophysical abstracts and Oceanic index.

Bibliographie des sciences de la terre. Orleans, France: Bureau de recherches géologique et minières, 1968- . Monthly.
Computer-prepared bibliography of geosciences issued in 9 sections, each with author, KWIC (Key Word In Context), and geographic indexes. Continuation of index card service previously published by the Bureau. Worldwide.

France. Centre national de la recherche scientifique. Bulletin signalétique. Paris, 1956- . Monthly.
A monthly abstracting service of worldwide scientific literature, issued in sections since 1961. Since 1969, the earth science sections have been: 210. Sciences de la terre I--Minéralogie, géochimie, géologie extraterrestre, pétrographie; 214. Sciences de la terre II--Géologie appliquée; 216. Sciences de la terre III--Géologie, paléontologie.

Geographical abstracts, Section A: Geomorphology. Norwich, England: Geo Abstracts, 1960- . 6/yr.
From 1960-1965, entitled Geomorphological abstracts. Worldwide in scope. Author and regional indexes in last issue of year, but comprehensive subject indexes as follows: Geomorphological abstracts, 1960-1965, Geographical abstracts, 1966 and 1967 (published 1970).

Geological Society of America. Bibliography and index of geology. Boulder, Colorado, 1934- . Monthly.
The principal English language bibliography of world geology. Prior to 1969 (v. 33), the bibliography was entitled the Bibliography and index of geology exclusive of North America (v. 1-32, 1934-1968) and served to complement the U.S. Geological Survey's Bibliography of North American geology. In 1969 (beginning with v. 33) the bibliography was expanded to include world geology, by incorporating the literature file (tape) on North America furnished by the U.S.G.S. and material supplied by cooperating agencies in other countries--thus far, the New Zealand Geological Survey and the United Kingdom Cooperative Abstracts group. Issued annually from 1934-1967, the bibliography became monthly in 1968, with each issue divided into 21 intradisciplinary geologic categories. Bibliographic citations are accompanied by groups of descriptors or keywords, and each issue has subject and author indexes. Annual index.

Geotitles weekly; current awareness service for geoscience. London: Geosystems, 1969- . Weekly.
Classifies titles, in English, of new geoscience publications, scanned weekly by regional editors. Worldwide. Coverage includes journals and serials, trade magazines, patents and standards, conference publications, books, reports and theses. Sent airmail, Geotitles also includes a news section announcing meetings and new books.

Mineralogical abstracts. London: Mineralogical Society of Great Britain and the Mineralogical Society of America, 1920- . Quarterly.
Covers mineralogy, petrology, geochemistry, and mineral deposits. Worldwide. From 1920-1958, published as a section of Mineralogical magazine.

Oceanic citation journal, with abstracts. La Jolla, California: Oceanic Research Institute, 1968- . 9/yr.
Covers the literature of the marine sciences. Annual keyword index is entitled the Oceanic index (which began in 1964). Oceanic index is the hardbound cumulative of the journal's quarterly indexes.

Referativnyi zhurnal. Moscow: Akademiia Nauk SSSR, 1956- . Monthly.
Perhaps this service abstracts the greatest total number

of geological articles. Worldwide. Following a classed arrangement (in Russian), the citations are given in the language of the publication, with a Russian abstract. Annual indexes. Issued in sections. Of interest to geologists are "Geologiya," "Geofizika" and "Gornoe Delo (Mining)."

U.S. Geological Survey. Abstracts of North American geology. Washington: Government Printing Office, 1966- . Monthly.
Abstracts "technical papers and books and also citations of maps on the geology of North America including Greenland, the West Indies, as well as the state of Hawaii, Guam and other island possessions of the United States." - Introd. North America is defined to include Mexico and Central America. Arrangement alphabetical by author, with subject indexes in each issue. Cumulates annually as:

----------. Bibliography of North American geology. Washington: Government Printing Office, 1923- . Annual.
The volumes of this bibliography are issued in the Survey's Bulletin series under the title Bibliography of North American geology. Arrangement by author, with subject index. Abstracts not repeated in the annual bibliography. BNAG cumulates decennially. The following U.S.G.S. Bulletins make up the bibliography to date: 746-747 (cover 1785-1918); 823 (1919-1928); 937 (1929-1939); 1049 (1940-1949); 1195 (1950-1959); 1196 (1960); 1197 (1961); 1232 (1962); 1233 (1963); 1234 (1964); 1235 (1965); 1266 (1966); 1267 (1967). BNAG complements the Geological Society of America's Bibliography and index of geology exclusive of North America until 1969, at which time the latter, as the Bibliography and index of geology, began to include North American literature as well as foreign.

----------. Geophysical abstracts. Washington: Government Printing Office, 1929- . Monthly.
Includes abstracts "of technical papers and books on the physics of the solid earth, the application of physical methods and techniques to geologic problems, and geophysical exploration."--Introd. Worldwide. Annual indexes. From 1929 until 1963, published variously as Bulletins of the U.S. Geological Survey and Information Circulars of the U.S. Bureau of Mines.

Zentralblatt für Geologie und Paläontologie. Stuttgart: E. Schweizerbart'sche, 1960- . Irregular.
Teil I. Allgemeine und angewandt Geologie. Regionale Geologie; teil II. Historische Geologie. Paläontologie. Important abstracting serial with long and complicated history. In 1943, the Zentralblatt superceded the "Referate" component of the Neues Jahrbuch für Mineralogie, Geologie und Paläontologie. Worldwide in scope.

Zentralblatt für Mineralogie. Stuttgart: E. Schweizerbart'sche, 1950- . Irregular.
Teil 1. Kristallographie und Mineralogie; teil II. Petrographie. Technische Mineralogie. Geochemie und Lagerstättenkunde.

THESIS BIBLIOGRAPHY

Bibliography of theses in geology

A bibliography of theses completed for advanced degrees in the geological sciences as reported by the departments listed in the Directory of geoscience departments, United States and Canada, published biennially by the American Geological Institute. The individual bibliographies making up this series to date are:

Chronic, B. John and Halka Chronic. Bibliography of theses written for advanced degrees in geology and related sciences at universities and colleges in the United States and Canada through 1957. Boulder, Colorado: Pruett Press, 1958. 1 v. Available from: American Geological Institute, Washington, D. C.

----------. Bibliography of theses in geology, 1958-1963. Washington: American Geological Institute, 1965. 1 v.

Ward, Dederick C. "Bibliography of theses in geology, 1964." In: Geoscience abstracts. v. 7, no. 12, pt. 1, 1965. p. 103-129.

----------, and T. C. O'Callaghan. Bibliography of theses in geology, 1965-1966. Prepared in cooperation with the Geoscience Information Society. Washington: American Geological Institute, 1969. 255 p.

Dissertation abstracts international. Sec. B: Physical sciences and technology. Ann Arbor, Michigan: University Microfilms, Inc., 1938- . Monthly.
Abstracts of doctoral dissertations from cooperating institutions in the U. S. and Canada. University Microfilms, Inc. also sells xerox and microfilm copies of the dissertations listed in Dissertation abstracts.

Dissertation abstracts international, retrospective index 1938-1969. Ann Arbor, Michigan: University Microfilms, Inc., 1970. 9 v.

National Advisory Committee on Research in the Geological Sciences. "Geological abstracts (of graduate Canadian theses)." An irregular feature in the Canadian mining journal, 1950- .

Purdue University. Thermophysical Properties Research Center. Masters theses in the pure and applied sciences. Edited by Beth M. Schick. Ann Arbor, Michigan: University Microfilms, Inc., 1957- . Annual.
Covers about 170 institutions in the United States.

INDEXES AND CATALOGS

An index can cover a group of periodicals and books, such as those of a society, or a separate periodical. Examples of the former type are as follows:

American Association of Petroleum Geologists. Comprehensive index of the publications of the American Association of Petroleum Geologists, 1917-1945. Daisy W. Heath. Tulsa, 1947. 603 p.

----------. 1946-1955. Daisy W. Heath and June McFarland. Tulsa, 1957. 301 p.

----------. 1956-1965. June McFarland and Mary E. Gieselman. Tulsa, 1967. 682 p.

Cumulative index to the publications of the Society of Exploration Geophysicists from 1931 through 1961. Tulsa: Society of Exploration Geophysicists, 1963. 534 p.

Indexes for separate periodicals or journals generally appear annually. They may cumulate either regularly or irregularly as follows:

Geological Society of London. Quarterly journal. Index, v. 1-50 (1845-1894); Index, v. 51-90 (1895-1934); Index, v. 91-118 (1935-1962).

Geological magazine consolidated index, 1904-1963. E. H. Pamphilon. Hertford, Herts, England: Austin, 1964. 235 p.

Journal of paleontology, Index v. 1-25, 1927-1951. Michael S. Chappars and others. Tulsa: Society of Economic Paleontologists and Mineralogists, 1961. 364 p. (Special Publication no. 8.)

Society of Exploration Geophysicists. KWIC index of Geophysics (1936-1965) and Geophysical Prospecting (1953-1965). n.p., 1966. 202 p.

Catalogs represent a systematic listing of the publications which an organization has published. Some useful catalogs and indexes of U.S. and Canadian governmental agencies are given below.

Canada. Geological Survey. Index of publications, Geological Survey of Canada, 1845-1958. A. G. Johnson. Ottawa: Department of Mines and Technical Surveys, 1961. 378 p.

----------. 1959-1964. H. M. A. Rice. 1965. 163 p. Kept up-to-date by annual supplements in the Survey's "Paper" series.

Canada. Geological Survey. Index to reports of the Geological Survey of Canada from 1927-1950. W. E. Cockfield and others. Ottawa: Department of Mines and Technical Surveys, 1962. 723 p.
This index is the sixth in the Index to reports series issued by the survey. A brief outline of earlier indexes: 1st covers reports from 1863-1884; 2nd covers reports from 1884-1904; 3rd covers Separate Reports from 1906-1910 and Summary Reports from 1905-1916; 4th covers Palaeontology in reports from 1847-1916; and 5th covers Memoirs from 1910-1926, Bulletins from 1913-1926 and Summary Reports from 1917-1926.

Canadian government publications. Mines Branch and Mineral Resources. Bibliography no. 12. Ottawa: Queen's Printer, 1967. 401 p. (Sectional Catalogue no. 12.)
Periodically revised.

U. S. Bureau of Mines. List of publications issued by the Bureau of Mines from July 1, 1910 to January 1, 1960 with subject and author index. Washington: Government Printing Office, 1960. 826 p.
Kept up-to-date by annual supplements, which cumulate every 5 years.

U. S. Geological Survey. Publications of the Geological Survey, 1879-1961. Washington: Government Printing Office, 1964. 457 p.
Complete catalog of the books, maps and charts published by the Survey. Revision of the catalog is not intended, but it is kept up-to-date by annual supplements (which will cumulate) and monthly notices. The catalog and annual supplements are indexed by subject, area and author. An index to U. S. G. S. publications by latitude and longitude is Albertson, G. H. Geologic index to the publications of the U. S. Geological Survey. Denver: Geological Publishing Co., 1963. Kept up-to-date by monthly, quarterly and annual supplements.

U. S. Superintendent of Documents. Monthly catalog of United States Government publications. Washington: Government Printing Office. 1895- . Monthly.
A current bibliography of publications issued by all branches of the federal government, including department and bureau publications. Each monthly issue contains an author-subject index. Annual index with the December issue.

LIBRARY CATALOGS

Library catalogs of important research libraries are useful to geologists, because they can be considered a bibliography of publications in the library's specialty and indicate a location for at least one copy of each title listed. Further, they are very useful for verifying titles found in the literature. A catalog of the largest geological library in the world is:

U. S. Geological Survey. Library. Catalog of the United States Geological Survey Library. Boston: G. K. Hall,

1964. 25 v.
The catalog indexes by author, title and subject approximately half a million publications on geology, paleontoloby, petrology, mineralogy, hydrology, cartography and mineral resources, as well as considerable material in allied sciences.

California. University at San Diego. Library. Catalogs of the Scripps Institution of Oceanography. Boston: G. K. Hall, 1970. 12 v.
Library's specialty is oceanography, marine biology, marine technology and oceanographic expeditions, but there are also substantial holdings in geology, physics, chemistry, biology and the atmospheric sciences.

ENCYCLOPEDIAS

Bertin, Leon. Larousse encyclopedia of the earth. New York: Prometheus Press, 1961. 419 p.

Fairbridge, Rhodes W., ed. Encyclopedia of earth sciences series. New York: Reinhold, 1966- .
V. 1. The encyclopedia of oceanography. 1966. 1021 p; v. 2. The encyclopedia of atmospheric sciences and astrogeology. 1967. 1200 p; v. 3. The encyclopedia of geochemistry. In preparation.

International dictionary of geophysics; seismology, geomagnetism, aeronomy, oceanography, geodesy, gravity, marine geophysics, meteorology, the earth as a planet and its evolution. Edited by S. K. Runcorn and others. New York: Pergamon, 1967. 3 v.
Encyclopedic work.

McGraw-Hill encyclopedia of science and technology. 3rd ed. New York: McGraw-Hill, 1970. 15 v.
Up-dated by yearbooks.

DICTIONARIES AND GLOSSARIES

American Geological Institute. *Dictionary of geological terms*. New York: Dolphin Books, 1962. 545 p. An abridged and revised edition of the *Glossary of geology and related sciences*, 2nd ed., 1960.

----------. *Glossary of geology and related sciences*. 2nd ed., with supplement. Washington, 1960. 325 p., 72 p. Many times more comprehensive than this glossary is a major glossary for the geological sciences now in preparation by the American Geological Institute.

Challinor, John. *A dictionary of geology*. 3rd ed. Cardiff: University of Wales Press, 1967. 298 p.

Chamber's mineralogical dictionary. New York: Chemical Publishing Co., 1960. 47 p.

Dennis, John G. *International tectonic dictionary*. Published for the International Geological Congress, Commission for the Geological Map of the World. Tulsa: American Association of Petroleum Geologists, 1967. 196 p. (A. A. P. G. Memoir no. 7.)

Himus, Godfrey W. *Dictionary of geology*. Baltimore: Penquin Books, 1954. 153 p.

Nelson, Archibald and K. D. Nelson. *Dictionary of applied geology; mining and civil engineering*. New York: Philosophical Library, 1967. 421 p.

Pfannkuch, Hans-Olaf. *Elsevier's dictionary of hydrogeology*. Amsterdam: Elsevier, 1969. 168 p.

Rice, Clara M. *Dictionary of geological terms, exclusive of stratigraphic formations and paleontologic genera and species*. Ann Arbor, Michigan: Edwards Brothers, 1955. 464 p.

Stokes, William L. and David J. Varnes. *Glossary of selected geologic terms*. Denver: Colorado Scientific Society, 1955. 165 p. (Proceedings, v. 16.)

U. S. Bureau of Mines. *A dictionary of mining, mineral and related terms*. Compiled by Paul W. Thrush and others.

Washington: Government Printing Office, 1968. 1269 p.

LANGUAGE DICTIONARIES

Baulig, Henri. Vocabulaire français-anglo-allemande de geomorphologie. Paris: Belles-Lettres, 1956. 230 p. (Strasbourg. Université. Publication no. 130.) Also covers some Scandinavian terms. Glossary of Arabic terms.

Bradley, J. E. S. and A. C. Barnes. Chinese-English glossary of mineral names. New York: Consultants Bureau, 1963. 120 p.

Cagnacci-Schwicker, Angelo. International dictionary of metallurgy, mineralogy, geology, mining and oil industries: English, French, German, Italian. Milan: Technoprint International, 1968; reprinted by McGraw-Hill, 1970. 1530 p.

Cooper, S. A. A concise international dictionary of mechanics and geology; English, French, German and Spanish. London: Cassell, 1949. 400 p.

Davies, George M. French-English vocabulary in geology and physical geography. London: J. Mann, 1960, c1932. 140 p.

De Vries, Louis. French-English science dictionary. 3rd ed. New York: McGraw-Hill, 1959. 655 p.

----------. German-English science dictionary. 3rd ed. New York: McGraw-Hill, 1959. 592 p.

Huebner, Walter. Geology and allied sciences; a thesaurus and a coordination of English and German specific and general terms. Pt. 1: German-English. New York: Veritas Press, 1939. 405 p.

Jones, William R. and Arnold Cissarz. German-English geological terminology. London: Murby, 1931. 250 p.

Leinz, Viktor and Josué C. Mendes. Vocabulário geológico (com a correspondente terminologia em inglês, alemão e francês). 3rd ed. São Paulo: Editora Nacional, 1963. 198 p. (Iniciação científica, v. 26.)

Manzoni, Marcello. Dizionario di geologia con appendice di riferimento inglese-italiano. Bologna: Zanichelli, 1968. 234 p.

Nederlandsch Geologisch Mijnbouwkundig Genootschap. Geological nomenclature. Edited by A. A. G. Schieferdecker. Gorinchen: J. Noorduijn, 1959. 523 p.
English-French-Dutch-German.

Novitsky, Alejandro. Diccionario minero-metalurgico-geologico-mineralogico-petrografico y de petroleo. Buenos Aires, 1951. 369 p.
English, Spanish, French, German, Russian.

Sofiano, T. A. Russko-angliiskii geologicheskii slovar'. Moscow: Glavnaya Redaktsiya Inostrannykh Nauchno-Tekhnicheskikh Slovarei Fitzmatgiza, 1960. 559 p.

----------. English-Russian geological dictionary. Moscow: publisher as above, 1961. 335 p.

Telberg, V. G. Russian-English petrographic dictionary. New York: Telberg, 1967. 250 p.

----------. Russian-English dictionary of paleontological terms. New York: Telberg, 1966. 168 p.

Windhausen, Heriberto. Diccionario y nomenclatura geológica en castellano, aleman, ingles. La Plata, Argentina, 1945. 200 p. (Publicaciónes didácticas y de divulgación científica, no. 3.)

Zeman, O. and K. Beněs. English-Czech geological dictionary. Prague: Československé Akademie Ved, 1963. 367 p.
Index of Czech terms.

LEXICONS OF GEOLOGIC FORMATION NAMES

Wilmarth, M. Grace. Lexicon of geologic names of the United States including Alaska. Washington: Government Printing Office, 1938. 2 v. (U.S. Geological Survey. Bulletin no. 896.)

Contains more than 9,000 stratigraphic units described in the literature before 1936.

Keroher, Grace C. and others. Lexicon of geologic names of the United States for 1936-1960. Washington: Government Printing Office, 1966. 3 v. (U.S. Geological Survey. Bulletin no. 1200.)
Lists 14,625 names including cross-references.

----------. Lexicon of geologic names of the United States for 1961-1967. Washington: Government Printing Office, 1970. 848 p. (U.S. Geological Survey. Bulletin no. 1350.)

Cohee, G. V. and others. Changes in stratigraphic nomenclature by the U.S. Geological Survey. Washington: Government Printing Office, 1965- . Annual.
Supplements the Lexicon by providing stratigraphic names which have been adopted, revised, reinstated or abandoned. Appears annually in the Bulletin series as follows: changes for 1963 (Bulletin no. 1194-A); 1964 (Bulletin no. 1224-A); 1965 (Bulletin no. 1244-A); 1966 (Bulletin no. 1254-A); 1967 (Bulletin no. 1274-A); 1968 (Bulletin no. 1294-A); 1969 (Bulletin no. 1324-A); 1970 (Bulletin no. 1254-A).

Lexique stratigraphique international. Publie sur l'initative de la Commission de stratigraphie du Congress géologique international. Paris: Centre national de la recherche scientifique (CNRS), 1956- .
A fundamental reference work still in progress and intended to be complete when all parts of its projected eight volumes are published. Each part includes the description, type-locality, age, distribution and references for the stratigraphical units of each country or group of countries. The stratigraphical formations are listed alphabetically, and at the end of each part there is an author index, a bibliography of cited references, a stratigraphical index and a general index. Many of the parts are prefaced by a summary of the geology of the region. Folded maps accompany each part. The general plan of the Lexique follows:

Volume I. Europe. Pt. 1. Greenland, Iceland, Spitsberg, Faeroe Islands; pt. 2. Finland, Norway, Sweden, Denmark; pt. 3. England, Wales, Scotland, Ireland; pt. 4. France, Belgium, Lowlands, Luxembourg; pt. 5.

Germany; pt. 6. Poland, Czechoslovakia; pt. 7. Switzerland; pt. 8. Austria; pt. 9. Hungary; pt. 10. Iberian Peninsula; pt. 11. Italy; pt. 12. Yugoslavia, Greece, Albania; pt. 13. Rumania, Bulgaria; pt. 14. General tables.

Volume II. U. S. S. R. (4 pts.).

Volume III. Asia. Pt. 1. China; pt. 2. Korea, Manchuria; pt. 3. Japan, Ryu-Kyu; pt. 4. Taiwan; pt. 5. Philippines; pt. 6. Indochina, Malaya, Thailand, Burma; pt. 7. Indonesia, British Borneo, Malayan Archipelago; pt. 8. India, Pakistan, Nepal, Bhutan, Burma, Ceylon; pt. 9. Afghanistan, Iran, Turkey, Cyprus; pt. 10. Arabian Peninsula; pt. 11. General tables.

Volume IV. Africa. Pt. 1. Morocco, Algeria, Tunisia; pt. 2. French West Africa, Sahara; pt. 3. British West Africa; pt. 4. Libya, Egypt, Sudan; pt. 5. Ethiopia, Somalies; pt. 6. French Equitorial Africa; pt. 7. Congo, Ruanda-Urundi, Angola; pt. 8. Kenya, Uganda, Tanganyika; pt. 9. Nyasaland, Rhodesias; pt. 10. Mozambique, Union of South Africa, South West Africa; pt. 11. Madagascar.

Volume V. Latin America. Pt. 1. Mexico; pt. 2. Central America, Antilles, Cuba; pt. 3. Venezuela; pt. 4. Columbia; pt. 5. Equador, Peru; pt. 6. Bolivia; pt. 7. Chile; pt. 8. Argentina; pt. 9. Uruguay, Paraguay, Falklands; pt. 10. Brazil, Guianas; pt. 11. General tables.

Volume VI. Oceania. Pt. 1. General; pt. 2. Oceania (Central); pt. 3. Oceania (Marginal); pt. 4. New Zealand; pt. 5. Australia; pt. 6. Antartica.

Volume VII. North America. Pt. 1. United States (Published also as U. S. Geological Survey Bulletin no. 1200, 1966. 3 v.); pt. 2. Canada.

Volume VIII. Major stratigraphic systems. Issued thus far: Infracambrian, Pennsylvanian, and Aptian.

HANDBOOKS

Handbooks specific to a discipline are listed in the appropriate subject section.

Clark, Sydney P., Jr., ed. Handbook of physical constants. Rev. ed. New York: Geological Society of America, 1966. 587 p. (Memoir no. 97.)
Compilation of a wide variety of physical constants needed by geologists and geophysicists.

Handbook of chemistry and physics; a ready reference book of chemical and physical data. Cleveland: Chemical Rubber Co., 1913- . Annual.
An important reference tool, useful to geologists.

TRANSLATIONS

National Translations Center. Consolidated index of translations into English. New York: Special Libraries Association, 1969. 948 p.
A list of serial publications which have been translated into English, consolidated from a number of standard and select sources published in previous years (to 1966). For an example, selected sources for geology include Geochemistry international (1964-65) and International geology review (1959-66). This volume is supplemented by:

Translations register-index. Chicago: National Translations Center, 1967- . Semi-monthly.
The National Translations Center is the principal U.S. depository for unpublished translations into English from world scientific literature. The center currently services about 150,000 translations prepared by government agencies, industry, professional societies, academic institutions and individuals. Translations register-index cites new acquisitions of the center. NTC has just established international exchange with the European Translation Centre in Delft, the Netherlands (publisher of the World list of scientific translations. Quarterly from 1967) and the National Lending Library of Great Britain

(publisher of the NLL translations bulletin. 1960-).

TRANSLATION JOURNALS IN THE GEOLOGICAL SCIENCES

Cover-to-cover translation journals and journals which contain translated selected papers in the geological sciences are published by the American Geological Instititute and the American Geophysical Union.

1. International geology review. Washington: American Geological Institute, 1959- . Monthly. "Papers selected from a large number of Russian geological periodicals, and sometimes from the Chinese, Japanese and other geological literature are published in English. Non-serial works are reviewed and on occasion significant portions translated and published in IGR. Pre-publication announcement of geologically important Soviet literature is a regular feature."--Preface.

2. Geochemistry international. Washington: American Geological Institute in cooperation with the Geochemical Society, 1964- . Bi-monthly. Publications in English of selected papers from Geokhimiya and other topical papers from world literature.

3. Paleontological journal. Washington: American Geological Institute, 1967- . Quarterly. Translation of the Russian Paleontologiskii zhurnal and distributed with 6 months of the original.

4. Akademiia Nauk SSSR. Doklady. Earth sciences section. Washington: American Geological Institute, 1959- . Bi-monthly. Reports of basic research in geology, geochemistry, geophysics, lithology, mineralogy, petrography, hydrogeology, oceanography and paleontology.

5. Akademiia Nauk SSSR. Izvestiya. Physics of the solid earth. Washington: American Geophysical Union, 1957- . Monthly. From 1957-1964 translated as Izvestiya. Geophysics series. From 1965- , as above.

6. Akademiia Nauk SSSR. Izvestiya. Atmospheric and ocean physics. Washington: American Geophysical Union, 1957- . Monthly. From 1957-1964 translated as Izvestiya. Geophysics series. From 1965- , as above.

7. Oceanology. Washington: American Geophysical Union, 1965- . Monthly. Translation of the leading Russian oceanographic journal, Okeanologiya.

8. Soviet hydrology: selected papers. Washington: American Geophysical Union, 1961- . 6/yr. Important articles from Soviet hydrological journals.

9. Geotectonics. Washington: American Geophysical Union, 1967- . Bi-monthly. Cover-to-cover translation of Geotektonika, articles on tectonics and structural geology.

TECHNICAL WRITING

There are many books about scientific and technical writing. Two aimed specifically at geologists are:

Low, Julian W. and Jules Braunstein. "Preparation of a technical article." In: American Association of Petroleum Geologists. Bulletin. v. 48, no. 11, 1964. p. 1837-1846.

U.S. Geological Survey. Suggestions to authors of the reports of the United States Geological Survey. 5th ed. Washington: Government Printing Office, 1958. 255 p.

SPECIAL SUBJECTS

GEOLOGICAL EDUCATION

American Geological Institute. Directory of geoscience departments, United States and Canada. Washington: American Geological Institute, 1949- . Biennial.

Heller, Robert L., ed. Geology and earth sciences sourcebook for elementary and secondary schools. New York: Holt, Rinehart and Winston, 1962. 496 p.
Being revised.

Journal of geological education. Washington: National Association of Geology Teachers, 1953- . 5/yr.
Index for years 1960-1969 in v. 18, no. 5, 1970.

Pangborn, Mark W., Jr. A buying list of 100 good geology books for the high school library. Published by the Geoscience Information Society and the National Association of Geology Teachers and distributed in cooperation with the National Science Teachers Association and Lamar State College of Technology, 1967. 7 p.
Reprinted, with revision, in the Journal of geological education. v. 16, no. 2, 1968. p. 65-68.

Proctor, Cleo V., Jr. "Paperback books for earth science teachers." In: Journal of geological education. v. 15, no. 1, 1967. p. 29-55.

U.S. National Council on Marine Resources and Engineering. University curricula in the marine sciences and related fields, academic years 1969-70 and 1970-71. Prepared by the Marine Sciences Affairs Staff, Office of the Oceanographer, U.S. Navy. Washington, 1969. 273 p.

Wall, Charles A. and Janet E. Wall. "Where can I find ... resources and references in earth sciences." In: Journal of geological education. v. 18, no. 4, 1970. p. 178-183.
First of a continuing series of installments to help teachers find sources of information on curriculum improve-

ment, curriculum projects, periodicals, texts and classroom materials.

COUNCIL ON EDUCATION IN THE GEOLOGICAL SCIENCES (CEGS)

Address: American Geological Institute, 2201 M Street, N.W., Washington, D.C. 20037.

One of several NSF sponsored college commissions, CEGS applies its work to the whole spectrum of undergraduate geological education. Not only does CEGS concentrate on the examination and development of the college geology curriculum, but it also strives to upgrade the professional standards of the college professor. To accomplish the latter, CEGS sponsors a coordinated series of published papers, symposiums and short courses which present the results of current scientific work. CEGS publications program supports the work of the council and consists of:

1. CEGS Contribution series. Articles on undergraduate geological instruction published irregularly in the Journal of geological education.

2. CEGS Short Review series. A continuing series of review articles in the Journal of geological education, each by a specialist, which cover selected geologic topics. An example: Morgan, James P. "Deltas--a resume." In: Journal of geological education. v. 18, no. 3, 1970. p. 107-119.

3. CEGS Program Publication series. Issued thus far: 1. Problems in physical geology; 2. Annotated bibliography of statistical applications in geology; 3. Fund sources for undergraduate geology; 4. Audio-tutorial instruction, a strategy for teaching introductory college geology; 5. Requirements in the field of geology.

4. CEGS Newsletter. 10/yr.

5. Short course lecture notes. An example: New concepts of continental margin sedimentation. Washington: American Geological Institute, 1969. 1 v.

6. Concepts in introductory geology, a series of undergraduate instructional units published by McGraw-Hill under the guidance of CEGS. The four modules, which will be

published initially are: 1. Formation of magmatic igneous rocks. Richard Berry; 2. Dimensions of a planet. O. T. Hayward; 3. Slope stability studies. Raymond Pestrong; 4. Seismology and the study of the earth's interior. Henry Pollack.

EARTH SCIENCE CURRICULUM PROJECT (ESCP)

Address: P.O. Box 1559, Boulder, Colorado 80302.

As a major federally-funded project from 1963-1970, ESCP carried out its mission to improve secondary school education in the earth and environmental sciences by developing an interdisciplinary and investigative approach to teaching earth science. The effectiveness of the program has been evaluated in: Champlin, Robert F. "A review of the research related to ESCP." In: Journal of geological education. v. 18, no. 1, 1970. p. 31-39.

Beginning in 1970, two new projects developed by the ESCP staff, will, in effect, supercede ESCP in Boulder: 1. Earth Science Teacher Preparation Project (pre-service and in-service preparation of earth science teachers) and 2. Environmental Studies Project (developing materials and techniques for helping urban students study their environment).

Publications of ESCP are:

1. Earth Science Curriculum Project. Reference series. Edited by William H. Matthews, III. Washington: distributed by the American Geological Institute, 1964- . RS-1. Sources of earth science information. William H. Matthews, III. 48 p.; RS-2. Selected references for earth science courses. William H. Matthews, III. 34 p.; RS-3. Selected earth science films. Wakefield Dort, Jr. 34 p.; RS-4. Selected maps and earth science publications for the states and provinces of North America. William H. Matthews, III. 42 p.; RS-5. Free materials for earth science teachers. William H. Matthews, III and Rolland B. Bartholomew. 25 p.; RS-6. Planetariums, observatories and earth science exhibits. William H. Matthews, III. 33 p.; RS-7. Topographic maps and how to use them. Malcolm P. Weiss. 20 p.; RS-8. Basic data and water budget computation for selected cities in North America. Douglas B. Carter. 36 p.; RS-9. Selected guides for geologic field study in Canada and the United States of America. Donald H. Lokke. 56 p.; RS-10. Weather maps--how they are made and used. In

press; RS-11. Selected earth science references. In preparation; RS-12. Geologic maps. In preparation.

2. Earth Science Curriculum Project. Investigating the earth. Boston: Houghton Mifflin Co., 1967. 594 p. Teacher's guide. 2 v. 1967. The textbook/laboratory manual developed for secondary schools. Being revised.

3. Earth Science Curriculum Project. Pamphlet series. Edited by Robert E. Boyer. Boston: Houghton Mifflin Co., 1971. PS-1. Field guide to rock weathering. Robert E. Boyer; PS-2. Field guide to soils. Henry Foth and Hyde S. Jacobs; PS-3. Field guide to layered rocks. Tom Freeman; PS-4. Field guide to fossils. James R. Beerbower; PS-5. Field guide to plutonic and metamorphic rocks. William D. Romey; PS-6. Color of minerals. George Rapp, Jr.; PS-7. Field guide to beaches. John H. Hoyt; PS-8. Field guide to lakes. Jacob Verduin; PS-9. Field guide to astronomy, without a telescope. William A. Dexter; PS-10. Meteorites. Carleton B. Moore.

4. The Environmental Studies Project, successor to the ESCP, will issue (1971) a pamphlet series, numbers 1-5 entitled, Change, Mapping, Counting, Judging and an overview number. They will be distributed from the ESP, Boulder.

5. The Earth Science Teachers Preparation Project, successor to the ESCP, will publish a pamphlet series on teacher education.

MATHEMATICAL GEOLOGY, COMPUTERS, MODELS

Agterberg, F. P. "Computer techniques in geology." In: Earth Science reviews. v. 3, no. 1, 1967. p. 47-77.

Computer applications in the earth sciences; an international symposium. Edited by Daniel F. Merriam. New York: Plenum Press, 1969. 281 p.

Computer contributions. Lawrence, Kansas: Kansas Geological Survey, no. 1, 1966- . Irregular.
Cover computer applications to geology, simulation and provide actual programs.

Geocom bulletin; international abstracts and current information on mathematical geoscience and computer technology.

London: Geosystems, v. 1, 1968- . Monthly.

Griffiths, J. C. "Current trends in geomathematics." In: Earth science reviews. v. 6, no. 2, 1970. p. 121-140.

Harbaugh, John W. and Daniel F. Merriam. Computer applications in stratigraphic analysis. New York: Wiley, 1968. 282 p.

---------- and Graeme Bonham-Carter. Computer simulation in geology. New York: Wiley, 1970. 575 p.

Howard, James C. Bibliography of statistical applications in geology. Washington: Council on Education in the Geological Sciences, American Geological Institute, 1968. 24 p. (C.E.G.S. Programs Publication no. 2.)

International Association for Mathematical Geology. Journal. New York: Plenum, v. 1, 1969- . Frequency to be determined.

Koch, George S., Jr. and Richard F. Link. Statistical analysis of geological data. New York: Wiley, v. 1, 1970. 375 p.

Krumbein, W. C. and F. A. Graybill. Introduction to statistical models in geology. New York: McGraw-Hill, 1965. 512 p.

Loudon, T. Victor. "Computing techniques in geology." In: Endeavour. v. 29, no. 108, 1970. p. 125-128.

Miller, Robert L. and James S. Kahn. Statistical analysis in the geological sciences. New York: Wiley, 1962. 483 p.

Models of geologic processes; an introduction to mathematical geology. AGI/CEGS Short Course Lecture Notes. Edited by Peter Fenner. Washington: American Geological Institute, 1969. 1 v.

Romanova, M. A. and O. V. Saramanov, eds. Topics in mathematical geology. Translated by J. P. Fitzsimmons. New York: Consultants Bureau, 1970. 281 p.

Smith, Frederick G. Geological data processing using Fortran IV. New York: Harper and Row, 1966. 284 p.

Vistelius, Andrew B. Studies in mathematical geology. New York: Consultants Bureau, 1967. 294 p.

FIELD GEOLOGY

Bradley, E. F. and O. T. Denmead, eds. The collection and processing of field data. New York: Wiley, 1967. 597 p.
Sections on hydrology, oceanography, use of computers and data processing, and acquisitions systems.

Compton, Robert R. Manual of field geology. New York: Wiley, 1962. 378 p.

Forrester, James D. Principles of field and mining geology. New York: Wiley, 1946. 647 p.

Geikie, James. Structural and field geology. 6th ed. Edinburgh: Oliver and Boyd, 1953. 397 p.

Geologic field techniques. New York: Holt, Rinehart and Winston, 1965- . Issued to date in this series: Lattman, Laurence H. and Richard G. Ray. Aerial photographs in field geology. 1965. 221 p.; Kottlowski, Frank E. Measuring stratigraphic sections. 1965. 253 p.

Lahee, Frederic H. Field geology. 6th ed. New York: McGraw-Hill, 1961. 926 p.

Low, Julian W. Geologic field methods. New York: Harper, 1957. 489 p.

----------. Plane table mapping. New York: Harper, 1952. 365 p.

REMOTE SENSING, PHOTOGRAMMETRY, PHOTOGEOLOGY

Allum, J. A. E. Photogeology and regional mapping. New York: Pergamon, 1966. 107 p.

American Society of Photogrammetry. Manual of color aerial photography. Washington, 1960. 868 p.

Avery, Thomas E. Interpretation of aerial photographs. 2nd ed. Minneapolis: Burgess, 1968. 324 p.

Bostock, H. S. A catalogue of selected photographs. Ottawa: Queen's Printer, 1968. 163 p. (Canada. Geological Survey. Paper no. 67-48.)
Identifies photographs, available from the National Air Photographic Library and Geological Survey, which are representative of geomorphologic features in Canada.

Colwell, R. N. "Remote sensing of natural resources." In: Scientific American. v. 218, no. 1, 1968. p. 54-69.

Crone, D. R. Elementary photogrammetry. New York: Ungar, 1968. 197 p.

Eitel, Dean F. and others. Selected bibliography of terrain sciences. 4th ed. Alexandria, Virginia: Raytheon Co., 1970. 96 p.
Stresses photo-interpretive techniques as applied to geology, geomorphology, soils, vegetation and land use.

Fezer, F. "Photo interpretation applied to geomorphology; a review." In: Photogrammetria. v. 27, no. 1, 1971. p. 7-53.

Introduction to electromagnetic remote sensing, with emphasis on applications to geology and hydrology. AGI/CEGS Short Course Lecture Notes. Washington: American Geological Institute, 1968. 1 v.

Journal of remote sensing. Sacramento: International Remote Sensing Institute, v. 1, 1970- . Bi-monthly.
In v. 1: "Remote sensing of the earth's surface" by George Rabchevsky, in three parts.

Lattman, Laurence H. and Richard G. Ray. Aerial photographs in field geology. New York: Holt, Rinehart and Winston, 1965. 221 p. (Geologic field techniques.)

Llaverias, R. K. Remote sensing bibliography for earth resources. Springfield, Va.: National Technical Information Service, 1970. 136 p. (PB 192 863.)

Lowman, Paul D. "Space photography, a review." In: Photogrammetric engineering. v. 31, no. 1, 1965. p. 76-86.

---------- and others. "Orbital photography for geological studies." A special issue of Photogrammetria. v. 24, no. 3/4, 1969. p. 77-174.

Leuder, Donald R. Aerial photographic interpretation: principles and applications. New York: McGraw-Hill, 1959. 462 p.

Miller, Victor C. "Current trends in photogeology and in the use of other remote sensing methods in geological interpretation." In: Earth science reviews. v. 4, no. 2, 1968. p. 135-152.

----------. Photogeology. New York: McGraw-Hill, 1961. 248 p.

Moffitt, Francis H. Photogrammetry. 2nd ed. Scranton, Pa.: International Textbook Co., 1967. 540 p.

National Aeronautics and Space Administration. Science and Technology Information Facility. Remote sensing of earth resources. Springfield, Va.: National Technical Information Service, 1970. 1221 p. (NASA SP-7036.)

Paylore, Patricia. "Eye in the sky." In: Special Libraries Association. Geography and Map Division. Bulletin. No. 80, June, 1970. p. 2-15.
State-of-the-art bibliography of remote sensing publications since 1966.

Ray, Richard G. Aerial photographs in geologic interpretation and mapping. Washington: Government Printing Office, 1960. 230 p. (U.S. Geological Survey. Professional Paper no. 373.)

Remote sensing of environment. New York: American Elsevier, v. 1, 1969- . Quarterly.

Seminar on New Horizons in Color Aerial Photography, New York, 1969. New horizons in color aerial photography; a seminar. Falls Church, Va.: American Society of Photogrammetry; Washington: Society of Photographic Scientists and Engineers, 1969. 423 p.

Sinha, Evelyn. Oceanography from space and aircraft, an annotated bibliography. La Jolla, California: Ocean Engineering Information Service, 1970. 79 p. (Ocean Engineering Information Series, v. 2.)

Smith, John T., Jr. and Abraham Anson. Manual of color aerial photography. Falls Church, Va.: American Society of Photogrammetry, 1968. 550 p.

Spurr, Stephen H. Photogrammetry and photo-interpretation, with a section on applications to forestry. 2nd ed. New York: Ronald, 1960. 472 p.

St. Joseph, J. K. S. The uses of air photography; nature and man in a new perspective. New York: John Day, 1966. 166 p.

Steiner, Dieter. "Annotated bibliography of bibliographies on photo interpretation and remote sensing." In: Photogrammetria. v. 26, no. 4, 1970. p. 143-161.

Strandberg, Carl H. Aerial discovery manual. New York: Wiley, 1967. 249 p.

Symposium on remote sensing of environment, University of Michigan. Proceedings. Ann Arbor; Center for Remote Sensing Information and Analysis, Institute of Science and Technology, 1962- . Irregular.
Important series of publications on remote sensing.

Thompson, Morris M. and others. Manual of photogrammetry. 3rd ed. Falls Church, Va.: American Society of Photogrammetry, 1966. 1199 p.

U.S. Geological Survey. A descriptive catalog of selected aerial photographs of geologic features in the United States. Charles S. Denny and others. Washington: Government Printing Office, 1968. 79 p. (Professional Paper no. 590.)
Stereoscopic pairs of photographs representative of geologic features available from the Survey.

----------. A descriptive catalog of selected aerial photographs of geologic features in areas outside the United States. Charles R. Warren and others. Washington: Government Printing Office, 1969. 23 p. (Professional Paper no. 591.)

Von Bandet, Horst F. Aerogeology. Houston: Gulf Publishing Co., 1962. 350 p.

Subject Section

EARTH SCIENCE, PHYSICAL GEOGRAPHY

This section is purposefully brief and only cites selected references which might be used in beginning courses.

REFERENCE WORKS

Bertin, Léon. Larousse encyclopedia of the earth. Translated by R. Bradshaw and M. M. Owen. New York: Prometheus, 1961. 419 p.

Columbia Lippincott gazeteer of the world. Edited by Leon Seltzer. Morningside Heights, N. Y.: Columbia University Press, 1962. 2148 p.

Earth Science Curriculum Project. Sources of earth science information. William H. Matthews, III. Englewood Cliffs, N. J.: Prentice-Hall, 1964. 48 p. (Reference Series, RS-1.) (Available from American Geological Institute, Washington, D. C.)

Geographical abstracts. Norwich, England: Geo Abstracts, 1966- . 6/yr.
Issued in 4 sections--Geomorphology; Climatology, biogeography, and cartography; Economic geography; and Social geography and cartography.

Harris, Chauncy D. Annotated world list of selected current geographical serials in English. 2nd ed. Chicago: University of Chicago, Department of Geography, 1964. 32 p. (Research Paper no. 96.)

Macmillan world gazeteer and geographical dictionary. Edited by T. C. Collocott and J. O. Thorne. New York: Macmillan, 1961. 796 p.

Monkhouse, F. J. A dictionary of geography. 2nd ed. Chicago: Aldine, 1970. 378 p.

Moore, Winifred G. Dictionary of geography. Revised ed. New York: Praeger, 1967. 246 p.

New geographical literature and maps. London: Royal Geographical Society, 1918- . Monthly.
Worldwide in scope.

Schneider, Allen A. and others. A dictionary of basic geography. Boston: Allyn and Bacon, 1970. 299 p.

Sealock, Richard B. and Pauline A. Seeley. Bibliography of place-name literature, United States and Canada. 2nd ed. Chicago: American Library Association, 1967. 352 p.

Stamp, L. Dudley. Dictionary of geography. London: Longmans, 1966. 429 p.
A selected bibliography of geographical books, p. 483-492.

United States. Board on Geographic Names. Gazetteers. Washington: Government Printing Office, 1955- .
A world-wide series of gazeteers of official standard names, their coordinates, identification and source.

Wall, Charles A. and J. E. Wall. "Where can I find resources and references in earth sciences." In: Journal of geological education. v. 18, 1970. p. 178-183.

Webster's geographical dictionary. Springfield, Massachusetts: G. and C. Merriam Co., 1963. 1293 p.

INTRODUCTORY TEXTS

Birot, Pierre. General physical geography. Translated by M. Ledesert. London: Harrap, 1966. 360 p.

Finch, Vernor C. and others. The earth and its resources. 3rd ed. New York: McGraw-Hill, 1959. 584 p.

Foundations of earth science series. Dubuque, Iowa: W. C. Brown, 1968- .
In progress. Series includes: 1. Fossils, paleontology and evolution. D. L. Clark; 2. Stratigraphy and geologic time. J. W. Harbaugh; 3. Astronomy and the origin of the earth. T. G. Mehlin; 4. Geophysics, geologic structures and tectonics. J. S. Sumner; 5. Applied

earth science. D. S. Turner; 6. The geosystem: dynamic integration of land, sea and air. G. R. Rumney; 7. Landforms and landscapes. S. D. Tuttle; 8. Historical geology of North America. M. Petersen and others; 9. Pleistocene glaciation and the coming of man. W. N. Melhorn; 10. Earth materials. H. Wenden.

Foundations of earth science series. Englewood Cliffs, N. J.: Prentice-Hall, 1968- .
In progress. Series includes: 1. Ancient environments. L. F. Laporte; 2. The atmosphere. R. M. Goody; 3. Earth materials. W. G. Ernst; 4. Geologic time. D. L. Eicher; 5. The history of life. A. L. McAlester; 6. Oceans. K. K. Turekian; 7. The solar system. J. A. Wood; 8. Structure of the earth. S. P. Clark; 9. The surface of the earth. A. L. Bloom.

Hodge, Paul W. Concepts of the universe. New York: McGraw-Hill, 1969. 125 p.

Hurley, Patrick M., ed. Advances in earth science; Contributions to the International Conference on the Earth Sciences, 1964. Cambridge: M.I.T. Press, 1966. 502 p.

Lake, Philip. Physical geography. 4th ed. Edited by J. A. Steers. Cambridge, England: Cambridge University Press, 1958. 483 p.

Lounsbury, John F. and L. Ogden. Earth science. New York: Harper and Row, 1969. 372 p.

Matthews, William H., III. Science probes the earth: new frontiers of geology. New York: Sterling, 1969. 176 p.

Monkhouse, F. J. Principles of physical geography. 6th ed. London: University of London, 1966. 576 p.

National Academy of Sciences/National Research Council. Committee on Resources and Man. Resources and man: a study and recommendations. San Francisco: Freeman, 1969. 259 p.
Eight authorities in different scientific fields summarize a two-year study of world resources.

Ordway, Richard J. Earth science. Princeton, N. J.: Van Nostrand, 1966. 705 p.

Patton, Clyde P. and others. Physical geography. Belmont, California: Wadsworth, 1970. 408 p.

Pearson, Ross N. Physical geography. New York: Barnes and Noble, 1968. 242 p.

Scientific American Resource Library. Readings in the earth sciences. San Francisco: Freeman, 1969. 2 v.

Skinner, Brian J. Earth resources. Englewood Cliffs, N. J.: Prentice-Hall, 1969. 150 p.

Spencer, Edgar W. Geology--a survey of earth science. New York: Crowell, 1965. 655 p.

Strahler, Arthur N. The earth sciences. New York: Holt, Rinehart and Winston, 1963. 681 p.

----------. Physical geography. 3rd ed. New York: Wiley, 1969. 733 p.

Trewartha, Glenn T., A. H. Robinson, and E. H. Hammond. Fundamentals of physical geography. New York: McGraw-Hill, 1968. 384 p.

METEOROLOGY, CLIMATOLOGY

Only a few books are cited here--typical of those which might be used in a beginning meteorology course.

REFERENCE WORKS

American Meteorological Society. Committee on the Compendium of Meteorology. Compendium of meteorology. Edited by Thomas F. Malone. Boston, 1951. 1334 p.

Fairbridge, Rhodes W., ed. The encyclopedia of atmospheric sciences and astrogeology. New York: Reinhold, 1967. 1200 p.

Glossary of meteorology. Edited by Ralph E. Huschke. Boston: American Meteorological Society, 1959. 638 p.

Meteorological and geoastrophysical abstracts. Boston: American Meteorological Society, 1950- . Monthly. Covers worldwide literature on meteorology, oceanography, hydrology, geophysics, astrophysics. Classed arrangement with monthly and annual author, subject, and geographical indexes. Until 1960 this publication was entitled Meteorological abstracts and bibliography.

"A Selective bibliography in meteorology." In: Weatherwise, v. 22, no. 4, 1969. p. 134-151.
A biennial feature of Weatherwise, arranged by topics as popular, elementary texts, history and paleoclimatology, advanced texts, special subjects, climatology, handbooks and workbooks, observing and periodicals.

TEXTS AND TREATISES

Ahlman, H. W. Glacier variations and climatic fluctuations. New York: American Geographical Society, 1953. 51 p.

Blair, Thomas A. and R. C. Fite. Weather elements. 5th ed. New York: Prentice-Hall, 1965. 364 p.
A text in elementary meteorology.

Cole, Franklyn W. Introduction to meteorology. New York: Wiley, 1970. 389 p.

Fairbridge, Rhodes W., ed. Solar variations, climatic change, and related geophysical problems. New York: New York Academy of Sciences, 1961. 739 p. (Annals v. 95, Article 1.)

Longley, Richmond W. Elements of meteorology. New York: Wiley, 1970. 310 p.
An excellent all inclusive text for those with only an introductory mathematics and physics background.

Lusignan, Bruce and J. Kiely, eds. Global weather prediction: the coming revolution. New York: Holt, Rinehart and Winston, 1970. 307 p.
Includes a brief history of meteorology and new methods of weather forecasting.

McIntosh, D. H. and A. S. Thom. Essentials of meteorology. London: Wykeham, 1969. 238 p. (Wykeham Science Series for Schools and Universities.)

Petterssen, Sverre. Introduction to meteorology. 3rd ed. New York: McGraw-Hill, 1969. 333 p.

Taylor, George F. Elementary meteorology. Englewood Cliffs, N.J.: Prentice-Hall, 1954, 8th printing 1965. 364 p.

Trewartha, Glenn T. The earth's problem climates. Madison, Wisconsin: University of Wisconsin Press, 1961. 334 p.

World survey of climatology. Edited by H. E. Landsberg. New York: American Elsevier, 1969- .
A set of 15 volumes composed of chapters by specialists will cover general climatology and climates of major geographic areas.

OCEANOGRAPHY

(See also Marine Geology, Mineral Resources of the Sea)

REFERENCE WORKS

American Society for Testing and Materials. Materials performance and the deep sea; a 1968 symposium. Philadelphia, 1969. 146 p. (STP no. 445.)

Deep sea research and oceanographic abstracts. New York: Pergamon, 1953- . Bimonthly.
Contains current papers on general oceanography and related subjects, and a separate oceanographic bibliography of titles and references to current papers on oceanography, entered under subject headings.

Diner, D. Computer programs in oceanography. Washington: National Oceanographic Data Center, 1970. 143 p. (Publication no. C-5.)

Fairbridge, Rhodes W., ed. The encyclopedia of oceanography. New York: Reinhold, 1966. 1021 p. (Encyclopedia of Earth Sciences Series, v. 1.)
Excellent articles usually with appended references.

Firth, F. E., ed. The encyclopedia of marine resources. New York: Van Nostrand-Reinhold, 1969. 740 p.

Forbes, Lynn. Oceanography in print; a selected list of educational resources. Falmouth, Mass.: Town of Falmouth Public Schools, 1968. 58 p. (Available from Oceanographic Education Center, Falmouth.)
An annotated bibliography of books, periodicals, films, maps, and catalogues, and bibliographies in the English language. The books are divided into sections based on degree of technicality, and these sections are in turn divided into general oceanography, physical oceanography, marine biology, marine geology, chemical oceanography, marine meteorology, and exploration and adventure.

Hunt, Lee M. and D. G. Groves, eds. A glossary of ocean science and undersea technology terms. Arlington, Va.:

Compass Publications, 1965. 172 p. (Appendix of charts and tables.)

Huxley, Anthony. Standard encyclopedia of the world's oceans and islands. New York: Putnam, 1962. 383 p.

Leenhardt, Oliver. Bibliography on marine seismics. Published for and with assistance of National Oceanographic Data Center. Washington, D.C.: American Meteorological Society, 1967. 95 p. (Special Bibliography on Oceanography Contribution, no. 3.)

Marine science affairs. Washington: Government Printing Office, 1967- . Annual.
National Council on Marine Research and Engineering Development's annual report, covering activities and accomplishments of all Federal agencies in the field of marine sciences.

National Council on Marine Resources and Engineering Development. University curricula in the marine sciences and related fields for the academic years 1969-70 and 1970-71. Washington: Government Printing Office, 1969. 273 p.

National Research Council. Committee on Oceanography. International directory of oceanographers. 4th ed. Washington: National Academy of Sciences, 1964.

----------. A directory of oceanographers in the United States. Compiled by Richard C. Vetter. Washington: National Academy of Sciences, 1969. 72 p.

----------. Oceanography information sources. Compiled by Richard C. Vetter. Washington: National Academy of Sciences-National Research Council, 1970. 51 p.

----------. Oceanography 1966. Washington: NAS-NRC, 1967. 183 p.
Review of current status of oceanography and recommendations for government support.

North, Jeannette P. Annotated acronyms and abbreviations of marine science related international organizations. Washington, D.C.: National Oceanographic Data Center, 1969. 115 p.

Oceanic index. La Jolla: Oceanic Research Institute, 1964-

This is an index by author and subject to Oceanic citation journal with abstracts which have been issued separately since 1968. The index is issued at irregular intervals and the citation journal is issued 9 times a year. In the latter there is a section on patents, contracts, organizational profiles, special publications, and news items.

Ocean research index: a guide to ocean and fresh water research, including fisheries research. Guernsey, England: Hodgson, 1970. 507 p.

Sea-Bed 1968. Indexed and annotated by Harry N. Winton. New York: Worldmark Press, 1970. 6 v.
250 United Nations documents concerning the ocean.

Underwater journal. Guilford, England: Iliffe Science and Technology Publications, 1971- . Bimonthly.
Includes a citations section and a patents abstracts section, covering all aspects of applied research and development in the underwater environment. From 1969-1971, the citations section was issued as Underwater bulletin, and the Underwater journal was issued separately.

U.S. Navy. Oceanographic Office. Glossary of oceanographic terms. 2nd ed. Washington: Government Printing Office, 1966. 204 p. (Special Publication no. SP-35.)

----------. Handbook of oceanographic tables, 1966. Washington: Government Printing Office, 1967. 427 p. (Special Publication no. 68.)

----------. Selected reading in the marine sciences. Washington, D.C.: Government Printing Office, 1969. 32 p. (Special Publication no. 129.)
Interest level varies from juvenile to technical.

Wiegel, Robert L. Waves, tides, currents, and beaches; glossary of terms and list of standard symbols. New York: The Engineering Foundation, Council on Wave Research, 1953. 113 p.

INTRODUCTORY TEXTS

Bascom, Willard. Waves and beaches; the dynamics of the ocean surface. Garden City, N. Y.: Anchor Books, 1964. 267 p.

Behrman, Daniel. The new world of the oceans: men and oceanography. Boston: Little, Brown, 1969. 436 p.

Bigelow, Henry B. Oceanography: its scope, problems, and economic importance. Boston: Houghton Mifflin, 1931. 263 p.

Clemons, Elizabeth. Waves, tides, and currents. New York: Alfred A. Knopf, 1967. 112 p.

Cotter, Charles H. The physical geography of the oceans. New York: Elsevier, 1965. 317 p.

Cox, Donald W. Explorers of the deep. Maplewood, N. J.: Hammond, 1968. 93 p.
Eighteen sketches of pioneers in oceanography.

Dietrich, Gunter. General oceanography, an introduction. Translated from German by F. Ostapoff. New York: Interscience, 1963. 588 p.

Ericson, David B. and G. Wollin. The ever-changing sea. New York: Knopf, 1967. 354 p.

Gaskell, T. F. World beneath the oceans. Garden City, N. J.: The Natural History Press, 1964. 154 p.

King, Cuchlaine A. M. An introduction to oceanography. New York: McGraw-Hill, 1963, c1962. 337 p.

MacMillan, D. H. Tides. New York: American Elsevier, 1966. 240 p.

McLellan, Hugh J. Elements of physical oceanography. New York: Pergamon, 1965. 150 p.

Maury, M. F. The physical geography of the sea. Edited by John Leighly. Cambridge, Mass.: Belknap Press, 1963, c1861.
Historic interest-written by the "Father of oceanography."

Miller, Robert C. The sea. New York: Random House, 1966. 315 p.

Ommanney, F. D. The ocean. 2nd ed. London: Oxford University Press, 1961. 238 p.

Pickard, George L. Descriptive physical oceanography. New York: Pergamon, 1964. 200 p.

Proudman, J. Dynamical oceanography. New York: Wiley, 1952. 409 p.

Ross, David A. Introduction to oceanography. New York: Appleton-Century-Crofts, 1970. 400 p.

Spar, Jerome. Earth, sea and air: a survey of the geophysical sciences. Reading, Mass.: Addison-Wesley, 1965. 152 p.

Von Arx, William S. An introduction to physical oceanography. Reading, Mass.: Addison-Wesley, 1962. 422 p.

Weyl, Peter K. Oceanography. New York: Wiley, 1970. 544 p.

Williams, Jerome. Oceanography: an introduction to the marine sciences. Boston: Little, 1962. 242 p.

---------- and others. Sea and air; the naval environment. Annapolis: United States Naval Institute, 1968. 338 p. Interdisciplinary approach to the processes and interrelations of the oceans and the atmosphere.

Zim, Herbert S. Waves. New York: William Morrow and Co., 1967. 63 p. Elementary discussion of waves with good diagrams and illustrations.

TREATISES

Capurro, Luis R. A. Oceanography for practicing engineers. New York: Barnes and Noble, 1970. 175 p.

Defant, Albert. Physical oceanography. New York: Pergamon, 1961. 2 v. Extensive bibliography in each volume.

Emery, K. O. A coastal pond; studied by oceanographic methods. New York: American Elsevier, 1969. 80 p. An account of an unusual ecological system.

Fomin, Luch M. Dynamic method in oceanography. Translated by Scripta Technica, Inc. Edited by T. Winterfeld. New York: Elsevier, 1964. 212 p.

Groen, Pier. The waters of the sea. Translated by M. Hollander. Princeton, N. J.: Van Nostrand, 1967. 328 p.

Harvey, H. W. Chemistry and fertility of sea waters. London: Cambridge University Press, 1966. 240 p.

Hidy, George M. The waves; the nature of sea in motion. New York: Van Nostrand Reinhold, 1971. 150 p.

Hill, M. N. and Arthur E. Maxwell, eds. The sea; ideas and observations on progress in the study of the seas. New York: Interscience, 1962-70. 4 v. A definitive reference work which updates the classic work by Sverdrup, The oceans. V. 1. Physical oceanography; v. 2. Composition of sea-water--comparative and descriptive oceanography; v. 3. The earth beneath the sea--history; v. 4. pts. 1 and 2. New concepts of sea floor evolution.

Horne, R. A. Marine geochemistry: the structure of water and the chemistry of the hydrosphere. New York: Wiley, 1969. 568 p.

International Oceanographic Congress, New York, 1959. Oceanography. Edited by Mary Sears. Washington: American Association for the Advancement of science, 1961. 654 p. (AAAS Publication no. 67.) Five parts: History of the oceans, Populations of the sea, The deep sea, Boundaries of the sea, Cycles of organic and inorganic substances in the ocean.

Ippen, A. T. Estuary and coastline hydrodynamics. New York: McGraw-Hill, 1966. 650 p.

Jerlov, Nils G. Optical oceanography. New York: Elsevier, 1968. 199 p.

King, Cuchlaine, A. M. Beaches and coasts. London: Arnold, 1959. 403 p.

Kinsman, Blair. Wind waves: their generation and propagation on the ocean surface. Englewood Cliffs, N. J.: Prentice-Hall, 1965. 676 p.

Lawrence, L. George. Electronics in oceanography. Indianapolis: Howard W. Sams, 1967. 288 p.

Martin, Dean F. Marine chemistry. New York: M. Dekker, 1968.
V. 1. Analytical methods. v. 2. Theory--in preparation.

Menard, Henry W. Anatomy of an expedition. New York: McGraw-Hill, 1969. 255 p.
"An account of an oceanographic expedition from its conception through the work at sea."

Murray, John and J. Hjort. The depths of the ocean: a general account of the modern science of oceanography based largely on the scientific researches of the Norwegian steamer Michael Sars in the North Atlantic. London: Macmillan, 1912. 821 p. (reprinted by Hafner). Classic work.

National Research Council. Committee on Oceanography and Committee on Ocean Engineering. An oceanic quest: the International Decade of Ocean Exploration. Washington: National Academy of Sciences, 1969. 115 p. (Publication no. 1709.)

Neumann, Gerhard. Ocean currents. New York: Elsevier, 1968. 351 p.

The ocean. San Francisco: Freeman, 1969. 140 p.
Papers originally appeared in September 1969 issue of Scientific American. A wide variety of topics covered, such as the origin of the oceans, geological structure, and food resources.

Oceanography and marine biology; an annual review. London: Allen and Unwin, 1963- .

Progress in oceanography. Edited by Mary Sears, R. W. Stewart, and T. Braarud. New York: Pergamon, 1963- .

Riley, J. P. and G. Skirrow, eds. Chemical oceanography. London: Academic, 1965. 2 v.
Comprehensive text for this field.

Robinson, Allan R. Wind-driven ocean circulation. Waltham, Mass.: Blaisdell, 1963. 161 p.

Russell, Robert C. H. and D. H. Macmillan. Waves and tides. New York: Philosophical Library, 1953. 348 p.

Stommel, Henry. The Gulf stream: a physical and dynamical description. 2nd ed. Berkeley: University of California Press, 1966. 248 p.

Sverdrup, H. U., M. W. Johnson, and R. H. Fleming. The oceans: their physics, chemistry, and general geology. New York: Prentice-Hall, 1962. 1087 p.
A classic reference book.

Thorne, Jim. The underwater world; a survey of oceanography today. New York: Crowell, 1969. 160 p.

Tricker, R. A. R. Bores, breakers, waves and wakes. New York: Elsevier, 1965. 250 p.

Worzel, J. Lamar, M. Ewing, and C. L. Pekeris. Propagation of sound in the ocean. Boulder, Colorado: Geological Society of America, 1963, c 1948. 229 p. (Memoir no. 27.)

INTRODUCTORY GEOLOGY

PHYSICAL GEOLOGY

Bates, Robert L. and Walter C. Sweet. Geology, an introduction. Boston: Heath, 1966. 367 p.

Cailleux, André. Anatomy of the earth. New York: McGraw-Hill, 1969. 255 p.

Eardley, Armand J. General college geology. New York: Harper and Row, 1965. 499 p.

Foster, Robert J. General geology. Columbus, Ohio: Merrill, 1969. 630 p.

Gilluly, James and others. Principles of geology. 3rd ed. San Francisco: Freeman, 1968. 687 p.

Gorshkov, G. and A. Yakushova. Physical geology. Translated by A. Gurevich. Moscow: Mir Publishers, 1967. 586 p.

Holmes, Arthur. Principles of physical geology. 2nd ed. New York: Ronald, 1965. 1288 p.
A textbook based on this treatise is Holmes, Doris L. Elements of physical geology. New York: Ronald Press, 1969, 288 p.

Johnson, Arvid M. Physical processes in geology. San Francisco: Freeman, 1970. 577 p.

Leet, Lewis Don and Sheldon Judson. Physical geology. 4th ed. Englewood Cliffs, N. J.: Prentice-Hall, 1971. 704 p.

Longwell, Chester R., Richard F. Flint and John E. Sanders. Physical geology. New York: Wiley, 1969. 685 p.

Putnam, William C. Geology. New York: Oxford University Press, 1964. 496 p.

Rogers, John J. W. and John A. S. Adams. Fundamentals of geology. New York: Harper and Row, 1966. 424 p.

Shelton, John S. Geology illustrated. San Francisco: Freeman, 1966. 434 p.

Shimer, John A. This changing earth; an introduction to geology. New York: Harper and Row, 1968. 233 p.

Stokes, William L. and Sheldon Judson. Introduction to geology, physical and historical. Englewood Cliffs, N. J.: Prentice-Hall, 1968. 530 p.

Termier, Henri and Geneviève Termier. Traité de géologie. Paris: Masson, 1952- .
Includes: Histoire géologique de la biosphère. 1952; L'évolution de la lithosphère. 1956-1961. 4 v.; Les temps fossilifères. 1964- .

Verhoogen, John and others. The earth; an introduction to physical geology. New York: Holt, Rinehart, and Winston, 1970. 760 p.
Analytical treatment with emphasis on geophysics and geochemistry.

White, J. F., ed. Study of the earth; readings in geological science. Englewood Cliffs, N. J.: Prentice-Hall, 1962. 408 p.

HISTORICAL GEOLOGY

Clark, Thomas H. and C. W. Stearn. Geological evolution of North America. 2nd ed. New York: Ronald Press, 1968. 570 p.
A physical and regional approach to historical geology.

Dott, Robert H. and R. L. Batten. Evolution of the earth. New York: McGraw-Hill, 1971. 649 p.

Dunbar, Carl O. and K. M. Waage. Historical geology. 3rd ed. New York: Wiley, 1969. 562 p.

Kummel, Bernhard. History of the earth; an introduction to historical geology. 2nd ed. San Francisco: Freeman, 1970. 707 p.

Miller, William J. An introduction to historical geology. 6th ed. Englewood Cliffs, N. J.: Van Nostrand, 1952. 555 p.

Moore, Raymond C. Introduction to historical geology. 2nd ed. New York: McGraw-Hill, 1958. 656 p.

Spencer, Edgar W. Basic concepts of historical geology. New York: Crowell, 1962. 472 p.

Stokes, William L. Essentials of earth history. 2nd ed. Englewood Cliffs, N. J.: Prentice-Hall, 1966. 468 p.

Stovall, J. W. and H. C. Brown. The principles of historical geology. Boston: Ginn, 1954. 471 p.

Woodford, Alfred O. Historical geology. San Francisco: Freeman, 1965. 512 p.

MINERALOGY
(See also Ores and Strategic Minerals, Gemology)

REFERENCE WORKS

Aballain, Madeleine and others. Index alphabétique de nomenclature mineralogique. Orleans, France: Bureau de Recherches Géologiques et Minières, 1968. 386 p.
A list of all mineral names, both valid and invalid, compiled while France was making a national inventory of museum mineral collections.

Barnard, Walther M. "Recently published guide books to mineral collecting localities in North America." In: American Mineralogist. v. 54, 1969, p. 1743-1746.

Bradley, J. E. S. and A. C. Barnes. Chinese-English glossary of mineral names. New York: Consultants Bureau, 1963. 120 p.

Chamber's mineralogical dictionary. New York: Chemical Publishing Co., 1957. 47 p.

Dietrich, Richard V. Mineral Tables; hand specimen properties of 1500 minerals. New York: McGraw-Hill, 1969. 237 p.

Ehlers, Curt. Nomina der Kristallographie und Minéralogie. Hamburg: Boysen-Maasch, 1952. 166 p.

Font-Altaba, M. World directory of mineralogists. 2nd ed. Barcelona: Editorial Jover, 1970. Published for the International Mineralogical Association.

France. Centre national de la recherche scientifique. Bulletin signalétique. Paris, 1956- . Monthly.
Section 210. Sciences de la terre I--Minéralogie, géochemie, géologie extraterrestre, pétrographie.

Harrison, R. K. and P. A. Sabine, eds. A petrological-mineralogical code for computer use. London: Her Majesty's Stationery Office, 1970. 134 p. (Institute of

Geological Sciences. Report no. 70/6.)

Hey, Max H. An index to mineral species and varieties arranged chemically, with an alphabetical index of accepted mineral names and synonyms. 2nd ed., revised. London: British Museum (Natural History), 1962. 728 p. Appendix, 1963. 135 p.

International Mineralogical Association. Commission on New Minerals and Mineral Names. "Report by the Commission on new names, rejections, and redefinitions voted by the commission." In: Mineralogical magazine. v. 33, 1962. p. 260-263. Supplement. v. 36, 1967. p. 131-136.

Kirsch, Helmut. Applied mineralogy for engineers, technologists and students. Translated by K. A. Jones. London: Chapman and Hall, 1968. 233 p.
A compendium of terms.

Leake, Bernard E. A catalog of analyzed calciferous and subcalciferous amphiboles together with their nomenclature and associated minerals. Boulder, Colorado: The Geological Society of America, 1968. 210 p. (G.S.A. Special Paper, no. 98.)

Lueschen, H. Die Namen der Steine. Munchen: Ott Verlag, 1968. 384 p.

Mineralogical abstracts. London: Mineralogical Society of Great Britain and Mineralogical Society of America. 1920- . Quarterly.
An abstracting journal, worldwide in scope, in the fields of mineralogy, petrology, sedimentary and structural petrology, clay mineralogy, chemical and physical, and geochemical techniques, and age determinations. From 1920-1958, published as a section of Mineralogical Magazine.

Waldbaum, D. R., ed. A bibliography of the feldspars. Ann Arbor, Michigan: University Microfilms, 1969. 145 p.
Two thousand annotated references on books, theses, papers and abstracts. Emphasis on important older papers and those published in last five years.

Zentralblatt für Mineralogie. Stuttgart: E. Schweizerbart-'sche, 1950- . Irregular.
Pt. 1. Kristallographie und Mineralogie; pt. 2. Petrographie. Technische Mineralogie. Geochemie und Lagerstättenkunde.

GENERAL AND SYSTEMATIC MINERALOGY

Barth, Tom F. W. Feldspars. New York: Wiley, 1969. 261 p.

Borner, Rudolph. Minerals, rocks, and gemstones. Translated by W. Mykura. Edinburgh: Oliver and Boyd, 1962. 250 p.

Cox, K. G., N. B. Price, and B. Harte. An introduction to the practical study of crystals, minerals and rocks. New York: McGraw-Hill, 1970. 256 p.

Dana, James D. The system of mineralogy of James Dwight Dana and Edward Salisbury Dana, Yale University, 1837-1892. 7th ed. Entirely rewritten and greatly enlarged by Charles Palache, H. Berman, and C. Frondel. New York: Wiley, 1944- .
Issued thus far: V. 1. Elements, sulfides, sulfosalts, oxides. 1944. 834 p.; v. 2. Halides, nitrates, borates, carbonates, sulfates, phosphates, arsenates, tungstates, molybdates, etc. 1951. 1124 p.; v. 3. Silica minerals. 1962. 334 p.

Deer, W. A., R. A. Howie, and J. Zussman. An introduction to the rock-forming minerals. New York: Wiley, 1966. 528 p.

----------. Rock-forming minerals. New York: Wiley, 1962-1963. 5 v.
V. 1. Ortho- and ring silicates. 1962. 333 p.; v. 2. Chain silicates. 1963. 379 p.; v. 3. Sheet silicates. 1962. 270 p.; v. 4. Framework silicates. 1963. 435 p.; v. 5. Nonsilicates. 1962. 371 p.

Dennen, William H. Principles of mineralogy. Revised printing with determinative tables. New York: Ronald Press, 1960. 453 p.

Desautels, Paul E. The mineral kingdom. New York: Grosset and Dunlap, 1968. 251 p.

Eitel, Wilhelm. Silicate science. New York: Academic, 1964-1966.
V. 1. Silicate structures. A revision of The physical chemistry of the silicates; v. 2. Glasses, enamels, slags; v. 3. Dry silicate systems; v. 4. Hydrothermal silicate systems; v. 5. Ceramics and hydraulic binders.

Ernst, Wallace G. Amphiboles; crystal chemistry, phase relation and occurrence. New York: Springer-Verlag, 1968. 125 p. (Minerals, Rocks, and Inorganic Materials, v. 1.)

Hurlbut, Cornelius S., Jr. Dana's manual of mineralogy. 18th ed. New York: Wiley, 1971. 579 p.

----------. Minerals and man. New York: Random House, 1968. 314 p.

Korzhinskii, D. S. Physiochemical basis of the analysis of the paragenesis of minerals. New York: Consultants Bureau, 1959. 142 p.

Kraus, Edward H., W. F. Hunt, and L. S. Ramsdell. Mineralogy; an introduction to the study of minerals and crystals. 5th ed. New York: McGraw-Hill, 1959. 686 p.

Loomis, Frederic B. Field book of common rocks and minerals. Revised ed. New York: Putnam, 1948. 352 p.

Marfunin, A. M. The feldspars--phase relations, optical properties, and geological distribution. Translated from the Russian by J. Kolodny. Jerusalem: Israel Program for Scientific Translation, 1966. 317 p. (IPST no. 1266.)

Mason, Brian H. and L. G. Berry. Elements of mineralogy. San Francisco: Freeman, 1968. 550 p.

Maxwell, John A. Rock and mineral analysis. New York: Wiley, 1969. 584 p. (Chemical Analysis Series, v. 28.)

Mineralogical Society of America. Special papers. Washington, 1963- .
1. International Mineralogical Association. 3rd, 1962. Proceedings. 1963; 2. Pyroxenes and amphiboles:

crystal chemistry and phase petrology. Edited by James J. Papike. 1969; 3. Fiftieth anniversary symposia. 1969.

Morrissey, C. J. Mineral specimens. New York: Elsevier, 1968. 202 p.

Pohopien, K. M. Field manual--an introduction to the megascopic study and determination of minerals and rocks. 3rd ed. Dubuque, Iowa: W. C. Brown, 1969. 144 p.
A book of tables.

Progress in the science and technology of the rare earths. Edited by Leroy Eyring. New York: Pergamon, 1964- . 3 v. to date.

Rutley, Frank. Elements of mineralogy. 25th ed. Revised by H. H. Read. London: Allen and Unwin, 1962. 525 p.

Schrocke, Helmut and Karl L. Weiner. Mineralien. New York: Francis Paul, 1967.
Part 1. 162 unbound color plates with mineral descriptions.

Schüller, Arno. Die Eigenschaften der Minerale; ein Lehr-- und Hilfsbuch zum Bestimmen von mineralischen Rohstoffen und Kristallformen. Berlin: Akademie-Verlag, 1954-1960.
V. 1. 5th ed., 1960; v. 2. 1954.

Simpson, Brian. Rocks and minerals. New York: Pergamon, 1966. 293 p.

Smith, O. C. Identification and qualitative chemical analysis of minerals. 2nd ed. Princeton, N. J.: Van Nostrand, 1953. 385 p.

Strunz, Hugo. Mineralogische Tabellen; eine Klassifizierung der Mineralien auf Kristallschemischer Grundlage. Mit einer Einfuhrung in die Kristallchemie. 4th ed. Leipzig: Geest and Portig, 1966. 448 p.

Vanders, Iris and Paul F. Kerr. Mineral recognition. New York: Wiley, 1967. 316 p.

Voskuil, Walter H. A geography of minerals. Dubuque, Iowa: W. C. Brown, 1969. 117 p.

CRYSTALLOGRAPHY

Donnay, J. D. H., ed. Crystal data: determinative tables. 2nd ed. Washington: American Crystallographic Association, 1963. 1302 p. (ACA Monograph no. 5.) Includes bibliography and bibliographic references.

International Union of Crystallography. World directory of crystallographers. 3rd ed. Utrecht, The Netherlands: A. Oosthoekds Uitgevers Mij N. V., 1965. A 4th edition is in preparation.

International Union of Crystallography. Commission on Crystallographic Computing. World list of crystallographic computer programs. Edited by David P. Shoemaker. 2nd ed. Cambridge: Massachusetts Institute of Technology, 1966. 52 p.

International Union of Crystallography. Commission on Crystallographic Teaching. Crystallographic book list. Edited by Helen Megaw. Brooklyn: Distributed by Polycrystal Book Service, 1965. 83 p.

* * *

Acta crystallographica. Copenhagen: Munksgaard (for the International Union of Crystallography), 1948- . Beginning 1968, Section A. Crystal physics, diffraction, theoretical and general crystallography. 6/yr.; Section B. Structural crystallography and crystal chemistry. 12/yr. A special issue dedicated to Paul P. Ewald on his eightieth birthday is v. A24, pt. 1 and v. B24, pt. 1, 1968.

Amorós, J. L. and P. Krishna. Polymorphism and polytupism in crystals. New York: Wiley, 1966. 341 p.

Amorós, Luis José and M. Amorós. Molecular crystals: their transforms and diffuse scattering. New York: Wiley, 1968. 479 p.

Azaroff, Leonid V. Introduction to solids. New York: McGraw-Hill, 1960. 460 p.

Bloss, F. Donald. Crystallography and crystal chemistry. New York: Holt, Rinehart and Winston, 1970. 256 p.

Bragg, Sir Wm. H. and W. L. Bragg, eds. The crystalline state. London: Bell, 1933-1965.
V. 1. A general survey. W. L. Bragg. 1933. 366 p.; v. 2. The optical principles of X-rays. R. W. James. 2nd ed. 1963. 640 p.; v. 3. Determination of crystal structure. H. S. Lipson and W. Cochran. 2nd ed. 1957. 356 p.; v. 4. Crystal structures of minerals. L. Bragg and G. F. Claringbull. 1965. 409 p.

Buckley, Harold E. Crystal growth. New York: Wiley, 1951. 571 p.

Buerger, Martin J. Contemporary crystallography. New York: McGraw-Hill, 1970. 380 p.

----------. Elementary crystallography; an introduction to the fundamental geometrical features of crystals. New York: Wiley, 1956. 528 p.

----------. Vector space, and its application in crystal-structure investigations. New York: Wiley, 1969. 347 p.

Burke, John G. Origins of the science of crystals. Berkeley, California: University of California Press, 1966. 198 p.
History of the study of crystallography.

Burns, Roger G. Mineralogical applications of crystal field theory. London: Cambridge University Press, 1970. 224 p.

Correns, Carl W. Introduction to mineralogy: crystallography and petrology. 2nd ed. Translated from the German edition by W. D. Jones. Berlin: Springer-Verlag, 1969. 485 p.

Evans, Robert C. An introduction to crystal chemistry. 2nd ed. Cambridge, England: Cambridge University Press, 1964. 410 p.

Frank F. C., J. B. Mullin, and H. S. Peiser, eds. Crystal growth 1968; Proceedings of the second International Conference on Crystal Growth, Birmingham, 1968. Amsterdam: North Holland, 1969. 862 p.

Higgs, Donald V. Angular relations of lines and planes. 2nd ed. San Francisco: Freeman, 1966. 43 p.
"Basic graphical and numerical methods for the solution of geologic and crystallographic problems."

Jong, Wieger F. de. General crystallography; a brief compendium. San Francisco: Freeman, 1959. 281 p.

Papike, J. J., ed. Pyroxenes and amphiboles: crystal chemistry and phase petrology. Washington: Mineralogical Society of America, 1969. 314 p. (Special Paper no. 2.)

Phillips, Frank C. An introduction to crystallography. 3rd ed. New York: Wiley, 1963. 340 p.

Ramachandran, G. N. and R. Srinivasan. Fourier methods in crystallography. New York: Wiley, in press.

Sands, Donald. Introduction to crystallography. New York: Benjamin, 1969. 165 p.

Schmid E. and W. Boas. Plasticity of crystals. London: Chapman and Hall, 1968. 353 p.
With particular reference to metals.

Sheftal, N. N., ed. Growth of crystals. New York: Consultants Bureau, 1958- .
Volumes 1 and 3 are the proceedings of the first conference on crystal growth held in 1956. 7 volumes have been published to date.

Wade, Franklin A. and R. B. Mattox. Elements of crystallography and mineralogy. New York: Harper, 1960. 332 p.

Wyckoff, Ralph W. G. Crystal structures. 2nd ed. New York: Wiley, 1963- .
6 v. to date.

Zeitschrift für kristallographie; kristallgeometrie, kristallphysik, kristallchemie. Frankfurt: Akademische Verlagsgesellschaft, 1877- . Bimonthly.

OPTICAL MINERALOGY

Bloss, F. Donald. An introduction to the methods of optical crystallography. New York: Holt, Rinehart and Winston, 1961. 294 p.

Hallimond, A. F. The polarizing microscope. 3rd ed. York, England: Vickers Instruments, 1970. 302 p.

Hartshorne, N. H. and S. Stuart. Crystals and the polarizing microscope. 4th ed. New York: American Elsevier, 1970. 614 p.

----------. Practical optical crystallography. 2nd ed. New York: American Elsevier, 1970. 335 p.
Includes a listing of some major manufacturers and suppliers of polarizing microscopes.

Heinrich, Eberhardt W. Microscopic identification of minerals. New York: McGraw-Hill, 1965. 414 p.

McCrone, W. C., R. G. Draftz, and J. G. Delly. The particle atlas: a photomicrographic reference for the identification of particulate substances. Ann Arbor, Michigan: Ann Arbor Science Publishing Co., 1967. 406 p.
A notable contribution to the compilation of inorganic and organic particles information which will be of particular interest to those working in the field of environmental geology.

McDougall, D. J., ed. Thermoluminescence of geological materials. New York: Academic, 1968. 678 p.
Pt. 4 includes a classified bibliography, subject and author indexes.

Novitsky, Alejandro. Tablas para la determinación microscopica y rayos X de minerales opacos. 2nd ed. Oruro, Bolivia, 1957. 268 p. (Universidad Técnica de Oruro. Biblioteca de Estudios Técnicos no. 2.)

Oelsner, O. Atlas of the most important ore mineral parageneses under the microscope. Translated by B. J. Hazzard. New York: Pergamon, 1966, c1961. 311 p.

Rath, R. Theoretische grundlagen der allgemeinen kristalldiagnose in durchfallenden licht. Berlin: Springer, 1969.

170 p.
Description of classic method of mineral recognition using the polarization microscope.

Schouten, Cornelis. Determination tables for ore microscopy. Amsterdam: Elsevier, 1962. 242 p.

Shubnikov, A. V. Principles of optical crystallography. New York: Consultants Bureau, 1960. 186 p.

Troger, Walter E. Optische Bestimmung der gesteinsbildenden Minerale. Stuttgart: E. Schweizerbart'sche, 1956-1967. 2 v.

U. S. Geological Survey. The microscopic determination of the nonopaque minerals. E. S. Larsen, Jr. and Harry Berman. 2nd ed. Washington: Government Printing Office, 1934. 266 p. (Bulletin no. 848.)

----------. Microscopic determination of the ore minerals. M. N. Short. 2nd ed. (reprinted). Washington: Government Printing Office, 1940. 314 p. (Bulletin no. 914.)

Uytenbogaardt, W. and E. A. J. Burke. Tables for microscopic identification of ore minerals. 2nd ed. New York: Elsevier, 1971. In press.

Wahlstrom, Ernest E. Optical crystallography. 4th ed. New York: Wiley, 1969. 489 p.

----------. Petrographic mineralogy. New York: Wiley, 1955. 408 p.

Winchell, Alexander N. Elements of optical mineralogy. New York: Wiley, 3 v.
V. 1. Principles and methods. 5th ed. 1937. 263 p.; v. 2. Descriptions of minerals. 4th ed. 1951. 551 p.; v. 3. Determinative tables. 2nd ed. 1939. 231 p.

---------- and Horace Winchell. The microscopical characters of artificial inorganic solid substances: optical properties of artificial minerals. 3rd ed. New York: Academic, 1964. 440 p.

Winchell, Horace. Optical properties of minerals: a determinative table. New York: Academic Press, 1965. 91 p.

METHODOLOGY, X-RAY MINERALOGY

Adams, John A. S. and P. Gasparini. Gamma-ray spectrometry of rocks. New York: Elsevier, 1970. 265 p. (Methods in Geochemistry and Geophysics, v. 10.)

Adler, Isidore. X-ray spectrography in geology. New York: Elsevier, 1966. 258 p. (Methods in Geochemistry and Geophysics, v. 4.)

American Society for Testing Materials. X-ray powder data file. Philadelphia, 1945- . Annual.
Card sets of data on organic and inorganic powder, issued with cumulative indexes.

Angino, Ernest E. and G. K. Billings. Atomic absorption spectrometry in geology. New York: American Elsevier, 1967. 154 p. (Methods in Geochemistry and Geophysics, v. 7.)

Azároff, Leonid V. Elements of X-ray crystallography. New York: McGraw-Hill, 1968. 610 p.

---------- and M. J. Buerger. The powder method in x-ray crystallography. New York: McGraw-Hill, 1958. 342 p.

---------- and R. J. Donahue. Laboratory experiments in x-ray crystallography. New York: McGraw-Hill, 1969. 136 p.

Berry, L. G. and R. M. Thompson. X-ray powder data for ore minerals; the peacock atlas. New York: Geological Society of America, 1962. 281 p. (Memoir no. 85.)

Birks, L. S. X-ray spectrochemical analysis. 2nd ed. New York: Wiley, 1969. 143 p.

Borg, I. Y. and D. K. Smith. Calculated x-ray powder patterns for silicate minerals. Boulder, Colorado: Geological Society of America, 1969. 896 p. (Memoir no. 122.)

Buerger, Martin J. The precession method in x-ray crystallography. New York: Wiley, 1964. 276 p.

----------. X-ray crystallography. New York: Wiley, 1942. 531 p.

Bunn, Charles W. Chemical crystallography; an introduction to optical and x-ray methods. 2nd ed. Oxford, England: Oxford University Press, 1961. 524 p.

Donnay, Gabrielle and others. "Fifty years of x-ray crystallography at the Geophysical Laboratory, 1919-1969." In: Carnegie Institution. Annual report of the Director Geophysical Laboratory. Washington, 1968-1969. p. 278-283.

Fischer, Alfred G., S. Honjo, and R. E. Garrison. Electron micrographs of limestones and their nannofossils. Princeton, N. J.: Princeton University Press, 1967. 141 p.
Application of electron microscopy to the study of mineral grains and fossils in limestones.

Guinier, A. X-ray diffraction in crystals, imperfect crystals, and amorphous bodies. Translated by P. Lorrain and D. St. M. Lorrain. San Francisco: Freeman, 1963. 373 p.

Henry, Norman F. M., H. Lipson and W. A. Wooster. The interpretation of x-ray diffraction photographs. 2nd ed. London: Macmillan, 1961. 282 p.

International Union of Crystallography. International tables for x-ray crystallography. Birmingham, England: Kynoch Press, 1952-1962. 3 v.
V. 1. Symmetry groups; v. 2. Mathematical tables; v. 3. Physical and chemical tables.

Jenkins, L. and J. L. de Vries. Practical x-ray spectrometry. 2nd ed., revised. New York: Springer, 1970. 181 p.

Kortum, G. Reflectance spectroscopy. Translated from the German by J. E. Lohr. New York: Springer, 1969. 366 p.

Robie, Richard A., P. M. Bethka, and K. M. Beardsley. Selected x-ray crystallographic data, molar volumes, and densities of minerals and related substances. Washington: Government Printing Office, 1967. 87 p. (U.S.

Geological Survey Bulletin, no. 1248.)

White, F. A. Mass spectrometry in science and technology. New York: Wiley, 1968. 352 p.

Wilson, A. J. C. Elements of x-ray crystallography. Reading, Mass.: Addison Wesley, 1969. 256 p.

Woolfson, M. M. An introduction to x-ray crystallography. Cambridge, England: Cambridge University Press, 1970. 380 p.

Zussman, J., ed. Physical methods in determinative mineralogy. New York: Academic, 1967. 514 p.

CLAY AND CLAY MINERALS (see also Sedimentology)

Beutelspacher, H. and H. W. van der Marel. Atlas of electron microscopy of clay minerals and their admixtures, a picture atlas. Amsterdam: Elsevier, 1968. 333 p. Selected micrographs of common clay minerals from the sediments of many countries and classical localities. English and German. Reviews current research.

Brown, George, ed. The x-ray identification and crystal structures of clay minerals. 2nd ed. London: Mineralogical Society, 1961. 544 p.

Caillère, Simonne and S. Hénin. Minéralogie des argiles. Paris: Masson, 1963. 355 p.

Carroll, Dorothy. Clay minerals: a guide to their x-ray identification. Boulder, Colo.: Geological Society of America, 1970. 80 p. (Special Paper no. 126.)

Clay and clay minerals. New York: Pergamon, 1968- . Bimonthly.
Formerly: National Conference on Clays and Clay Minerals. Clays and clay minerals; Proceedings. 1952-1968. Annual.

Grim, Ralph E. Clay mineralogy. 2nd ed. New York: McGraw-Hill, 1968. 596 p.

International Clay Conference. Proceedings. 1, 1963- . Proceedings of first conference were published by Pergamon, 1963-65; second and third conference published by Israel Universities Press, Jerusalem, in 1966 and 1969 respectively. The 4th Conference to be held in Madrid 1972.

Kerr, Paul F. and P. K. Hamilton. *Glossary of clay mineral names.* New York: Columbia University Press, 1949. 66 p. (American Petroleum Institute Project 49. Preliminary Report no. 1.)

Levin, Ernest M., C. R. Robbins, and H. F. McMumurdie. *Phase diagrams for ceramists.* Columbus, Ohio: American Ceramic Society, 1964. 601 p.
General discussion of phase diagrams, accompanied by an annotated bibliography, pp. 32-36, on phase equilibria. Continues with compilation of phase diagrams, in particular the oxide-silicate systems. Supplement, 1969.

Millot, George. *Geology of clays; weathering, sedimentology, geochemistry.* Translated by W. R. Farrand and H. Paquet. New York: Springer, 1970.

Olphen, Hendrik van. *An introduction to clay colloid chemistry, for clay technologists, geologists, and soil scientists.* New York: Interscience, 1963. 301 p.

PETROLOGY, PETROGRAPHY

REFERENCE WORKS AND TREATISES

Holmes, Arthur. The nomenclature of petrology, with references to selected literature. 2nd ed. London: Murby, 1928. (1971 Hafner reprint.) 284 p.

Telberg, V. G. Russian-English petrographic dictionary. New York: Telberg Book Corp., 1967. 250 p.

* * *

Barth, Thomas F. W. and others, eds. Die Entstehung der Gesteine, ein Lehrbuch der Petrogenese. Berlin: Springer, 1960, c1939. 422 p.

Bayly, Brian. Introduction to petrology. Englewood Cliffs, N. J.: Prentice Hall, 1968. 320 p.

Fairbairn, Harold W. Structural petrology of deformed rocks. 2nd ed. Cambridge, Mass.: Addison-Wesley, 1949. 344 p.

Geological Society of America. Petrologic studies; a volume in honor of A. F. Buddington. Edited by A. E. J. Engel, H. L. James, and B. F. Leonard. New York: Geological Society of America, 1962. 660 p.

Harker, Alfred. Petrology for students; an introduction to the study of rocks under the microscope. 8th ed., revised by C. E. Tilley and others. Cambridge, England: Cambridge University Press, 1954. 283 p. Paperback edition published in 1960.

Heinrich, Eberhardt W. Microscopic petrography. New York: McGraw-Hill, 1956. 296 p.

Huang, Walter T. Petrology. New York: McGraw-Hill, 1962. 480 p.

Jackson, Kern C. Textbook of lithology. New York: McGraw-Hill, 1970. 552 p.

Jeffrey, P. G. Chemical methods of rock analysis. New York: Pergamon, 1970. 507 p.

Johannsen, Albert. Manual of petrographic methods. New York: Hafner, 1968. c1918. 649 p.

Journal of petrology. Oxford: Clarendon Press, 1960- 3/yr.

Jung, Jean. Précis de pétrographie; roches sédimentaires métamorphique et éruptive. 3rd ed. Paris: Masson, 1969. 332 p.

Knopf, Eleanora B. and Earl Ingerson. Structural petrology. New York: Geological Society of America, 1938. 270 p. (Memoir no. 6.)

Loewinson-Lessing, F. Y. A historical survey of petrology. Translated by S. I. Tomkeieff. Edinburgh: Oliver and Boyd, 1954. 112 p.

Moorhouse, Walter W. The study of rocks in thin section. New York: Harper, 1959. 514 p.

Niggli, Paul. Rocks and mineral deposits. English translation by R. L. Parker. San Francisco: Freeman, 1954. 559 p.

Parkhomenko, Eleonora I. Electrical properties of rocks. Translated by G. V. Keller. New York: Plenum Press, 1967. 314 p.

Pirsson, Louis V. and Adolf Knopf. Rocks and rock minerals. 3rd ed. New York: Wiley, 1947. 349 p.

Ronner, Felix. Systematische Klassifikation der Massengesteine. Wein: Springer, 1963. 380 p.
Excellent bibliography.

Sander, Bruno. Gefügekunde der Gesteine. Berlin: Springer, 1930. 352 p.

----------. An introduction to the study of fabrics of geological bodies. Translated from German by F. C.

Phillips and G. Windsor. New York: Pergamon, 1970. 641 p.
Pt. 1 concerns theory and pt. 2, laboratory investigations and interpretations of grain fabrics. A translation of Sander's 1948 and 1950 texts supplemented by a bibliography updated to 1967.

Spock, Leslie E. Guide to the study of rocks. 2nd ed. New York: Harper, 1962. 298 p.

Travis, Russell B. Classification of rocks. Golden, Colorado: Colorado School of Mines, 1955. 98 p. (C.S.M. Quarterly. v. 50, no. 1.)

Williams, Howel, F. J. Turner, and C. M. Gilbert. Petrography; an introduction to the study of rocks in thin section. San Francisco: Freeman, 1954. 406 p.

IGNEOUS ROCKS, MIGMATITES, ORIGINS

Balk, Robert. Structural behavior of igneous rocks. New York: Geological Society of America, 1937. 177 p. (Memoir no. 5.)

Basalts: the Poldervaart treatise on rocks of basaltic composition. Edited by H. H. Hess and Arie Poldervaart. New York: Interscience, 1967-68. 2 v.

Bowen, Norman L. Evolution of the igneous rocks. Princeton, N. J.: Princeton University Press, 1928. 332 p. (reprinted by Dover, 1956).

Daly, Reginald A. Igneous rocks and the depths of the earth. New York: McGraw-Hill, 1933. 598 p. (reprinted by Hafner, 1968).

Gilluly, James and others. Origin of granite. Ann Arbor, Michigan: Edwards, 1948. 139 p. (Memoir no. 28.)

Hamilton, Warren and W. Bradley Myers. The nature of batholiths. Washington: Government Printing Office, 1967. 30 p. (U.S. Geological Survey Professional Paper no. 554-C.)

Hatch, Frederick H., A. K. Wells and M. K. Wells. The petrology of the igneous rocks. 12th ed. London: Murby, 1961. 515 p.

Heinrich, Eberhardt W. *The geology of carbonatites.* Chicago: Rand McNally, 1966. 555 p.

Isachsen, Yngvar W., ed. *Origin of anorthosite and related rocks; Proceedings of a symposium in Plattsburgh, N.Y., 1966.* Albany, N. Y.: Univ. of State of New York, 1969. 470 p. (N.Y. State Museum and Science Service. Memoir no. 18.)

Johanssen, Albert. *A descriptive petrography of the igneous rocks.* Chicago: University of Chicago Press, 1931-1939. 4 v.
V. 1. Introduction, textures, classifications, and glossary. 1939. (Revised ed.) 318 p.; v. 2. The quartz-bearing rocks. 1932. 428 p.; v. 3. The intermediate rocks. 1937. 360 p.; v. 4. pt. 1. The feldspathoid rocks. pt. 2. The periodotites and perknites. 1938. 551 p.

Larsen, L. H., M. Prinz, and V. Manson, eds. *Igneous and metamorphic geology. A volume in honor of Arie Poldervaart.* Boulder, Colorado: Geological Society of America, 1968. 561 p. (Memoir no. 115.)

McBirnery, A. R., ed. *Proceedings of the andesite conference.* Portland, Oregon: Oregon Dept. of Geology and Mineral Industries, 1969. 193 p. (Bulletin no. 65.)

Marmo, V. *Granite petrology.* New York: Elsevier, in preparation. (Developments in petrology.)

Mehnert, K. R. *Migmatites and the origin of granite rocks.* New York: Elsevier, 1968. 393 p.

Newall, Geoffrey and N. Rast, eds. *Mechanism of igneous intrusion.* Liverpool: Gallery, 1970. 380 p. (*Geological journal.* Special issue no. 2.)

Raguin, E. *Pétrographie des riches plutoniques dans leur Cadre géologique.* Paris: Masson, 1970. 240 p.

Sederholm, Jakob J. *Granites and migmatites: selected papers.* New York: Wiley, 1967. c1935. 608 p.

Shand, Samuel J. *Eruptive rocks; their genesis, composition, classification, and their relation to ore-deposits with a chapter on meteorites.* New York: Hafner, 1969, c1947. 488 p.

Turner, Francis J. and John Verhoogen. Igneous and metamorphic petrology. 2nd ed. New York: McGraw-Hill, 1960. 694 p.

Tuttle, Orville F. and J. Gittings. Carbonatites. New York: Interscience, 1966. 591 p.

---------- and N. L. Bowen. Origin of granite in the light of experimental studies in the system $NaAlSi_3O_8$-$KAlSi_3O_8$-SiO_2-H_2O. Baltimore, Maryland: Waverly Press, 1958. 153 p. (Geological Society of America. Memoir no. 74.)

Wager, Laurence R. and G. M. Brown. Layered igneous rocks. San Francisco: Freeman, 1967. 588 p.

Wahlstrom, Ernest E. Introduction to theoretical igneous petrology. New York: Wiley, 1961. c1950. 365 p.

Wyllie, Peter J. Ultramafic and related rocks. New York: Wiley, 1967. 464 p.

Zavaritskii, A. N. and V. S. Sobolev. The physiochemical principles of igneous petrology. Jerusalem: Israel Program for Scientific Translations, 1964. 414 p.

VOLCANOLOGY

International Volcanological Association. Catalogue of the active volcanoes of the world including Solfatara fields. Naples, 1951- .

----------. Bulletin volcanologique. Naples, 1924- . Quarterly.

"Volcanology, a directory of names and addresses." In: Tectonophysics, v. 5, no. 2, 1968. p. 161-176.

* * *

Bullard, Fred M. Volcanoes in history, in theory, in eruption. Austin: University of Texas Press, 1962. 441 p.

Coats, Robert R., R. L. Hay, and C. A. Anderson, eds. Studies in volcanology. A memoir in honor of Howel

Williams. Boulder, Colorado: Geological Society of America, 1968. 678 p. (Memoir no. 116.)

Gorshkov, Georgii S. Volcanism and the Upper Mantle; investigations in the Kurile Island Arc. Translated from Russian by C. P. Thornton. New York: Plenum, 1970. 385 p.

Green, Jack and N. M. Short. Volcanic landforms and surface features. New York: Springer, 1971. 522 p.

König, M. A. Vulkane und erdbeden. Thun: Ott, 1969. 192 p.

MacDonald, Gordon A. and A. T. Abbott. Volcanoes in the sea; the geology of Hawaii. Honolulu: University of Hawaii Press, 1970. 441 p.
Glossary p. 417-420.

McBirney, Alexander R. "Some current aspects of volcanology." In: Earth science reviews. v. 6, no. 5, 1970. p. 337-352.

Ollier, Clifford D. Volcanoes. Cambridge, Mass.: MIT Press, 1970. 177 p.
(Introduction to Systematic Geomorphology, v. 6.)

Rittman, Alfred. Volcanoes and their activity. Translated from the 2nd German ed. by E. A. Vincent. New York: Interscience, 1962. 305 p.

Thorarinsson, Sigurdur. Surtsey: the new island in the North Atlantic. New York: Viking, 1967. 47 p.

SEDIMENTARY ROCKS (see also Sedimentology)

LaMoreaux, Philip E. and others. Annotated bibliography of carbonate rocks. University, Alabama: Geological Survey of Alabama, 1970. 242 p. (Bulletin no. 94, pt. A.)

Potter, Paul E. "A selective annotated bibliography on carbonate rocks." In: Bulletin of Canadian Petroleum Geology, v. 16, 1968. p. 87-103.

* * *

American Association of Petroleum Geologists. Carbonate Rock Subcommittee. Symposium of Carbonate Rocks. Proceedings.
1st. Ham, W. E., ed. Classification of carbonate rocks. Tulsa: American Association of Petroleum Geologists, 1962. 279 p. (Memoir no. 1); 2nd. Pray, Lloyd C. and R. C. Murray, eds. Dolomitization and limestone diagenesis. Tulsa: Society of Economic Paleontologists and Mineralogists, 1965. 180 p. (Special Publication no. 13); 3rd. Friedman, Gerald, ed. Depositional environments in carbonate rocks. Tulsa: Society of Economic Paleontologists and Mineralogists, 1969. 209 p. (Special Publication no. 14).

Carver, Robert E., ed. Procedures in sedimentary petrology. New York: Wiley, 1971. 653 p.

Carozzi, Albert V. Microscopic sedimentary petrography. New York: Wiley, 1960. 485 p.

Cayeux, Lucien. Sedimentary rocks of France: carbonate rocks (limestones and dolomites). Translated and updated by A. V. Carozzi. Darien, Conn.: Hafner, 1970. 506 p.

Engelhardt, W. von, H. Füchtbauer, and G. Müller, eds. Sediment-Petrologie. Stuttgart: E. Schweizerbart'sche, 1964- .
V. 1. Methoden der Sediment-Untersuchung. G. Müller. 1964. 303 p. Translated as Methods in sedimentary petrology by Hans-Ulrich Schmincke in 1967. 283 p.; v. 2. Sediment und Sedimentgesteine. H. Füchtbauer and G. Müller. 1970. 726 p.

Folk, Robert L. Petrology of sedimentary rocks. Rev. ed. Austin, Texas: Hemphill's, 1968. 170 p.

Garrels, Robert M. and F. T. Mackenzie. Evolution of sedimentary rocks. New York: Norton, 1971. 416 p.

Hatch, Frederick H. and R. H. Rastall. The petrology of the sedimentary rocks. 4th ed. London: Allen and Unwin, 1965. 395 p.

Journal of sedimentary petrology. Tulsa: Society of Economic Paleontologists and Mineralogists, 1931- . Quarterly.

Milner, Henry B. Sedimentary petrography. 4th ed. New York: Macmillan, 1962. 2 v.

Tickell, Frederick G. The techniques of sedimentary mineralogy. Amsterdam: Elsevier, 1965. 220 p. (Developments in Sedimentology, v. 4.)

METAMORPHIC ROCKS

Fyfe, W. S., F. J. Turner and J. Verhoogen. Metamorphic reactions and metamorphic facies. New York: Geological Society of America, 1958. 259 p. (Memoir no. 73.)

Harker, Alfred. Metamorphism, a study of the transformation of rock-masses. 3rd ed. London: Methuen, 1950. 362 p.

Pitcher, Wallace S. and Glenys W. Flinn, eds. Controls of metamorphism. New York: Wiley, 1965. 368 p.
Results of a symposium devoted to a review of recent researches on experimental data as concerns metamorphic crystallization.

Ramberg, Hans. The origin of metamorphic and metasomatic rocks; a treatise on recrystallization and replacement in the earth's crust. Chicago: University of Chicago Press, 1952. 317 p.

Spry, Alan. Metamorphic textures. New York: Pergamon, 1969. 368 p.
Emphasis on recent advances.

Turner, Francis J. Metamorphic petrology; mineralogical and field aspects. New York: McGraw-Hill, 1968. 403 p.

----------. Mineralogical and structural evolution of the metamorphic rocks. New York: Geological Society of America, 1948. 342 p. (Memoir no. 30.)

---------- and John Verhoogen. Igneous and metamorphic petrology. 2nd ed. New York: McGraw-Hill, 1960. 694 p.

Winkler, H. G. E. Petrogenesis of metamorphic rocks. 2nd ed. Translated by N. D. Chatterjee and E. Froese. New York: Springer, 1967. 237 p.

Wynne-Edwards, H. R., ed. Age relations in high-grade metamorphic terrains. Ottawa: Geological Association of Canada, 1969. 228 p. (Special Paper no. 5.) Papers presented at symposium in 1967. New ideas and data to those interested in evolution of the crust.

STRUCTURAL GEOLOGY, TECTONICS

REFERENCE WORKS

Bederke, Erich and H-G. Wunderlich. Atlas zur geologie. Germany: Bibliographisches Institut Mannheim, 1968. 50 p.
Only compilation of its kind--areal geological maps and structural features of the world.

Cloos, Ernst. Lineation, a critical review and annotated bibliography. New York: Geological Society of America, 1946. 122 p. (Memoir no. 18.)
Lineation, review of literature 1942-1952. G. S. A. Memoir no. 18 Supplement, 1953. Reprinted in 1962 as one volume.

Dennis, John G., ed. International tectonic dictionary. English terminology. Tulsa: American Association of Petroleum Geologists, 1967. 196 p. (Memoir no. 7.)
History of usage and meaning of structural terms.

TEXTS AND TREATISES

Akademiia Nauk SSSR. Institut Fiziki Zemili. Folded deformations in the earth's crust, their types and origins. Edited by V. V. Beloussov and A. A. Sorskii. Translated by Z. Garfunkel. Jerusalem: Israel Program for Scientific Publications, 1965. 366 p.
A collection of papers. Available from NTIS, Springfield, Virginia.

Anderson, E. M. The dynamics of faulting and dyke formation with application to Britain. 2nd ed. Edinburgh: Oliver and Boyd, 1951. 216 p.

Aubouin, Jean. Geosynclines. New York: Elsevier, 1965. 335 p. (Developments in Geotectonics, v. 1.)

Badgley, Peter C. Structural and tectonic principles. New York: Harper, 1965. 521 p.

An important reference work synthesizing the study of both structural geology and tectonics in a single text.

----------. *Structural methods for the exploration geologist and a series of problems for structural geology students.* New York: Harper, 1959. 280 p.

Bemmelen, Reinout W. van. *Mountain building; a study primarily based on Indonesia, region of the world's most active crustal deformations.* The Hague: Nijhoff, 1954. 177 p.

Billings, Marland P. *Structural geology.* 3rd ed. New York: Prentice-Hall, 1972. 672 p.

Braunstein, J. and G. D. O'Brien. *Diapirs and diapirism.* Tulsa: American Association of Petroleum Geologists, 1968. 444 p. (Memoir no. 8.)

Bucher, Walter H. *Deformation of the earth's crust; an inductive approach to the problem of diastrophism. . .* . Princeton, N. J.: Princeton University Press, 1933. 518 p. (reprinted by Hafner, 1963).

Daly, Reginald A. *Architecture of the earth.* New York: Appleton-Century, 1938. 211 p.

Dennison, John M. *Analysis of geologic structures.* New York: Norton, 1968. 209 p.

Donn, W. L. and J. A. Shimer. *Graphic methods in structural geology.* New York: Appleton-Century-Crofts, 1958. 180 p.

Goguel, Jean. *Tectonics.* English translation of the 1952 French ed. San Francisco: Freeman, 1962. 384 p. The 2nd French ed. was published in 1965 by Masson, Paris.

Hansen, Edward. *Strain facies.* New York: Springer, 1970. 220 p. (Minerals, Rocks, and Inorganic Materials, v. 2.)
Study of outcrop folds to understand large scale structures.

Hills, Edwin S. *Elements of structural geology.* London: Methuen, 1963. 483 p.

Second edition in preparation.

Illies, J. H. and St. Mueller, eds. Graben problems. Stuttgart: E. Schweizerbart'sche, 1970. 316 p. (International Upper Mantle Scientific Report no. 27.)

Jaeger, John C. Elasticity, fracture and flow. 3rd ed. London: Methuen, 1970. 268 p.

Kent, P. E., G. E. Satterthwaite, and A. M. Spencer, eds. Time and place in orogeny. London: Geological Society of London, 1969. 311 p. (Special Publication no. 3.)

Kober, Leopold. Der Bau der Erde; eine Einführung in die Geotektonik. 2nd ed. Berlin: Borntraeger, 1928. 499 p.

Moore, Carl A. Handbook of subsurface geology. New York: Harper and Row, 1963. 235 p.

Nevin, Charles M. Principles of structural geology. 4th ed. New York: Wiley, 1949. 410 p.

Phillips, Frank C. The use of stereographic projection in structural geology. 2nd ed. London: Arnold, 1960. 86 p.

Platt, John I. and J. Challinor. Single geologic structures. 4th ed. London: Murby, 1968. 56 p.

Price, Neville J. Fault and joint development in brittle and semi-brittle rock. New York: Pergamon, 1966. 192 p.

Ragan, Donal M. Structural geology; an introduction to geometrical techniques. New York: Wiley, 1968. 166 p.

Ramsay, John G. Folding and fracturing of rocks. New York: McGraw-Hill, 1967. 568 p.

Sander, Bruno. Einführung in die Gefügekunde der geologischen Körper. Berlin: Springer, 1948-1950. 2 v.

Scheidegger, Adrian E. Principles of geodynamics. 2nd ed. New York: Academic Press, 1963. 362 p.

Sitter, Lamoraal Ulbo de. Structural geology. 2nd ed. New York: McGraw-Hill, 1964. 551 p.

Spencer, Edgar W. Introduction to the structure of the earth. New York: McGraw-Hill, 1969. 597 p.

Whitten, E. H. T. Structural geology of folded rocks. Chicago: Rand McNally, 1966. 663 p.

ROCK MECHANICS (See Also Engineering Geology)

REFERENCE WORKS

"Bibliography of Canadian contributions in the field of rock mechanics." In: Canadian mining and metallurgical bulletin. v. 57, 1964. p. 540-545; Supplement. v. 58, 1965. p. 320-321; 2nd supplement. v. 59, 1966. p. 1198-1200; 3rd supplement. v. 60, 1967. p. 655-656; 4th supplement. v. 61, 1968. p. 758-759; 5th supplement. v. 62, 1969. p. 647-648.

Burk, C. R. Rock mechanics: a bibliography. Santa Barbara, California: General Motors Corp., 1969. 144 p. 1200 references to books, articles and reports on rock mechanics and its application to mining.

KWIC index of rock mechanics literature, 1870-1969. Compiled by the Rock Mechanics Section, Imperial College of Science and Technology, London. New York: American Institute of Mining, Metallurgical and Petroleum Engineers, 1970. 2 v.

Mitra, Devi and J. W. Willson. A bibliography on geomechanics. Salt Lake City: University of Utah, 1962. 210 p. (Utah. University. Bulletin 53, no. 22.) Covers soil and rock mechanics.

Rock mechanics abstracts. New York: American Institute of Mining, Metallurgical and Petroleum Engineers, 1970- . Quarterly.

U.S. Air Force. Office of Aerospace Research. Cambridge Research Laboratories. Terrestrial Research Laboratories. Bibliography of experimental rock deformation. 2nd ed. R. E. Riecker and others. Bedford, Mass., 1965. 104 p. (AFCRL-65-740.)

----------. Supplement. 1966. 10 p. (AFCRL-65-740-2.)

TEXTS AND TREATISES

Baidyuk, Bronislav V. Mechanical properties of rocks at high temperatures and pressures. Translated from the Russian. New York: Plenum, 1967. 75 p.

Coates, D. F. Rock mechanics principles. Ottawa: Mines Branch, 1967. 363 p. (Monograph no. 874, rev. 1967.)

Griggs, David and John Handin, eds. Rock deformation, a symposium. New York: Geological Society of America, 1960. 382 p. (Memoir no. 79.)

International Conference on State of Stress in the Earth's Crust. State of Stress in the earth's crust. Proceedings. Edited by W. R. Judd. New York: Elsevier, 1964. 732 p.

International journal of rock mechanics and mining sciences. New York: Pergamon, 1964- . Bimonthly.

Jaeger, John and N. G. W. Cook. Fundamentals of rock mechanics. London: Methuen, 1969. 513 p. Extensive references p. 472-502.

National Academy of Sciences. National Research Council. Rock mechanics research. Washington: National Academy of Sciences, 1966. 820 p. (Publication no. 1466.) A survey of U.S. research to 1965 with a partial survey of Canadian universities.

Obert, Leonard and Wilbur I. Duvall. Rock mechanics and the design of structures in rock. New York: Wiley, 1967. 650 p.

Rock mechanics. Fels mechanik. Mécanique des roches; Journal of the International Society for Rock Mechanics. Vienna: Springer, 1963- . Quarterly.

Symposium on rock mechanics. Proceedings. 1st, 1956- . Annual.
Imprint varies.

GLOBAL HYPOTHESES

Blackett, Patrick M. S., E. S. Bullard, and S. K. Runcorn, eds. A symposium on continental drift. London: Royal

Society of London, 1965. 328 p. (Philosophical Transactions. Series A. v. 258, no. 1088.)

Continental drift, a symposium... . Hobart, Tasmania: University of Tasmania, Department of Geology, 1958. 375 p.

Drake, Charles L. The geological revolution. Eugene: Oregon State System of Higher Education, 1970. 55 p. (Condon Lecture Publications.)
Excellent introduction to the new global tectonics.

DuToit, Alexander I. Our wandering continents; an hypothesis of continental drifting... . Edinburgh: Oliver and Boyd, 1937. 366 p. (reprinted by Hafner, 1957).

Garland, George D., ed. Continental drift. Toronto: University of Toronto Press, 1966. 140 p. . (Royal Society of Canada. Special Publication no. 9.)

Hammond, Allen L. "Plate tectonics: the geophysics of the earth's interior." In: Science. v. 173, July 2, 1971. p. 40-41.
A companion article is scheduled.

International Upper Mantle Committee. Continental margins and island arcs. Ottawa: Geological Survey of Canada, 1966. (CGS Paper no. 66-15.)

----------. The world rift system. Ottawa: Geological Survey of Canada, 1966. (CGS Paper no. 66-14.)

Isacks, B. L., J. Oliver and L. R. Sykes. "Seismology and the new global tectonics." In: Journal of geophysical research. v. 73, 1968. p. 5855-5900.

Johnson, Helgi and B. L. Smith, eds. The megatectonics of continents and oceans. New Brunswick, New Jersey: Rutgers University Press, 1970. 282 p.
Papers from 1966 symposium held at Rutgers University.

Kraus, Ernst. Die Entwicklungsgeschichte der Kontinente und Ozeane. Berlin: Akademie-Verlag, 1959. 285 p.

Markowitz, William and B. Guinot, eds. Continental drift, secular motion of the pole and rotation of the Earth. Berlin: Springer, 1968. 107 p.

Meyerhoff, Arthur A. "Continental drift: implications of paleomagnetism studies, meteorology, physical oceanography, and climatology." In: Journal of Geology. v. 78, 1970. p. 1-51.

----------. "Continental drift, II: high latitude evaporite deposits and geologic history of Arctic and North Atlantic oceans." In: Journal of Geology. v. 78, 1970. p. 406-444.
Extensive bibliography. These two papers are the first of a proposed continuing series.

Morgan, W. Jason. "Rises, trenches, great faults and crustal blocks." In: Journal of geophysical research. v. 73, 1968. p. 1959-1982.

Runcorn, S. K., ed. Continental drift. New York: Academic Press, 1962. 338 p.

Society of Economic Paleontologists and Mineralogists. Polar wandering and continental drift, a symposium. Edited by A. C. Munyan. Tulsa, 1963. 169 p. (Special Publication no. 10.)

Sutton, J. Continental drift. Guernsey, England: Hodgson, 1970. 22 p. (Sigma Library of Science Surveys.)

Takeuchi, Hitosni, S. Uyeda, and H. Kanamori. Debate about the earth; approach to geophysics through analysis of continental drift. Revised ed. San Francisco: Freeman, 1970. 282 p.

Wegener, Alfred. The origin of continents and oceans. Translated from the 4th rev. German ed. by John Biran. New York: Dutton, 1924. 246 p. (reprinted by Dover, 1965 and McGraw-Hill, 1970.)

"The world rift system." A special issue of Tectonophysics. v. 8, no. 4-6, 1969. p. 265-586. (International Upper Mantle Committee. Scientific Report no. 19.)

GEOMORPHOLOGY
(For Weathering, See Also Soil Science)

REFERENCE WORKS

Baulig, Henri. Vocabulaire francais-anglo-allemand de géomorphologie. Nouveau tirage. Paris: Les Belles Lettres, 1966. 230 p. (Strasbourg. Université. Fondation Baulig. Publications, v. 4.)

Clayton, Keith M., ed. A bibliography of British geomorphology. London: George Philip, 1964. 200 p. (British Geomorphological Research Group. Occasional Publication no. 1.)
Publications of a general nature published in British journals also included. Covers principally 1945-1962. Titles after 1962 in Geographical abstracts. A. Geomorphology.

Dolan, Robert and J. McCloy. Selected bibliography on beach features and related nearshore processes. Baton Rouge: Louisiana State University Press, 1965. 59 p.

Fairbridge, Rhodes W., ed. The encyclopedia of geomorphology. New York: Reinhold, 1968. 1295 p. (Encyclopedia of Earth Sciences Series, v. 3.)

Freth, John H. Review and annotated bibliography of ancient lake deposits (Precambrian to Pleistocene) in the western states. Washington: Government Printing Office, 1964. 119 p. (U.S. Geological Survey. Bulletin no. 1080.)

Geographical Abstracts. A. Geomorphology. London: London School of Economics, Department of Geography, 1962- . 6/yr.
Worldwide in scope. Arranged by author and region, and indexed in last issue of year. Formerly Geomorphological abstracts, 1960-1962.

McGill, John T. Selected bibliography of coastal geomorphology of the world. Los Angeles, 1960. 50 p.

This bibliography is complemented by those in: (1) Guilcher, André. Coastal and submarine morphology, and (2) Valentin, Harmut. Die Küsten der Erde, which contain numerous references to European literature.

Monroe, Watson H. A glossary of karst terminology. Washington: Government Printing Office, 1970. 36 p. (U.S. Geological Survey. Water Supply Paper no. 1899-K.)

Richards, Horace G. and Rhodes W. Fairbridge. Annotated bibliography of Quaternary shorelines (1945-1964). Philadelphia: Academy of Natural Sciences, 1965. 280 p. (Special Publication no. 6.)

----------. 1965-1969. Philadelphia: Academy of Natural Sciences, 1970. 240 p. (Special Publication no. 10.)

Russell, Richard J. Glossary of terms used in fluvial, deltaic and coastal morphology and processes. Baton Rouge: Louisiana State University Press, 1969. 97 p. (Coastal Studies Series, no. 23.)

Stone, R. O. "A desert glossary." In: Earth-Science Reviews. v. 3, 1967. p. 211-268.

U.S. Department of the Army. Corps of Engineers. Beach Erosion Board. (Publications.)
This agency issues several series of interest to the geomorphologist--technical reports, memorandums, and bulletins pertaining to beaches and shallow water processes.

TEXTS AND TREATISES

American Association for the Advancement of Science. Arid lands in perspective. Tucson: University of Arizona Press, 1969. 421 p.

Bagnold, Ralph A. The physics of blown sand and desert dunes. London: Methuen, 1954, c1942. 265 p.

Bloom, Arthur L. The surface of the earth. Englewood Cliffs, N.J.: Prentice-Hall, 1969. 152 p.

Chorley, Richard J., A. J. Dunnan, and R. P. Beckinsale. The history of the study of landforms, or the development of geomorphology. London: Methuen, 1964- .
V. 1. Geomorphology before Davis. 1964. 678 p.

Cooper, William S. Coastal dunes of California. Boulder, Colorado: Geological Society of America, 1967. 131 p. (Memoir no. 104.)
Together with Cooper's Coastal sand dunes of Oregon and Washington. (Memoir no. 72, reprinted in 1967), the geomorphic history of coastal sand dunes along western coast of United States is presented.

Cotton, Charles A. Geomorphology; an introduction to the study of landforms. 7th ed. London: Whitcombe and Tombs, 1958. 505 p.

Davies, G. L. The earth in decay. New York: Elsevier, 1970. 406 p.
History of geomorphology in a study of British thought on the origin of the earth's topography, 1578-1878.

Derrau, Max. Les formes du relief terrestre. (Notions de géomorphologie). Paris: Masson, 1969. 112 p.

----------. Précis de géomorphologie. 5th ed. Paris: Masson, 1967. 416 p.

Dregne, Harold E. Arid lands in transition. Washington: American Association for the Advancement of Science, 1970. 524 p.

Dury, George H. Essays in geomorphology. New York: Elsevier, 1966. 404 p.

----------. Face of the earth. Baltimore: Penguin Books, 1959. 223 p.

----------. Rivers and river terraces. London: Macmillan, 1970. 283 p.

Flemal, Ronald C. "The attack on the Davisian system of geomorphology: a synopsis." In: Journal of geological education. v. 19, no. 1, 1971. p. 3-13.

Guilcher, André. Coastal and submarine morphology. Translated by B. W. Sparks and R. H. W. Kneese. London: Methuen, 1958. 274 p.

Harvey, A. M., ed. Geomorphology in a tropical environment. Nottingham, England: British Geomorphological Research Group, 1969. 96 p. (Occasional Paper no. 5.)

Hunt, Charles B. *Physiography of the United States*. San Francisco: Freeman, 1967. 480 p.

International Association for Quaternary Research. *Loess and related eolian deposits of the world*. Edited by C. Bertrand Schultz and J. C. Frye. Lincoln: University of Nebraska Press, 1968. 369 p. (INQUA Congress VII. Proceedings, v. 10.)

Introduction to systematic geomorphology series. Canberra, Australia: Australia National University Press, 1969- .

This series is also being published by the M.I.T. Press in Cambridge, Massachusetts. Series includes: 1. *Humid landforms*. I. Douglas. 1971; 2. *Desert and savanna landforms*. J. A. Mabbutt. 1970; 3. *Landforms of cold climates*. J. L. Davies. 1969; 4. *Coasts*. E. C. F. Bird. 1969; 5. *Structural landforms*. C. R. Twidale. 1970; 6. *Volcanoes*. Cliff Ollier. 1969; 7. *Karst*. J. N. Jennings. 1970.

Johnson, Douglas W. *Shore processes and shoreline development*. New York: Hafner, 1965, c1919. 584 p.
A classic work.

King, Cuchlaine A. M. *Techniques in geomorphology*. New York: St. Martin's Press, 1966. 342 p.

King, Lester C. *The morphology of the earth; a study and synthesis of world scenery*. 2nd ed. Edinburgh: Oliver and Boyd, 1967. 726 p.
Bibliography p. 683-709. The evolution of the earth's crust principally from the Mesozoic, its landscapes and tectonics, and continental drift. Three maps are included on the morphology of the Atlantic, Pacific, and Indian Ocean basins.

----------. *South African scenery; a textbook of geomorphology*. 3rd ed. Edinburgh: Oliver and Boyd, 1963. 308 p.

Leopold, Luna B., M. G. Wolman, and J. P. Miller. *Fluvial processes in geomorphology*. San Francisco: Freeman, 1964. 522 p.

McGinnies, William G., B. J. Goldman, and P. Paylore, eds. *Deserts of the world; an appraisal of research in-*

to their physical and biological environments. Tucson: University of Arizona Press, 1968. 788 p. Excellent bibliography.

Machatschek, Fritz. Geomorphology. Edited by H. Graul and C. Rathjens. Translation of the 9th edition by D. J. Davis. New York: Elsevier, 1969. 212 p.

----------. Das Relief der Erde. 2nd ed. Berlin: Borntraeger, 1955. 2 v.

Neal, James T., ed. Playa surface morphology: miscellaneous investigations. AFCRL Environmental Research Paper, no. 283, 1968. 150 p. Available from NTIS, Springfield, Va. (AD669 074.)

Penck, Walther. Morphological analysis of land forms. London: Macmillan, 1953. 429 p.

Scheidegger, Adrian E. Theoretical geomorphology. 2nd ed., revised. New York: Springer, 1970. 435 p.

Scovel, James and others. Atlas of landforms. New York: Wiley, 1966. 164 p.

Shepard, Francis P. Our changing coastlines. New York: McGraw-Hill, 1971. 512 p.

Small, R. J. Study of landforms; a textbook of geomorphology. London: Cambridge University Press, 1970. 486 p.

Sparks, B. W. Geomorphology. New York: Wiley, 1960. 371 p.

Steers, James A. Coasts and beaches. Edinburgh: Oliver and Boyd, 1969. 136 p.

Suess, Eduard. The face of the earth. (Das Antlitz der Erde.) Translated by Hertha B. C. Sollas. Oxford: Clarendon Press, 1904-1924. 5 v.

Thornbury, W. D. Principles of geomorphology. 2nd ed. New York: Wiley, 1969. 594 p.

----------. Regional geomorphology of the United States. New York: Wiley, 1965. 609 p.

Tricart, Jean. Principes et methodes de la géomorphologie. Paris: Masson, 1965. 496 p.

Upton, William B., Jr. Landforms and topographic maps. New York: Wiley, 1970. 148 p.
Illustrates the landforms of the United States through the use of three-color, contoured topographic maps.

Valentin, Harmut. Die Küsten der Erde; Beiträge zur allgemeinen und regionalen Küstenmorphologie. 2nd ed. Gotha: VEB GeographischKartographische Anstalt, 1954. 118 p. (Petermann Geographischen Mitteilungen. Ergänzungsheft, no. 246, 2. aufl.)

Wagret, Paul. Polderlands. New York: Barnes and Noble, 1968. 288 p.

Zenkovich, V. P. Processes of coastal development. Edited by J. A. Steers. Translated by D. G. Fry. New York: Interscience, 1967. 738 p.

GLACIOLOGY, PERIGLACIAL FEATURES

American Geographic Society, New York. Department of Exploration and Field Research. Glaciology collection. Boston: G. K. Hall, 1971. 3 v.

Armstrong, Terence E., B. Roberts, and C. Sithinbank. Illustrated glossary of snow and ice. Cambridge, England: Scott Polar Research Institute, 1966. 60 p. (Special Publication no. 4.)

Hamelin, Louis Edmond and Frank A. Cook. Le périglaciaire par l'image. Quebec: Presses de l'Université Laval, 1967. 237 p. (Travaux et Documents du Centre d'Études Nordiques, no. 4.)
An illustrated glossary of periglacial phenomena in French and English.

"Reports of the Commission on Periglacial Morphology of I. G. U." In: Biuletyn Peryglacjalny. no. 14, 1964. p. 1-110.
The reports contain excellent bibliographies to recent studies in periglacial morphology, as well as summations of current research. The following countries are covered: United States, 1959-63; Canada, 1954-63; Poland; Italy, 1956-63; Belgium, 1956-63; Scandinavia, 1960-63; North

Africa; Czechoslovakia, 1955-63; Great Britain. Periglacial nomenclature is discussed in the same bulletin.

U.S. Department of the Army. Corps of Engineers. Cold Regions Research and Engineering Laboratory. Bibliography on snow, ice, and frozen ground with abstracts. Hanover, N. H., 1951- . Annual.
Worldwide in scope. Available from the NTIS, Springfield, Virginia. The many publications (research reports, technical reports, etc.) issued by this agency make its recognition a valuable asset in the field of technology.

U.S. Department of the Navy. Hydrographic Office. A functional glossary of ice terminology. Washington, 1952. (Publication no. 609.)

* * *

Agassiz, Louis. Studies on glaciers. Translated and edited by Albert V. Carozzi. New York: Hafner, 1967. 213 p.

Ahlmann, H. W. Glacier variations and climatic fluctuations. New York: American Geographical Society, 1953. 51 p.

Coleman, Arthur Philemon. Ice ages, recent and ancient. New York: AMS Press, 1969, c1926. 269 p.

Daly, Reginald. The changing world of the ice age. New York: Hafner, 1963, c1934. 271 p.

Dyson, James L. The world of ice. New York: Knopf, 1962. 292 p.
Good popular treatment.

Embleton, Clifford and C. A. M. King. Glacial and periglacial geomorphology. New York: St. Martin's Press, 1968. 386 p.
Emphasis on landforms and processes of erosion and deposition by ice.

Fletcher, N. H. The chemical physics of ice. London: Cambridge University Press, 1970. 271 p.

Flint, Richard F. Glacial and Quaternary geology. New York: Wiley, 1971.

Journal of glaciology. Cambridge, England: The Glaciological Society, 1947- . 3-4/yr.
Contains as a regular feature "Glaciological literature," divided into sections on conferences, general glaciology, glaciological instruments and methods, physics of ice, land ice, icebergs, glacial geology, frost action on rocks and soil, meteorological and climatological glaciology, and snow.

Kingery, W. D., ed. Ice and snow; properties, processes, and applications. Cambridge, Mass.: M.I.T. Press, 1963. 684 p.

Klebelsberg, R. von. Handbuch der Gletscherkunde und Glazialgeologie. Vienna: Springer, 1948-1949. 2 v.

Lilboutry, L. Traité de glaciologie. Paris: Masson, 1964- .
V. 1. Glacé, neige, hydrologie nivale; v. 2. Glaciers, variations du climat, sols gelés.

Paterson, W. S. B. The physics of glaciers. New York: Pergamon, 1969. 250 p.

Péwé, Troy L., ed. The perigalcial environment: past and present. Montreal: McGill University Press, 1969. 287 p.

Sharp, Robert P. "Glacier flow: a review." In: Geological Society of America. Bulletin. v. 65, 1954. p. 821-838.

----------. Glaciers. Eugene: Oregon State System of Higher Education, 1960. 78 p.

Tricart, Jean. Geomorphology of cold environments. Translated by E. Watson from the revised, abridged 1963 edition of Traité de géomorphologie. New York: St. Martin's Press, 1970. 320 p.
Divided into 1. existing glaciers and cold environments; 2. periglacial processes and landforms; and 3. glacial processes and landforms.

MARINE GEOLOGY
(See also Geomorphology, Oceanography and Sedimentology)

REFERENCE WORKS

California. University at San Diego. Library. Catalogs of the Scripps Institution of Oceanography. Boston: G. K. Hall, 1970. 12 v.
Library's specialty is oceanography, marine biology, marine technology, and oceanographic expeditions, but there are also substantial holdings in geology, physics, chemistry, biology, and the atmospheric sciences.

Fairbridge, Rhodes W., ed. The encyclopedia of oceanography. New York: Reinhold, 1966. 1021 p.
Excellent articles with appended references.

TEXTS AND TREATISES

Clements, Thomas, R. E. Stevenson and D. M. Haimos, eds. Essays on marine geology in honor of K. O. Emery. Los Angeles: University of Southern California Press, 1963. 210 p.

Daly, Reginald. Floor of the ocean. Chapel Hill: University of North Carolina Press, 1942. 177 p.

Dill, R. F. Submarine erosion. New York: Elsevier, in preparation.

Emery, Kenneth O. Sea off southern California, a modern habitat of petroleum. New York: Wiley, 1960. 366 p.

Gorsline, D. S., ed. "Submarine canyons." A special issue of Marine geology. v. 8, no. 3/4, 1970. p. 181-291.

Heezen, Bruce C. and C. D. Hollister. The face of the deep. New York: Oxford University Press, 1971. 700 p.
A study of the ocean floor which discusses subjects of

sea floor spreading, continental drift, Mid-Oceanic Ridge, turbidity currents, Atlantis, and development of submarine canyons.

Ingle, J. C., Jr. The movement of beach sand. Amsterdam: Elsevier, 1966. 221 p. (Developments in Sedimentology, no. 5.)

Johnson, Douglas W. The origin of submarine canyons; a critical review of hypotheses. New York: Hafner, 1967, c1939. 126 p.

Keen, Michael J. An introduction to marine geology. Elmsford, N. Y.: Pergamon, 1968. 226 p.

Kuenen, Philip H. Marine geology. New York: Wiley, 1950. 568 p.

Louisiana State University. Coastal Studies Institute. (Publications.)
The institute publishes two series (Contributions, and Coastal Studies Series) which reflect recent work on coastal geology.

Manten, A. A. and Talwani, M. "The origin of marine geology." and "A review of marine geophysics." In: Marine geology. v. 2, no. 1/2, 1964. p. 1-28, 29-80.

Marine geology. Amsterdam, The Netherlands: Elsevier, 1964- . Monthly.

Marine geophysical researches; an international journal for the study of the earth beneath the sea. Dordrecht, Holland: Reidel, 1970- . Quarterly.

Menard, H. W. Marine geology of the Pacific. New York: McGraw-Hill, 1964. 271 p.

Miller, Robert L., ed. Papers in marine geology; Shepard commemorative volume. New York: Macmillan, 1964. 531 p.

Ottmann, F. Introduction à la géologie marine et littorale. Paris: Masson, 1965. 260 p.

Shepard, Francis P. The earth beneath the sea. Revised ed. Baltimore: Johns Hopkins Press, 1967. 242 p.

Essentially a popularization of author's Submarine geology.

----------. Submarine geology. 2nd ed. New York: Harper, 1963. 557 p.
Excellent text in marine geology.

---------- and R. F. Dill. Submarine canyons and other sea valleys. Chicago: Rand McNally, 1966. 381 p.

Trask, Parker D., ed. Recent marine sediments; a symposium. New York: Dover, 1968. 736 p.

U.S. Geological Survey. Atlantic continental shelf and slope of the United States. Washington: Government Printing Office, 1966. (U.S. Geological Survey. Professional Paper no. 529.)

Whittard, W. F. and R. Bradshaw, eds. Submarine geology and geophysics. London: Butterworths, 1965. 464 p. (Colston Papers no. 17.)

PALEONTOLOGY
(See Also Stratigraphy)

REFERENCE WORKS

Biological abstracts. Philadelphia, 1927- . Semimonthly. Classified abstracting service covering over 5, 000 world journals in the biological sciences. Annual author and subject indexes. Useful to paleontologists are the sections on ecology, botany, genetics, and zoology including paleozoology.

Fossilium catalogus. The Hague, Netherlands: W. Junk, 1913- .
Pt. 1. Animalia; pt. 2. Plantae. A series of literature surveys which the publisher plans to reprint beginning 1970.

Guelpa, Paulette-Marie. "Science de la terre-Paleontologie." In: Malcles, L-N. Les sources du travail bibliographique. v. 3. Paris: Minard, 1958. p. 248-260.
Catalogs, bibliographies, and reference sources for vertebrate and invertebrate paleontology, paleobotany, paleoecology, and evolution.

International code of zoological nomenclature adopted by the XV International Congress of Zoology. Published for the International Commission on Zoological Nomenclature. London: International Trust for Zoological Nomenclature, 1964. 176 p.

Kummel, Bernhard and David M. Raup, eds. Handbook of paleontological techniques. Prepared under the auspices of the Paleontological Society. San Francisco: Freeman, 1965. 852 p.
A handbook of field and laboratory techniques, ranging from fossil collecting to preparation of illustrations. "Bibliography on paleontological techniques," p. 707-764. "Compilation of bibliographies of use to paleontologists and stratigraphers," p. 765-832.

Lehmann, Ulrich. Palaontologische Wörterbuch. Stuttgart: Enke, 1964. 335 p.

Moore, Raymond C. and others. "Developments, trends, and outlooks in paleontology." In: Journal of Paleontology. v. 42, 1968. p. 1327-1377.

Russell, R. Dana. "SEPM history 1926-1946." In: Journal of paleontology. v. 44, 1970. p. 173-194.

Schenk, Edward T. and J. H. McMasters. Procedure in taxonomy. 3rd ed. Enlarged by A. M. Keen and S. W. Muller. Stanford, California: Stanford University Press, 1959. 149 p.

Telberg, V. G. Russian-English dictionary of paleontological terms. New York: Telberg, 1966. 168 p.

Westermann, G. E. C. Directory of paleontologists of the world. 2nd ed. Sponsored by the International Paleontological Union. Hamilton, Ontario: McMasters University Bookstore, 1968. 250 p.
Exclusive of Soviet Union and China.

Zentralblatt für Geologie und Paläntologie. Stuttgart: E. Schweizerbart'sche, 1950- . Irregular.
Teil II: Historische Geologie. Paläontologie.

Zoological record. London: Zoological Society of London, 1864- . Annual.
Contains roughly 15, 000 references a year from world-wide literature. Useful to paleontologists. Issued in sections (Protozoa, Porifera, etc.), each section with a subject index.

GENERAL AND THEORETICAL PALEONTOLOGY, EVOLUTION

American Association for the Advancement of Science. The species problem; a symposium... . Edited by Ernst Mayr. Washington, 1957. 395 p. (Publication no. 50.)

Beerbower, James R. Search for the past; an introduction to paleontology. 2nd ed. Englewood Cliffs, N. J.: Prentice-Hall, 1968. 572 p.

Bernal, John Desmond. The origin of life. Cleveland: World Publishing Co., 1967. 345 p.

Brouwer, Aart. General Paleontology. Translated from the Dutch by R. H. Kaye. Edinburgh: Oliver and Boyd, 1967. 216 p.

Campbell, K. S. W., ed. Stratigraphy and paleontology--essays in honor of Dorothy Hill. Canberra: Australian National University, 1969. 392 p.
Essays on biological interpretation of fossils, biostratigraphy and biogeography, descriptive paleontology, and marine sedimentation and geomorphology.

Casanova, Richard L. An illustrated guide to fossil collecting. Rev. ed. Healdsburg, California: Naturegraph, 1970. 128 p.

Chavan, André and André Cailleux. Détermination pratique des fossiles. Paris: Masson, 1957. 387 p.

Darwin, Charles. On the origin of species by means of natural selection, or the preservation of favoured races in the struggle for life. New York: Appleton, 1905. 2 v. in 1.
First edition was published in 1859. Reprinted by Mentor Books in 1958.

Davies, Arthur M. An introduction to paleontology. 3rd ed. London: Murby, 1961. 322 p.

Dobzhansky, T. G. Genetics and the origin of species. 3rd ed., revised. New York: Columbia University Press, 1951. 364 p. (paperback edition, 1964).

Gregory, W. K. Evolution emerging. New York: Macmillan, 1951. 2 v.

Hotton, Nicholas, III. The evidence of evolution. Princeton, N.J.: Van Nostrand, 1968. 160 p.

Horowitz, A. S. and P. E. Potter. Introductory petrography of fossils. New York: Springer-Verlag, 1971. 300 p.

Huxley, Julian S. Evolution the modern synthesis. New York: Harper, 1943. 645 p. (paperback edition, 1964).

----------, A. C. Hardy, and E. B. Ford, eds. Evolution as a process. 2nd ed. London: Allen and Unwin, 1958. 367 p.

International Symposium on the Origin of Life on the Earth, Moscow, 1957. Proceedings. Edited for the Academy of Sciences of the U.S.S.R. by A. I. Oparin and others. New York: Pergamon, 1959. 691 p.

Jepsen, G. L., E. Mayr, and G. G. Simpson, eds. Genetics, paleontology and evolution. Princeton, N. J.: Princeton University Press, 1949. 474 p. (paperback edition, Atheneum, 1963).

Journal of paleontology. Tulsa: Society of Economic Paleontologists and Mineralogists, 1927- . Bimonthly. The SEPM also has begun a memoir series, 1968- .

Matthews, William H., III. Fossils; an introduction to prehistoric life. New York: Barnes and Noble, 1962. 337 p.

Middlemiss, F. A. and P. F. Rawson, eds. Faunal provinces in space and time. Liverpool: Geological Journal, 1971. Special issue no. 4. In press.

Moret, Léon. Manuel de paleontologie animale. 5th ed. Paris: Masson, 1966. 781 p.

Müller, A. H. Lehrbuch der Palaozoologie. Jena: Fischer, 1963- .
V. 1. Allgemeine Gründlagen. 2nd ed. 1963. 387 p.; v. 2. Invertebraten. Pt. 1. Protozoa--Mollusca 1. 2nd ed. 1963. 574 p., Pt. 2. Mollusca 2--Arthropoda 1. 2nd ed. 1965. 448 p., Pt. 3. Arthropoda 2--Stomochorda. 1963. 698 p.; v. 3. Vertebraten. Pt. 1. 1965. Pt. 2. 1968. Pt. 3. 1970.

North American Paleontological Convention. Chicago, 1969. Proceedings. Edited by Ellis L. Yochelson. Lawrence, Kansas: Allen Press, 1970-71. 2 v.
Pt. A. Teaching Paleontology; Pt. B. Computers in paleontology; Pt. C. The genus: a basic concept in paleontology; Pt. D. Paleoclimatology; Pt. E. Evolution of communities; Pt. F. Correlation by fossils; Pt. G. The Ultra microplankton; Pt. H. Evolution of higher categories; Pt. I. Extraordinary fossils; Pt. J. Reef organisms through time; Pt. K. Phosphate in fossils; Pt. L. Cretaceous biogeography.

Oparin, A. I. Origin of life on the earth. New York: Academic Press, 1967. 495 p.

Orlov, Yu A., ed. Osnovy paleontologii. Moscow: Akademiia Nauk. 1959-1963. 15 v.
Several volumes of Osnovy paleontologii are available in English translation (Israel Program for Scientific Translation) from the National Technical Information Service (NTIS), Arlington, Virginia: V. 1. (OTS 61-11419); v. 11. (TT66-51147); v. 13. (TT67-51341).

Piveteau, Jean, ed. Traité de paleontologie. Paris: Masson, 1952-1957. 7 v. in 10.
V. 1. Les stades inférieurs d'organisation du Règne animale--Introduction. Généralités. Protistes. Spongiaires. Coelentères. Bryozoaires; v. 2. Problèmes d'adaption et de phylogenese--Brachiopodes. Chetognathes. Annelides. Gephyriens. Mollusques; v. 3. Les formes ultimes d'Invertebres; morphologie et évolution--Onychophores. Arthropodes. Echinodermes. Stomocordes; v. 4. L'origine des vertèbres. Leur expansion dans les eaux douces et le milieu marin--Agnathes. Placordermes. Elasmobranches. Actinopterygiens. Crossopterygiens. Dipneustes; v. 5. La sortie des eaux. Naissance de la Tetrapodie. L'exubérance de la vie vegétative. La conquête de l'air--Amphibiens. Reptiles. Oiseaux; v. 6. L'origine des Mammifères et les aspects fondamentaux de leur évolution--2 v.; v. 7. Vers la forme humaine. Le problème biologique de l'homme. Les époques de l'Intelligence--Primates. Homme.

Raup, David M. and S. M. Stanley. Principles of paleontology. San Francisco: Freeman, 1971. 388 p.

Raymond, Percy E. Prehistoric life. Cambridge, Mass.: Harvard University Press, 1947. 324 p.

Schindewolf, Otto H. Gründfragen der Palaeontologie; geologische Zeitmessung, organsche Stammensentwicklung, biologische Systematik. Stuttgart: E. Schweizerbart'sche, 1950. 505 p.

Shimer, Hervey W. and Robert R. Schrock. Index fossils of North America: based on revision and reillustration of Grabau and Shimer's North American index fossils. New York: Wiley, 1944. 837 p.

Simpson, George G. Life of the past; an introduction to paleontology. New Haven: Yale University Press, 1953. 198 p.

----------. The major features of evolution. New York: Columbia University Press, 1953. 434 p.

----------. The meaning of evolution. Revised ed. New Haven, Conn.: Yale University Press, 1967. 368 p.

Spencer, Jean M. "Geologic factors controlling mutation and evolution--a review." In: Baylor geological studies. Bulletin no. 7, 1964. 35 p.

Stirton, R. A. Time, life, and man; the fossil record. New York: Wiley, 1959. 558 p.

Systematics Association. The species concept in paleontology, a symposium. Edited by P. C. Sylvester-Bradley. London, 1956. 145 p. (Publication no. 2.)

Tax, Sol, ed. Evolution after Darwin. Chicago: University of Chicago Press, 1960. 3 v.

Teichert, Curt and E. L. Yochelson, eds. Essays in paleontology and stratigraphy; The R. C. Moore commemorative volume. Lawrence, Kansas: Kansas University Press, 1967. 626 p.

Traité de zoologie; Anatomie, systematique, biologie. Paris: Masson, 1948- .
A multi-volumed treatise by specialists in the field, and a valuable reference to paleontologists.

Wagstaffe, Reginald and J. Haveloc Fidler. The preservation of natural history specimens, v. 2. New York: Philosophical Library, 1968. 404 p.
Includes methods for geology, vertebrate paleozoology and paleobotany.

Watson, D. M. S. Paleontology and modern biology. New Haven, Conn.: Yale University Press, 1951. 216 p.

Weller, J. Marvin. The course of evolution. New York: McGraw-Hill, 1969. 696 p.

Zittel, Karl Alfred Ritter von. Textbook of paleontology. London: Macmillan, 1925-1937. (Reprinted by Hafner, 1964.)
The following three volumes are English editions of Zittel's Grundzüge der Palaeontologie: V. 1. Invertebrates, 2nd ed., 1937; v. 2. Vertebrates (excluding mammalia), 2nd ed., 1932; v. 3. Mammalia, 2nd ed., 1925.

INVERTEBRATE PALEONTOLOGY

Beznosova, G. A. and F. A. Zhuravaleva, eds. Paleontologicheskii Slovar. Moscow: Isdatel'stvo "Nauk," 1965. 615 p.

Carter, John L. and R. C. Carter. Bibliography and index of North American Carboniferous brachiopods (1898-1968). Boulder, Colorado: Geological Society of America, 1970. 382 p. (Memoir no. 128.)
"This bibliography and accompanying index supplements and brings up to date... Stuart Weller's indispensible Bibliographic index of North American Carboniferous invertebrates (U.S. Geological Survey. Bulletin no. 153, 1898)."--Preface.

Häntzschel, Walter, F. El-Baz, and G. C. Amstutz. Coprolites, an annotated bibliography. Boulder, Colorado: Geological Society of America, 1969. 132 p. (Memoir no. 108.)

Vokes, H. E. "Genera of the Bivalvia; a systematic and bibliographic catalog." In: Bulletins of American paleontology. v. 51, no. 232, 1967. 394 p.

* * *

Boardman, Richard S. and A. H. Cheetham. "Skeletal growth, intracolony variation, and evolution in Bryozoa: a review." In: Journal of paleontology. v. 43, 1969. p. 205-233.

Crimes, T. P. and J. C. Harper, eds. Trace fossils. Liverpool: Geological Journal, 1970. 547 p. Special issue no. 3.

Easton, William H. Invertebrate paleontology. New York: Harper, 1960. 701 p.

Hyman, Libbie H. *The invertebrates.* New York: McGraw-Hill, 1940-1967. 6 v.

Johnson, J. Harlan. "Geologic importance of calcareous algae with annotated bibliography." In: *Colorado School of Mines. Quarterly.* v. 38, no. 1, 1943; "Bibliography of fossil algae, 1942-1955." In: *C. S. M. Quarterly.* v. 52, no. 2, 1957; "Bibliography of fossil algae, algal limestone, and the geological work of algae, 1956-1965." In: *C. S. M. Quarterly.* v. 62, no. 4, 1967.

----------. *Limestone-building algae and algal limestones.* Golden, Colorado: Colorado School of Mines, 1961. 297 p.

---------- and others. *Fossil algae series.* Golden, Colorado. *Colorado School of Mines. Quarterly.* 1952- . Issued to date: Mississippian algae. v. 51, no. 4, 1956; Devonian algae. v. 53, no. 2, 1958; Silurian algae. v. 54, no. 1, 1959; Ordovician algae. v. 56, no. 2, 1961; Pennsylvanian and Permian algae. v. 58, no. 3, 1963; Jurassic algae. v. 59, no. 2, 1964; Cambrian algae. v. 61, no. 1, 1966.

Joint Committee on Invertebrate Paleontology. *Treatise on invertebrate paleontology.* Edited by Raymond C. Moore. Boulder, Colorado: Geological Society of America and Lawrence, Kansas: University of Kansas Press, 1953- . In progress.
"A series on the major groups of invertebrate animals. Its aim is to present the most comprehensive and authoritative, yet compact, statement of present knowledge about invertebrate fossil groups."--*GSA Catalog of Works in Print.* Volumes already published: Pt. C. Protista 2 (Sarcodina, chiefly "Thecamoebians" and Foraminiferida). 1964. 900 p.; pt. D. Protista 3 (chiefly Radiolaria, Tintinnina). 1954. 195 p.; pt. E. Archaeocyatha, Porifera. 1955. 122 p.; pt. F. Coelenterata. 1956. 498 p.; pt. G. Bryozoa. 1953. 253 p.; pt. H. Brachiopoda. 1965. 927 p.; pt. I. Mollusca 1 (Mollusca General Features Scaphopoda, Amphineura, Monoplacophora, Gastropoda General Features, Archaeogastropoda, mainly Paleozoic Caenogastropoda and Opisthobranchia). 1960. 351 p.; pt. K. Mollusca 3 (Cephalopoda General Features, Endoceratoidea, Actinoceratoidea, Nautiloidea, Bactritoidea). 1964. 519 p.; pt. L. Mollusca 4 (Ammonoidea). 1957. 490 p.; pt. N. Mollusca 6 (Bivalvia).

2 v. 1969; pt. O. Arthropoda 1 (Arthropoda General Features, Protarthropoda, Euarthropoda General Features, Trilobitomorpha). 1959. 560 p.; pt. P. Arthropoda 2 (Chelicerata, Pycnogonida, Palaeoisopus). 1955. 181 p.; pt. Q. Arthropoda 3 (Crustacea, Ostracoda). 1961. 442 p.; pt. R. Arthropoda 4 (Crustacea exclusive of Ostracoda, Myriapoda, Hexapoda). 1969. 651 p.; pt. S. Echinodermata 1 (Echinodermata General Features, Homalozoa, Crinozoa, exclusive of Crinoidea). 1967 (1968). 650 p.; pt. U. Echinodermata 3 (Asterozoans, Echinozoans). 1966. 695 p.; pt. V. Graptolithina. 1955. 101 p.; pt. W. Miscellanea (Conodonts, Conoidal Shells of Uncertain Affinities, Worms, Trace Fossils, Problematica). 1962. 259 p.
Volumes in preparation: Pt. A. Introduction; pt. B. Protista 1 (Chrysomonadida, Coccolithophorida, Charophyta, Diatomacea, etc.); pt. J. Mollusca 2 (Gastropoda, Streptoneura exclusive of Archaeogastropoda, Euthyneura); pt. M. Mollusca 5 (Coleoidea); pt. T. Echinodermata 2 (Crinoidea); pt. X. Addenda, Index.
Volumes in revision: Pt. E. Archaeocyatha, Porifera; pt. G. Bryozoa; pt. L. Ammonoidea; pt. V. Graptolithina.
Volume being supplemented: Pt. F. Coelenterata (Anthozoa Rugosa).

Middlemiss, F. A. A guide to invertebrate fossils. London: Hutchinson, 1968. 128 p.

Moore, Raymond C., C. G. Lalicker, and A. G. Fischer. Invertebrate fossils. New York: McGraw-Hill, 1952. 766 p.

Purnell, Louis R. Catalog of the type specimens of invertebrate fossils. Pt. 1. Paleozoic cephalopoda. Washington: Government Printing Office, 1968. 198 p. (U.S. National Museum. Bulletin no. 262.)
"First in a proposed series listing invertebrate fossil type specimens of the Division of Invertebrate Paleontology of the Smithsonian Institution."--Introd.

Twenhofel, William H. and Robert R. Shrock. Principles of invertebrate paleontology. 2nd ed. New York: McGraw-Hill, 1953. 816 p.

VERTEBRATE PALEONTOLOGY

Bibliography of fossil vertebrates. The monographs making up this series vary as to publisher and title. They are listed chronologically as follows:

Hay, Oliver P. Bibliography and catalogue of the fossil vertebrata of North America. Washington: Government Printing Office, 1902. 868 p. (U.S. Geological Survey. Bulletin no. 179.)

----------. Second bibliography and catalogue of the fossil vertebrata of North America. Washington: Carnegie Institute, 1929-1930. 2 v. (Publication no. 390.)

Romer, Alfred S. and others. Bibliography of fossil vertebrates exclusive of North America, 1509-1927. New York: Geological Society of America, 1963. 2 v. (Memoir no. 87.)
This publication, together with the two works by Hay, provide worldwide coverage from 1509-1927. Worldwide coverage from 1928- is provided by the following publications of the Geological Society of America under the title Bibliography of fossil vertebrates: 1928-1933. C. L. Camp and V. L. Vanderhoof. 1940. (Special Paper no. 27); 1934-1938. C. L. Camp, D. N. Taylor, and S. P. Welles. 1942. (Special Paper no. 42); 1939-1943. C. L. Camp, S. P. Welles, and M. Green. 1949. (Memoir no. 37); 1944-1948. C. L. Camp, S. P. Welles, and M. Green. 1953. (Memoir no. 57); 1949-1953. C. L. Camp and H. J. Allison. 1961. (Memoir no. 84); 1954-1958. C. L. Camp, H. J. Allison, and R. H. Nichols. 1964. (Memoir no. 92); and 1959-1963. C. L. Camp, and others. 1968. (Memoir no. 117).

Society of Vertebrate Paleontology. Bibliography of vertebrate paleontology and related subjects. Cambridge, Mass.: Museum of Comparative Zoology, Harvard University, 1946- . Annual.
Based on the periodicals received by the Library of the American Museum of Natural History in New York City.

* * *

Colbert, Edwin H. Evolution of the vertebrates. 2nd ed. New York: Wiley, 1969. 535 p.

Halstead, L. B. The pattern of vertebrate evolution. San Francisco: Freeman, 1969. 209 p.

Kühn, Oskar, ed. Handbuch der Palaoherpetologie; encyclopedia of palaoherpetology. Konigstein: Otto Koeltz, 1969- . Irregular.

Olson, Everett C. Vertebrate paleozoology. New York: Wiley, 1971, in press.

Romer, Alfred S. Osteology of the reptiles. Chicago: University of Chicago Press, 1956. 772 p.

----------. The procession of life. Cleveland: World Publishing Co., 1968. 323 p.

----------. Vertebrate paleontology. 3rd ed. Chicago: University of Chicago Press, 1966. 468 p. Supplemented by Notes and comments on vertebrate paleontology. Chicago Press, 1968. 304 p.

----------. The vertebrate story. 4th ed. Chicago: University of Chicago Press, 1959. 437 p.

Scott, William Berryman. A history of land mammals in the Western Hemisphere. Revised ed. New York: Hafner, 1962, c1937. 786 p.

Simpson, George G. The principles of classification and the classification of mammals. New York: American Museum of Natural History, 1945. 350 p. (Bulletin, v. 85.)

Swinton, W. E. The dinosaurs. New York: Wiley, 1970. 331 p.

Young, John Z. The life of the vertebrates. 2nd ed. New York: Oxford University Press, 1962. 820 p.

HOMINIDS

Fay, George E. A bibliography of fossil man of the old world. Greeley, Colorado: Museum of Anthropology, Northern Colorado State University, 1959- .
Pt. 1 covers 1845-1955. 1959. Appendix, 1964; pt. 2 covers 1955-1963. 1964. Supplement no. 1 covers 1963-1968. 1969.

* * *

Clark, Wilfrid E. Le Gros. *The fossil evidence for human evolution.* 2nd ed. Chicago: Chicago University Press, 1964. 200 p.

----------. *History of the primates; an introduction to the study of fossil man.* 9th ed. London: British Museum of Natural History, 1966. 119 p.

Day, Michael H. *Guide to fossil man; a handbook of human paleontology.* Cleveland: Meridian Books, 1968. 289 p.

Leakey, Louis S. B. *Adam, or ape: a sourcebook of discoveries about early man.* Cambridge, Mass.: Schenkman, 1968. 384 p.

---------- and V. M. Goodall. *Unveiling man's origins; ten decades of thought about human evolution.* Cambridge, Mass.: Schenkman, 1969. 220 p.

Symposium on Time and Stratigraphy in the Evolution of Man, Washington, 1965. *Proceedings.* Washington: National Academy of Sciences, National Research Coundil, 1967. 97 p.

PALEOBOTANY (See Also Palynology)

Andrews, Henry N., Jr. *Index of generic names of fossil plants, 1820-1965.* Washington: Government Printing Office, 1970. 354 p. (U.S. Geological Survey. Professional Paper no. 1300.)

Botanical Society of America. Paleobotanical Section. *Bibliography of American Paleobotany.* Publisher varies. Annual.
Compiled by the secretary of the Section for members of the society. Beginning 1969, available to the public.

Boureau, Edouard and others. "Bibliographie internationale de paléobotanique." Issued irregularly as part of a series *World report on paleobotany* in the handbook serial *Régnum végétabile.* Utrecht: International Union of Biological Sciences, Section on Paleobotany, 1956- . Issued thus far in *Régnum végétabile:* Pt. 1 in v. 7, 1956; pt. 2 in v. 11, 1958; pt. 3 in v. 19, 1960; pt. 4 in v. 24, 1962; pt. 5 in v. 35, 1964; pt. 6 in v. 42,

1966; pt. 7 in v. 57, 1968.

Grebenshchikov, Oleg S. Geobotanic dictionary. Moscow: Nauka, 1965. 227 p.
Russian, English, German, and French terms.

* * *

Alvin, K. L., P. D. W. Barnard and W. G. Chaloner. Studies in fossil plants. London: Academic Press, 1968. 226 p.
21 papers published in honor of T. M. Harris. Also available as the Journal of the Linnean Society of London (Botany). v. 61, no. 384.

Andrews, Henry N. Studies in paleobotany; with a chapter on palynology by C. J. Felix. New York: Wiley, 1961. 487 p.

Arnold, Chester A. An introduction to paleobotany. New York: McGraw-Hill, 1947. 433 p.

Boureau, Edouard, ed. Traité de paléobotanique. Paris: Masson, 1964- .
V. 1. Introduction. Généralités. Thallophyta; v. 2. Bryophyta. Psilophyta. Lycophyta; v. 3. Sphenophyta. Noeggerathiophyta; v. 4. Filicophyta; v. 5. Pteriodosphermaphyta; v. 6. Glossopteridophyta. Cycadophyta. Cordaitophyta. Ginkophyta. Coniferophyta; v. 7. Chlamydosphermophyta. Angiospermophyta; v. 8. Sporologie; v. 9. Paleophytogeographie; Conclusions générales. v. 1, 1964; v. 2, 1967; v. 4, 1970.

Brongniart, Adolphe T. Histoire des végétaux fossiles, ou recherches botaniques et géologiques. Paris: G. Dufour and Crochard, 1828-1837. 2 v. (reprinted by Asher, 1965).
Classification and distribution of fossil plants.

Darrah, William C. Principles of paleobotany. 2nd ed. New York: Ronald Press, 1960. 295 p.

Delevoryas, Theodore. Morphology and evolution of fossil plants. New York: Holt, Rinehart and Winston, 1962. 189 p.

Emberger, Louis. Les plantes fossiles dans leurs rapports avec les végétaux vivants. 2nd ed. Paris: Masson, 1968. 758 p.

Gothan, Walther and Hermann Weyland. Lehrbuch der Palaobotanik. 2nd ed. Berlin: Akademie-Verlag, 1964. 594 p.

"Major evolutionary events and the geological record of plants, a symposium arranged by H. P. Banks for the 1969 International Botanical Congress." In: Biological reviews. 1970. p. 317-351.

Moret, Leon. Manuel de paleontologie végétale. 3rd ed. Paris: Masson, 1964. 244 p.

Scott, Dukinfield H. Studies in fossil botany. 3rd ed. New York: Hafner, 1962. 2 v.
V. 1. Pteriodophyta; v. 2. Spermophyta.

Seward, Albert C. Fossil plants for students of botany and geology. New York: Hafner, 1963, c1898-1919. 4 v.

----------. Plant life through the ages; a geological and botanical retrospect. 2nd ed. New York: Hafner, 1959, c1933. 603 p.

Walton, J. An introduction to the study of fossil plants. 2nd ed. London: Black, 1953. 201 p.

MICROFAUNA

GENERAL REFERENCES

Hiltermann, Heinrich. Bibliographie stratigraphisch wichtiger mikropaläontologischer Publikationen von etwa 1830 bis 1958 mit Kurzreferaten. Stuttgart: E. Schweizerbart'sche, 1961. 403 p.
Continued annually in Paläontologische Zeitschrift.

* * *

Arbeitskreis Deutscher Mikropalaontologie. Leitfossilien der Mikropalaontologie; ein Abriss (von Helmut Bartenstein et al). Berlin: Borntraeger, 1962. 2 v.

Glaessner, Martin F. Principles of micropaleontology. New York: Hafner, 1963, c1945. 266 p.

Jones, Daniel J. Introduction to microfossils. New York: Harper, 1956. 406 p. (reprinted by Hafner, 1969).

Kuznetsov, Sergei I., ed. Geologic activity of microorganisms. New York: Consultants Bureau, 1962. 112 p. (Transactions of the Institute of Microbiology, no. 9).

---------- and others. Introduction to geological microbiology. Translated by Paul T. Broneer. New York: McGraw-Hill, 1963. 252 p.

McLean, James D., ed. The manual of micropaleontological techniques. Alexandria, Va.: McLean Paleontological Laboratory, 1959- .
Reprints of scattered small articles which appear in various geological and paleontological journals. The McLean Paleontological Laboratory publishes a card catalogue of American Foraminifera, a card catalogue of Ostracoda, and is responsible for reprinting a number of out-of-print publications on Foraminifera.

Pokorny, Vladimir. Grundzüge der zoologischen Mikropaläontologie. Berlin: Deutscher Verlag der Wissenschaften, 1958. 2 v.
V. 1 and v. 2 have been translated into English by K. A. Allen. Oxford, New York: Pergamon, 1963, 1965.

CONODONTS

Ash, Sidney R. "Bibliography and index of Conodonts, 1949-1958." In: Micropaleontology. v. 7, 1961. p. 213-244.

----------. 1959-1963. In: Brigham Young University geology studies. v. 10, 1963. p. 3-50.

Ellison, Samuel P., Jr. Annotated bibliography and index of Conodonts. Austin: Bureau of Economic Geology, 1962. 128 p. (Texas University. Publication no. 6210.)
"Supplement to the annotated bibliography and index of Conodonts." In: Texas journal of science. v. 15, p. 50-67; "Second supplement..." In: Texas journal of science. v. 16, 1964. p. 216-242; "Third supplement

. . ." In: Texas journal of science. v. 19, 1967. p. 5-40.

Fay, Robert O. Catalogue of Conodonts. Lawrence: University of Kansas, 1952. 206 p. (University of Kansas Paleontological Contributions--Vertebrata, Article no. 3.)

Lindström, Maurits. Conodonts. New York: Elsevier, 1964. 196 p.
Summary work-structure, morphology, occurrence, history, nomenclature, and taxonomy of conodonts.

FORAMINIFERA

Boltovskoy, E. Diccionario foraminiferologico plurilingue en cinco idiomas. Buenos Aires: Ministerio de Marina, 1956. 196 p. (Servicio de Hidrografia Naval. Miscellaneous Publication no. 1001.)
Terms in English, Spanish, German, French, and Russian.

Ellis, Brooks F. and A. R. Messina. Catalogue of Foraminifera. New York: American Museum of Natural History, 1940- .
Originally published in 30 numbered volumes. Now loose-leaf supplements are issued to be interfiled with these. A compilation of genera and species with illustrations of each. Provides translations of foreign literature.

----------. Catalogue of index Foraminifera. New York: American Museum of Natural History, 1965-67. 3 v.
Catalogue of larger index Foraminifera.

Ellis, Brooks F. and others. Catalogue of index smaller Foraminifera. New York: American Museum of Natural History, 1968-1969. 3 v.

Sherborn, Charles D. An index to the genera and species of the Foraminifera. Washington: Smithsonian Institution, 1893-1896. 2 v. (Miscellaneous Collection nos. 856 and 1031.)
A single volume of this index was reprinted by the Smithsonian Institution in 1955. 485 p.

Thalmann, Hans E. An index to the genera and species of the Foraminifera, 1890-1950. Stanford, Calif.: Stanford University, George Vanderbilt Foundation, 1960. 393 p.

----------. "Bibliography and index to new genera, species, and varieties of Foraminifera for the year 19--." In: Journal of paleontology. v. 7-33, 1933-1959.
During these years a bibliography was published with occasional supplements and corrections for specific periods.

Toomey, Donald F. and B. L. Mamet. "Annotated bibliography of Paleozoic non-Fusulinid Foraminifera." In: Contributions from the Cushman Foundation for Foraminiferal Research. v. 10, 1959. p. 71-105; Addendum, v. 12, 1961. p. 33-46; Addendum 1, v. 14, 1963. p. 77-94; Addendum 2, v. 16, 1965. p. 1-21; Addendum 3, v. 17, 1966. p. 46-66; Addendum 4, v. 18, 1967. p. 55-83; Addendum 5, v. 19, 1968. p. 41-69; Addendum 6, v. 20, 1969. p. 45-64; Addendum 7, v. 21, 1970. p. 50-70.

---------- and G. A. Sanderson. "A bibliography of the family Fusulinidae." In: Journal of paleontology. v. 28, 1954. p. 465-484. Supplements to this bibliography are published irregularly in the Journal of paleontology. Addendum, v. 30, 1956. p. 1360-1366; Addendum 2, v. 39, 1965. p. 1192-1206; Addendum 3, v. 40, 1966. p. 1402-1408; Addendum 5, v. 41, 1967. p. 1564-1568; Addendum 6, v. 43, 1969. p. 563-567; Addendum 7, v. 44, 1970. p. 770-775. Each bibliography in the series updates and amends previous ones.

* * *

Cushman, Joseph A. Foraminifera: their classification and economic use. 4th ed. Cambridge, Mass.: Harvard University Press, 1948. 605 p.

Journal of Foraminiferal research. Washington, D.C.: Cushman Foundation for Foraminiferal Research, U. S. National Museum, 1971- . Quarterly.
From 1950-1970, entitled Cushman Foundation for Foraminiferal Research. Contributions. Includes, as a regular feature, "Recent literature on the Foraminifera" by Ruth Todd.

Koenigswald. G. H. R. von, ed. Evolutionary trends in Foraminifera; a collection of papers... . Amsterdam:

Elsevier, 1963. 355 p.

Loeblich, Alfred R., Jr. and Helen Tappan. "Foraminiferal classification and evolution." In: *Geological Society of India. Journal.* v. 5, 1964. p. 5-40.

Neumann, Madeleine. *Manuel de micropaléontologie des Foraminiferes (Systematique stratigraphie).* Paris: Gauthier Villars, 1967- .

Phleger, Fred B. *Ecology and distribution of recent Foraminifera.* Baltimore: Johns Hopkins Press, 1960. 297 p.

OSTRACODS

Coryell, Horace N. *Bibliographic index and classification of the Mesozoic Ostracoda.* Dayton, Ohio: University of Dayton Press, 1963. 2 v.
A world catalog.

Ellis, Brooks F. and A. R. Messina. *Catalogue of Ostracoda.* New York: American Museum of Natural History, 1952-1964. 20 v.
Looseleaf up-dating pages issued at intervals.

Howe, Henry Van Wagenen. *Ostracod taxonomy.* Baton Rouge: Louisiana State University Press, 1962. 366 p.

Van Morkhoven, F. P. C. M. *Post-Paleozoic Ostracoda, their morphology, taxonomy, and economic use.* New York: Elsevier, 1962. 2 v.

FLAGELLATES

Downie, Charles and William A. S. Sarjeant. *Bibliography and index of fossil Dinoflagellates and Acritarchs.* New York: Geological Society of America, 1964. (Memoir no. 94.)

Eisenack, Alfred. *Katalog der fossilen Dinoflagellaten, Hystrichospharen und verwandten mikrofossilien.* Stuttgart: E. Schweizerbart'sche, 1964- .
V. 1. Dinoflagellaten. 888 p. First supplement published in 1967.

Loeblich, A. R., III and others. Annotated index of fossil and recent Silicoflagellates and Ebridians, with descriptions and illustrations of validly proposed taxa. Boulder, Colorado: Geological Society of America, 1968. 319 p. (Memoir no. 106.)

PALYNOLOGY (See Also Paleobotany)

Erdtman, Gunnar. "Literature on palynology." Irregularly in Geologiska Föreningens i Stockholm. Förhandlingar and Grana palynologica, 1927-1960.
From 1927-1957, published in G. F. F., vols. 49, 52, 54, 56, 57, 58, 62, 66, 67, 69, 70, 71, 72, 73, 74, 75, 76, 77, 79. Continued as "Basic literature on palynology" in Grana palynologica, v. 2, no. 1, 1959 and v. 2, no. 2, 1960.

Kremp, Gerhard O. W. Morphologic encyclopedia of palynology; an international collection of definitions and illustrations of spores and pollen. Tucson: University of Arizona Press, 1965. 185 p.

---------- and others. Catalog of fossil spores and pollen. University Park, Penn.: Palynological Laboratories, University of Pennsylvania, 1957- .

Manten, A. A., ed. "Bibliography of palaeopalynology 1836-1966." A special issue of Review of palaeobotany and palynology. v. 8, 1969. 57 p.

Van Campo, Madeleine and others. Bibliographie, supplement à pollen et spores. Paris: Muséum national d'histoire naturelle, 1959- . Annually, as a supplement to the journal, Pollen et spores.
Serves as a continuation of the Erdtman series "Literature on palynology."

* * *

American Association of Stratigraphic Palynologists. Proceedings. 1st, 1968- .
Proceedings of the first annual meeting published as Geoscience and man. Baton Rouge: Louisiana State University, 1970. 152 p.

Brown, Clair A. Palynological techniques. Baton Rouge, La., 1960. 188 p.

Erdtman, Gunnar. Handbook of palynology: morphology, taxonomy, ecology. Copenhagen: Munksgaard, 1969. 486 p.

----------. An introduction to pollen analysis. 2nd ed. Waltham, Mass.: Chronica Botanica, 1954. 239 p.

----------. Pollen morphology and plant taxonomy; an introduction to palynology. Waltham, Mass.: Chronica Botanica, 1952-1966. 3 v.

Faegri, Knut and J. Iversen. Textbook of pollen analysis. 2nd ed., revised. New York: Hafner, 1964. 237 p.

Hart, George F. The systematics and distribution of Permian miospores. Johannesburg: Witwatersrand University Press, 1965. 252 p.

Manten, A. A. "Half a century of modern palynology." In: Earth-Science Reviews. v. 2, 1966. p. 277-316.

Pollen et spores. Paris: Museum national d'histoire naturelle, 1959- . 3/yr.

Review of palaeobotany and palynology; an international journal. Amsterdam: Elsevier, 1967- . 4/yr.

Society of Economic Paleontologists and Mineralogists. Palynology in oil exploration, a symposium... . Edited by A. T. Cross. Tulsa, 1964. 212 p. (Special Publication no. 11.)

Tschudy, R. H. and R. A. Scott, eds. Aspects of palynology. New York: Wiley, 1969. 510 p.
A survey of fossil spores, fossil pollen, and other plant microfossils. Quaternary excluded.

Wodehouse, Rober P. Pollen grains... . New York: Hafner, 1959, c1935. 574 p.

PALEOCLIMATOLOGY, PALEOECOLOGY, PALEOGEOGRAPHY

Ager, Derek V. Principles of paleoecology. New York: McGraw-Hill, 1963. 371 p.

Bowen, R. Paleotemperature analysis. New York: Elsevier, 1966. 265 p. (Methods in Geochemistry and Geophysics, v. 2.)

Brooks, Charles E. P. Climate through the ages. 2nd ed. New York: Dover, 1970. 395 p.

Croizat, Leon C. M. Panbiogeography: or an introductory synthesis of zoogeography, phytogeography, and geology. Caracas: The author, 1958. 2 v. in 3.

Cushing, E. J. and H. E. Wright, Jr., eds. Quaternary paleoecology. New Haven, Conn.: Yale University Press, 1967. 433 p. (International Association for Quaternary Research. Congress VII. Proceedings, v. 7.)

Darlington, Philip J. Zoogeography: the geographical distribution of animals. New York: Wiley, 1957. 675 p.

Eckman, Swen P. Zoogeography of the sea. London: Sidgwick and Jackson, 1953. 417 p.

Fairbridge, Rhodes W., ed. Solar variations, climatic change, and related geophysical problems. New York: New York Academy of Sciences, 1961. 739 p. (Annals v. 95, article no. 1.)

Geological Society of America. Treatise on marine ecology and paleoecology. Edited by J. W. Hedgpeth and H. S. Ladd. New York: Geological Society of America, 1957. 2 v. (Memoir no. 67.)
A comprehensive reference source of marine paleoecology with extensive annotated bibliographies.

Godwin, Harry. The history of the British flora; a factual basis for phytogeography. Cambridge, England: Cambridge University Press, 1956. 383 p.

Hecker, R. F. Introduction to paleoecology. New York: Elsevier, 1965. 166 p.

Heusser, Calvin J. Late-Pleistocene environments of North Pacific North America; an elaboration of late-glacial and post-glacial climatic, physiographic, and biotic changes. New York: American Geographical Society, 1960. 308 p.

Hopkins, David M., ed. The Bering land bridge. Stanford: Stanford University Press, 1967. 495 p.

Imbrie, John and Norman D. Newell, eds. Approaches to paleoecology. New York: Wiley, 1964. 432 p.

International Association for Quaternary Research. Arctic and alpine environments. Edited by H. E. Wright, Jr. and W. H. S. Osburn. Bloomington: Indiana University Press, 1968. 308 p. (Congress VII. Proceedings, v. 10.)

----------. Pleistocene extinctions; the search for a cause. Edited by P. S. Martin and H. E. Wright, Jr. New Haven: Yale University Press, 1967. 453 p. (Congress VII. Proceedings, v. 6.)

Laporte, L. F. Ancient environments. New Jersey: Prentice Hall, 1968. 115 p.

Matthew, William D. Climate and evolution. 2nd ed. New York: New York Academy of Sciences, 1939. 223 p. (Special Publication no. 1.)

Mitchell, J. Murray, Jr., ed. Causes of climatic change. Boston: American Meteorological Association, 1968. 160 p. (International Association for Quaternary Research. Congress VII. Proceedings, v. 5.) (AMS Meteorological Monograph v. 8, no. 30.)

Nairn, Alan E. M., ed. Descriptive palaeoclimatology. New York: Interscience, 1961. 380 p.

NATO Palaeoclimates Conference. Newcastle-upon-Tyne and Durham, England, 1963. Proceedings: Problems in palaeoclimates. Edited by A. E. M. Nairn. New York: Interscience, 1961. 705 p.

Neaverson, Ernest. Stratigraphical paleontology; a study of ancient life-provinces. 2nd ed. Oxford: Clarendon Press, 1955. 806 p.

Palaeogeography, palaeoclimatology, palaeoecology; an international journal for the geosciences. Amsterdam: Elsevier, 1965- . Quarterly.

Pewe, Troy L., ed. The periglacial environment: past and present. Montreal: McGill University Press, 1969. 287 p.
Revisions and new papers added to those given in Alaska at the Symposium on Cold Climate Environments and Processes as part of the INQUA Congress, 1965.

Reeves, C. C., Jr. Introduction to paleolimnology. New York: Elsevier, 1968. 228 p. (Developments in Sedimentology, no. 1.)

Schwarzbach, Martin. Climates of the past. Translated by R. O. Muir. Princeton, N. J.: Van Nostrand, 1963. 328 p.

Shapley, Harlow, ed. Climatic change; evidence, causes, and effect. Cambridge, Mass.: Harvard University Press, 1953. 318 p.

World climate from 8000 to 0 B.C.; Proceedings of the International Symposium held at Imperial College, London, April 1966. London: Royal Meteorological Society, 1967. 230 p.

Wright, H. E., Jr., ed. Quaternary geology and climate. Washington: National Academy of Sciences, 1969. 168 p. (Publication no. 1701.)
Collection of papers by authors from many parts of the world.

Wulff, Ergenii R. An introduction to historical plant geography. Waltham, Mass.: Chronica Botanica Co., 1943. 223 p.

Zeuner, Friedrich E. The Pleistocene Period, its climate, chronology, and faunal successions. London: Hutchinson, 1959. 447 p.

Zoogeography. A symposium... . Edited by Carl L. Hubbs. Washington: American Association for the Advancement of Science, 1958. 509 p. (Publication no. 51.)

SEDIMENTOLOGY

(See Also Clay and Clay Minerals, Marine Geology, Sedimentary Rocks, and Soil Science)

REFERENCE WORKS

Annotated bibliography on hydrology and sedimentation, United States and Canada.
The following have been issued in this series:

U. S. Federal Interagency River Basin Committee. Annotated bibliography on sedimentation. Washington: Government Printing Office, 1950. 351 p. (Sedimentation Bulletin no. 2.)

U. S. Federal Interagency River Basin Committee. Annotated bibliography on hydrology, 1941-1950. Washington: American Geophysical Union, 1952. 408 p. (Notes on Hydrologic Activities. Bulletin no. 5.)

U. S. Interagency Committee on Water Resources. Annotated bibliography on hydrology, 1951-1954, and sedimentation, 1950-1954. Washington: American Geophysical Union, 1955. 207 p. (Joint Hydrology Sedimentation Bulletin, no. 7.)

U. S. Geological Survey. Annotated bibliography on hydrology and sedimentation, United States and Canada, 1955-1958. Compiled by H. C. Riggs. Washington: Government Printing Office, 1962. 236 p. (Water Supply Paper no. 1546.)

U. S. Interagency Committee on Water Resources. Annotated bibliography on hydrology and sedimentation, 1959-1962 (United States and Canada). Compiled by C. E. Bradbury and associates. Washington: Bureau of Reclamation, 1964. 323 p. (Joint Hydrology and Sedimentation Bulletin, no. 8.)

Annotated bibliography on hydrology and sedimentation 1963-1965 (United States and Canada). Compiled by the

Hydrology and Sedimentation Committees, Water Resources Council. Washington: Government Printing Office, 1969. (Joint Hydrology-Sedimentation Bulletin, no. 9.)

Annotated bibliography on hydrology and sedimentation 1966-1968 (United States and Canada). Compiled by the Hydrology and Sedimentation Committees. Washington: Government Printing Office, 1970. (Joint Hydrology and Sedimentation Bulletin, no. 10.)

Bonorino, F. G. and M. E. Teruggi. Lexico sedimentológico. Buenos Aires: Instituto Nacional Investigaciones Ciencias Naturales, 1952. 164 p.
A lexicon of terms in English, Spanish, and German.

Chambre syndicale de la recherche et de la production du pétrole et du gaz naturel. Sous-commission laboratoires de stratigraphie. Essai de nomenclature et caractérisation des principales structures sédimentaires. Paris: Editions Technip, 1966. 293 p.

----------. Essai de nomenclature des roches sédimentaires. Paris: Editions Technip, 1961 (v. 1, 3rd ed.), 1965 (v. 2).

Conybeare, C. E. B. and K. A. W. Crook. Manual of sedimentary structures. Canberra, Australia: Department of Natural Development, Bureau of Mineral Resources, Geology and Geophysics, 1968. 327 p. (Bulletin no. 102).

Haifa. Technion. Israel Institute of Technology. Sedimentation: annotated bibliography of foreign literature for 1959 to 1964. Edited by E. Goldberg. Springfield, Va.: NTIS, 1965. 192 p. (TT-66-51153.)
Covers literature published outside the English speaking world. Also available from the NTIS are the following: 1965-1967. 1968. 299 p. (TT-67-51412); 1966-1967. 1968. 271 p. (TT-68-50451); and 1967-1968. 1969. 395 p. (TT-69-55007).

Houston Geological Society. Deltas in their geologic framework. Houston, 1966. 251 p.

----------. Deltas of the world, modern and ancient--a bibliography. Houston, 1971.

International Geographical Union. Commission on Coastal Sedimentation. Bibliography, 1955-1958. Baton Rouge: Louisiana State University, 1958. 148 p. (Coastal Studies Institute. Contribution no. 60.)

Pettijohn, Francis J. and P. E. Potter. Atlas and glossary of primary sedimentary structures. Berlin: Springer, 1964. 370 p.
A guide containing 117 plates of ancient and modern depositional structures, each with explanatory matter. The glossary contains over 300 entries referring to the original or other comprehensive source of the definition. Equivalent terms are given for German, French, and Spanish.

Russell, Richard J. Glossary of terms used in fluvial, deltaic and coastal morphology and processes. Baton Rouge: Louisiana State University Press, 1969. 97 p. (Coastal Studies Series, no. 23.)

TEXTS AND TREATISES

Bouma, Arnold H. Methods for the study of sedimentary structures. New York: Wiley, 1969. 458 p.
Methods and techniques of sediment analysis. Appendices include shipboard radiography, comparative features of X-ray films, tables of conversion units, and an international directory of manufacturers. Bibliography p. 427-446.

Developments in sedimentology. New York: Elsevier, 1964- .
Titles in this series issued to date: 1. International Sedimentological Congress, 1963. Deltaic and shallow water marine deposits. Edited by L. M. J. U. van Straaten. 1964. 464 p.; 2. Sedimentology and ore genesis. Edited by Gerhardt C. Amstutz. 1964. 184 p. (Proceedings of a symposium of 6th International Sedimentological Congress, 1963); 3. Turbidites. Edited by Arnold H. Bouma and A. Brouwer. 1964. 264 p.; 4. The techniques of sedimentary mineralogy. Edited by Frederick G. Tickell. 1965. 220 p.; 5. The movement of beach sand; an analysis using fluorescent grains. J. C. Ingle, Jr. 1966. 221 p.; 6. The identification of detrital feldspars. Leendert van der Plas, Jr. 1966. 305 p.; 7. Sedimentary features of flysch and greywackes. Stanislaw Dzulynski and E. K. Walton.

1965. 274 p.; 8. Diagenesis in sediments. Edited by Gunnar Larsen and G. V. Chilingar. 1967. 551 p.; 9. Carbonate rocks. Edited by George V. Chilingar, H. J. Bissell, and R. W. Fairbridge. 1967. 2 v.; 10. Cyclic sedimentation. Peter M. D. Duff, A. Hallam, and E. K. Walton. 1967. 280 p.; 11. Introduction to palaeolimnology. C. C. Reeves, Jr. 1968. 228 p.; 13. Palaeozoic reefs of Gotland. A. A. Manten. 1970; 14. Desert sedimentary environments. Glennie, K. S. 1971.

Griffiths, John C. Scientific methods in analysis of sediments. New York: McGraw-Hill, 1967. 508 p.

Krumbein, William C. and L. L. Sloss. Stratigraphy and sedimentation. 2nd ed. San Francisco: Freeman, 1963. 660 p.

Kukal, Zdenek. Geology of recent sediments. New York: Academic, 1971. 490 p.

Pettijohn, Francis J. Sedimentary rocks. 2nd ed. New York: Harper, 1957. 718 p.

Ruchin, L. B. Grundzüge der lithologie; lehre von den sedimentgesteinen. Berlin: Akademie-Verlag, 1958. 806 p.

Sedimentology. Amsterdam: Elsevier, 1962- . 8/yr.

Selley, R. C. Ancient sedimentary environments: a brief survey. Ithaca: Cornell University Press, 1970. 240 p.

"Sedimentology of recent and old sediments; a symposium." In: Geologie en Mijnbouw. v. 21, no. 7 (new series), 1959. p. 185-230.

Trask, P. D., ed. Applied sedimentation. New York: Wiley, 1950. 707 p.

Twenhofel, William H. Principles of sedimentation. 2nd ed. New York: McGraw-Hill, 1950. 673 p.

---------- and S. A. Tyler. Methods of study of sediments. New York: McGraw-Hill, 1941. 183 p.

PROBLEMS AND PROCESSES

Allen, John R. L. Current ripples; their relation to patterns of water and sediment motion. Amsterdam: North Holland, 1968. 433 p.

----------. Physical processes of sedimentation. London: Allen and Unwin, 1970. (Earth Science Series, no. 1.) 248 p.

American Association of Petroleum Geologists. Research Committee. Geometry of sandstone bodies, a symposium.... Edited by J. A. Peterson and J. C. Osmond. Tulsa, 1961. 240 p.

American Institute of Mining and Metallurgical Engineers. Problems of clay and laterite genesis; symposium.... New York, 1952. 244 p.

Amstutz, G. C. Sedimentology and ore genesis; Proceedings of the 6th International Sedimentological Congress, Delft, 1963. New York: Elsevier, 1964. 184 p. (Developments in Sedimentology, v. 2.)

Boswell, Percy G. H. Muddy sediments; some geotechnical studies for geologists, engineers, and soil scientists. Cambridge, England: W. Heffer, 1961. 140 p.

De Segonzac, G. Dunoyer. "The birth and development of the concept of diagenesis (1866-1966)." In: Earth-Science Reviews. v. 4, 1968. p. 153-201.

Duff, P. M. D., A. Hallam, and E. K. Walton. Cyclic sedimentation. New York: Elsevier, 1967. 280 p.

Kansas. Geological Survey. Symposium on cyclic sedimentation. Lawrence, 1966. 2 v. (Bulletin no. 169.)

Klein, George D., ed. Late Paleozoic and Mesozoic continental sedimentation, northeastern North America. Boulder, Colorado: Geological Society of America, 1968. 309 p. (Special Paper no. 106.)

Lajoie, J., ed. Flysch sedimentology in North America. Waterloo, Ontario: Geological Association of Canada, 1970. 300 p. (Special Paper no. 7.)

Lauff, George H., ed. Estuaries. Washington: American Association for the Advancement of Science, 1967. 757 p. (Publication no. 83.)

Lebedev, Lev M. Metacolloids in endogenic deposits. Translated from the Russian by J. B. Southland. New York: Plenum, 1967. 298 p.

Lombard, Augustin. Geologie sédimentaire; les series marines. Paris: Masson, 1956. 722 p.

Morisawa, Marie. Streams; their dynamics and morphology. New York: McGraw-Hill, 1968. 175 p.

Müller, G. and G. M. Friedman, eds. Recent developments in carbonate sedimentology in Central Europe; a symposium. Berlin: Springer, 1968. 255 p.

Persons, Benjamin S. Laterite--genesis, location, use. New York: Plenum, 1970. 110 p.

Pettijohn, Francis J., P. E. Potter, and R. Siever. Geology of sand and sandstone. Bloomington, Ind.: Indiana University, 1965. 205 p.
Excellent source for selected references on sedimentation.

Potter, Paul E. and F. J. Pettijohn. Paleocurrents and basin analysis. New York: Academic Press, 1963. 296 p.

Richards, Adrian F., ed. Marine geotechnique. Urbana: University of Illinois, 1967. 327 p.
Study of marine sediments with emphasis on engineering properties. Papers presented at the 1st International Research Conference on Marine Geotechnique.

Shirley, M. L. and J. A. Ragsdale, eds. Deltas in their geologic framework. Houston: Houston Geological Society, 1966. 251 p.

Society of Economic Paleontologists and Mineralogists, Tulsa. Symposiums. Turbidity currents. 1951. (Special Publication no. 2); Finding ancient shorelines. 1956. (Special Publication no. 3); Recent marine sediments. 1955. (Special Publication no. 4); Regional aspects of carbonate deposition. 1957. (Special Publication no. 5);

Silica in sediments. 1959. (Special Publication no. 7); Dolomitization and limestone diagenesis. 1965. (Special Publication no. 13); Primary sedimentary structures and their hydrodynamic interpretations. 1965. (Special Publication no. 12); Depositional environment in carbonate rocks. 1969. (Special Publication no. 14); Deltaic sedimentation. 1970. (Special Publication no. 15).

Strakhov, Nikolai. Principles of lithogenesis. Translated by J. P. Fitzsimmons. New York: Consultants Bureau, 1967-1969. 3 v.

Sundborg, Ake. The river Klaralven, a study of fluvial processes. Stockholm: Royal Institute of Technology, Institution of Hydraulics, 1960. 316 p. (Bulletin no. 52.)

Termier, Henri and Genevieve Termier. Erosion and sedimentation. Princeton, N. J.: Van Nostrand, 1963. 433 p.

Trask, Parker D., ed. Recent marine sediments; a symposium. New York: Dover, 1968, c1955. 736 p.

STRATIGRAPHY
(See Also Paleontology, Sedimentology)

GEOCHRONOLOGY, RADIOMETRIC DATING

American journal of science. Radiocarbon supplement. New Haven, Conn., v. 1, 1959- . Semi-annual.
Laboratories are listed in v. 11, no. 2, 1969. p. 659-667.

Baadsgaard, H. "Geochronology." In: Dansk Geologisk Forening. Meddeleser. v. 16, no. 1, 1965. p. 1-48.

Bandy, Orville L., ed. Radiometric dating and paleontologic zonation. Boulder, Colorado: Geological Society of America, 1970. 247 p. (Special Paper no. 124.)

Berry, William B. N. Growth of a prehistoric time scale. San Francisco: Freeman, 1968. 158 p.
Development of time scale based on organic evolution.

Bowen, R. N. C. The exploration of time. London: G. Newnes, 1958. 143 p.

Changes in climate; Proceedings of the Rome Symposium, October 2-7, 1961. Paris: UNESCO, 1963. 488 p. (Arid Zone Research, no. 20.)

Dalrymple, G. Brent and M. A. Lanphere. Potassium-argon dating. San Francisco: Freeman, 1969. 258 p.

Eicher, Don L. Geologic time. Englewood Cliffs, N. J.: Prentice-Hall, 1968. 149 p. (Foundations of Earth Science Series.)

Faul, Henry. Ages of rocks, planets, and stars. New York: McGraw-Hill, 1966. 109 p.

----------, ed. Nuclear geology; a symposium on nuclear phenomena in the earth sciences. New York: Wiley, 1954. 414 p.

Franke, H. W. Methoden der géochronologie. Berlin: Springer, 1969. 136 p.

Geochron Laboratories, Inc. Bibliography of geochronology. Cambridge, Mass., n.d., 24 p. Periodically updated.

Hamilton, E. I. Applied geochronology. With a chapter on Comparative geochemistry by L. H. Ahrens. New York: Academic Press, 1965. 287 p.

---------- and R. M. Farquhar, eds. Radiometric dating for geologists. New York: Interscience, 1968. 506 p.

Libby, W. F. Radiocarbon dating. 2nd ed. Chicago: University of Chicago Press, 1955. 175 p.

National Research Council. Committee on Nuclear Science. Geochronology of North America. Washington, 1965. 313 p. (NAS-NRC Publication, no. 1276.)
Compilation of previously published data on the North American continent.

New York Academy of Science. Geochronology of rock systems. Edited by Franklin N. Furness. New York, 1961. 594 p. (Annals. v. 91, art. 2. p. 159-594.)

Schaeffer, O. A. and J. Zahringer, eds. Potassium argon dating. New York: Springer, 1966. 234 p.

Smiley, Terah L., ed. Geochronology, with special reference to southwestern United States. Tucson: University of Arizona Press, 1955. 200 p. (University of Arizona. Bulletin. v. 26, no. 2.)

Symposium on Radioactive Dating, Athens, 1962; Proceedings: Radioactive dating. Vienna: International Atomic Energy Agency, 1963. 440 p.

Zeller, E. J. "Modern methods for measurement of geologic time." In: California. Division of Mines and Geology. Mineral information service. v. 18, no. 1, 1965. p. 9-15.

Zeuner, Friedrich E. Dating the past; an introduction to geochronology. 5th ed. London: Methuen, 1958. 516 p.

PRINCIPLES AND METHODOLOGY

Donovan, Desmond T. Stratigraphy; an introduction to principles. London: Murby, 1966. 199 p.
History of development of stratigraphy included.

Dunbar, Carl O. and John Rodgers. Principles of stratigraphy. New York: Wiley, 1957. 356 p.

Gignoux, Maurice. Stratigraphic geology. English translation from the 4th French ed., 1950, by G. G. Woodford. San Francisco: Freeman, 1955. 682 p.

Grabau, Amadeus W. Principles of stratigraphy. Prefatory note by Marshall Kay. New York: Dover, 1960, c1924. 2 v.

Harbaugh, John W. and D. F. Merriam. Computer applications in stratigraphic analysis. New York: Wiley, 1968. 282 p.

Kay, Marshall and E. H. Colbert. Stratigraphy and life history. New York: Wiley, 1965. 736 p.

Kottlowski, Frank E. Measuring stratigraphic sections. New York: Holt, Rinehart and Winston, 1965. 253 p. (Geologic Field Techniques Series.)

Krumbein, W. C. and L. L. Sloss. Stratigraphy and sedimentation. 2nd ed. San Francisco: Freeman, 1963. 660 p. ("Stratigraphic maps p. 432-500.")

Stamp, L. D. Introduction to stratigraphy, British Isles. 3rd ed., revised. London: Murby, 1957. 381 p.

Termier, Henri and G. Termier. Paleontologie stratigraphique. Paris: Masson, 1959-1960. 4 v. in 1 (515 p.).

Weller, James M. Stratigraphic principles and practice. New York: Harper, 1960. 725 p.

NOMENCLATURE, STRATIGRAPHIC UNITS AND SEQUENCES

REFERENCE WORKS

American Commission on Stratigraphic Nomenclature. "Code of stratigraphic nomenclature." In: American Association of Petroleum Geologists. Bulletin. v. 45, 1961. p. 645-665.
Also reprinted in: Krumbein, W. C. and L. L. Sloss. Stratigraphy and sedimentation. 2nd ed. San Francisco: Freeman, 1963. Appendix. A revision in 1970 includes all post-1961 amendments. 22 p.

The geologic systems. Edited by Kalervo Rankama. New York: Interscience, 1963- .
A series of books, each a collection of regional surveys, on the geologic systems of the world. Published to date: The Precambrian. v. 1, 1963. 279 p.; v. 2, 1965. 454 p.; v. 3, 1968. 325 p.; v. 4, 1970. 288 p.; The Quaternary. v. 1, 1965. 300 p.; v. 2, 1967. 477 p.

Lexicon of geologic names of the United States.
For the breakdown of the lexicon see under "Lexicon of geologic formation names" in the general section of this guide.

Lexique stratigraphique international. Publie sur l'initative de la Commission de stratigraphie du Congress géologique international. Paris: Centre national de la recherche scientifique, 1956- .
For the breakdown of this important work, see under "Lexicon of geologic formation names" in the general section of this guide. Termes stratigraphiques majeurs is a series of articles on the principal geologic systems making up v. 8 of Lexique stratigraphique international.

Lotze, Franz, ed. Handbuch der stratigraphischen Geologie. Stuttgart: Enke, 1959- .
Issued thus far: v. 2. Quatär. Paul Woldstedt. 1969. 263 p.; v. 3. Tertiar. pt. 1. Grundzüge regionaler Stratigraphie. Adolf Papp. 1959. 411 p. pt. 2. Wirbeltierfaunen. Erich Thenius. 1959. 316 p.; v. 4. Jura. Helmut Hölder. 1964. 603 p.; v. 13. Präkambriam. pt. 1. Nördliche Halbkugel. P. Bankwitz and others. 1966. 388 p. pt. 2. Südliche Halbkugel. P. Bankwitz and others. 1968. 702 p.

U.S. Geological Survey. Paleotectonic maps. Edited by Edwin McKee. Washington: Government Printing Office, 1956- .
Issued to date: Jurassic system, with a separate section on paleogeography by R. W. Inmay. 1956. (Miscellaneous Geologic Investigations Map, 1-175); Triassic system. 1959. (Miscellaneous Geologic Investigations Map, 1-300); Permian system. 1967. (Miscellaneous Geologic Investigations Map, 1-450, supplemented by Professional Paper no. 515, Paleotectonic investigations of the Permian System in the United States).

Van Eysinga, F. W. B. "Stratigraphic terminology and nomenclature; a guide for editors and authors." In: Earth-science reviews, v. 6, no. 4, 1970. p. 267-288.

SPECIFIC TIME/ROCK UNITS (See Also Breakdowns of Series in the Preceding Section)

Note: the citations in this section are listed in ascending order, Precambrian to Quaternary.

Royal Society of Canada. The Proterozoic in Canada. Edited by J. E. Gill. Toronto: University of Toronto Press, 1957. 119 p. (Special Publication no. 2.)

* * *

Geological Society of London. The Phanerozoic time-scale; a symposium dedicated to Professor Arthur Holmes. Edited by W. B. Harland. London, 1964. 458 p. (Geological Society of London. Quarterly journal, v. 120S.)

Holland, Charles H., ed. Cambrian of the new world. New York: Wiley, 1971. 456 p.

International Geological Congress. 20th, Mexico, 1956. El sistema Cambrico, su paleogeografia y el problema de su base: symposium. Edited by John Rogers. Mexico, 1956, 2 v.

U.S. Geological Survey. Taconic stratigraphic names: definitions and synonymies. Edited by E-an Zen. Washington: Government Printing Office, 1964. 95 p. (Bulletin no. 1174.)

Berry, W. B. N. and A. J. Boucot, eds. Correlation of the North American Silurian rocks. Boulder, Colorado: Geological Society of America, 1970. 289 p. (Special Paper no. 102.)

International Symposium on the Devonian System, Calgary, Canada, 1967. Proceedings. Edited by D. H. Oswald. Calgary: Alberta Society of Petroleum Geologists, 1967. 2 v.

West Texas Geological Society. Lexicon Committee. Lexicon of pre-Pennsylvanian stratigraphic names of West Texas and southeast New Mexico. Midland, 1958. 153 p.

American Association of Petroleum Geologists. Pennsylvanian system in the United States. Edited by Carl C. Branson. Tulsa, 1962. 505 p.

Bibliography of Carboniferous geology. Heerlen: Netherlands Geologische Stichting, Afdeling Geologische Bureau voor het Mijngebied, v. 1, 1959- . 2/yr. Worldwide in scope.

Flawn, Peter T. and others. The Ouachita system. Austin: Bureau of Economic Geology, University of Texas, 1961. 401 p. (Texas. University. Publication no. 6120.)

McKee, E. D. and others. Paleotectonic investigations of the Permian System in the United States. Washington: Government Printing Office, 1967. 271 p. (U.S. Geological Survey. Professional Paper no. 515.)

* * *

Sherlock, Robert L. The Permo-Triassic formations, a world review. London: Hutchinson's, 1948. 367 p.

American Association of Petroleum Geologists. Jurassic and Carboniferous of western Canada. Edited by A. J. Goodman. Tulsa, 1958. 514 p.

Arkell, William J. Jurassic geology of the world. Edinburgh: Oliver and Boyd, 1956. 806 p.

----------. Jurassic system in Great Britain. Oxford: Oxford Press, 1970, c1933. 681 p.

International Geological Congress. 20th, Mexico, 1956. El sistema Cretácio; un symposium sobre el Cretácico en el Hemisferio occidental y su correlación mundial. Edited by Lewis B. Kellum. Mexico, 1959- . 3 v.

* * *

Eames, F. E. and others. Fundamentals of mid-Tertiary stratigraphical correlation. Cambridge, England: Cambridge University Press, 1962. 163 p.

Eocene Colloquium, 1968. Proceedings. Orleans, France: Bureau de recherches géologique et minières, 1969. 3 v. (Memoire nos. 58, 59, and 69.)

Pearson, Ronald. Animals and plants of the Cenozoic Era; some aspects of the faunal and floral history of the last sixty million years. London: Butterworths, 1964. 236 p.

Youngquist, Walter. Annotated lexicon of names applied to Tertiary stratigraphic units in Oregon and Washington west of the Cascade Mountains, with bibliography. Eugene, Oregon: The author, 1961. 92 p.

Charlesworth, John K. The Quaternary era, with special reference to its glaciation. London: Arnold, 1957. 2 v.

Flint, Richard F. Glacial and Quaternary geology. New York: Wiley, 1971. 892 p.
Supercedes Flint's Glacial and Pleistocene geology, 1957.

International Association for Quaternary Research. Means of correlation of Quaternary successions. Edited by Roger B. Morrison and H. E. Wright, Jr. Salt Lake City: University of Utah Press, 1968. 631 p. (International Association for Quaternary Research. Congress VII. Proceedings, v. 8.)

----------. Proceedings.
Imprint varies. 2nd, Leningrad, 1932; 3rd, Vienna, 1936; 4th, Rome, 1953; 5th, Madrid and Barcelona, 1957; 6th, Warsaw, 1961; 7th, Boulder, 1965; 8th, Paris, 1969. The 7th Congress proceedings covered a wide variety of subjects on the Quaternary. 19 volumes published to-date may be purchased from the Engineering Bookstore, University of Minnesota, Minneapolis, or from the individual

publishers. The 8th Congress proceedings (2 v.) may be purchased from INQUA, Paris.

International Studies on the Quaternary; Papers prepared on the occasion of the VII Congress of the International Association for Quaternary Research, Boulder, Colorado, 1965. Edited by H. E. Wright, Jr. and D. G. Frey. New York: The Geological Society of America, 1965. (Special Paper no. 84.)

Quaternary research. New York: Academic Press, 1970- . Quarterly.

Schumm, Stanley A. and W. C. Bradley, eds. United States contributions to Quaternary research; Papers prepared for the VIII Congress of the International Association for Quaternary Research, Paris, 1969. Boulder, Colorado: Geological Society of America, 1969. 305 p. (Special Paper no. 123.)

West, R. G. Pleistocene geology and biology. New York: Wiley, 1968. 377 p.
Emphasis on British Isles.

Woldstedt, Paul. Das Eiszeitalter; Grundlinien einer Geologie des Quartärs. Stuttgart: Enke, 1958- .
V. 1. Die allgemeinen Erscheinungen des Eiszeitalters. 3rd ed., 1961. 374 p.; v. 2. Europa, Vorderasien und Nordafrika im Eiszeitalter. 2nd ed., 1958. 438 p.; v. 3. Afrika, Asien, Australien und Amerika im Eiszeitalter. 1965. 328 p.

Wright, H. E., Jr., ed. Quaternary geology and climate. Washington: National Academy of Sciences, 1969. 162 p. (Publication no. 1701.) (International Association for Quaternary Research. Congress VII. Proceedings, v. 16.)

---------- and David G. Frey, eds. The Quaternary of the United States. Princeton, N. J.: Princeton University Press, 1965. 922 p.

Zeuner, Friedrich E. The Pleistocene Period, its climate, chronology, and faunal successions. London: Hutchinson, 1959. 447 p.

SOLID-EARTH GEOPHYSICS

REFERENCE WORKS

Clark, Sydney P., Jr. Handbook of physical constants. Revised ed. New York: Geological Society of America, 1966. 587 p. (Memoir no. 97.)

Eos; transactions of the American Geophysical Union. Washington: American Geophysical Union, 1920- . Monthly.
This periodical includes the following media pertinent to current awareness in the field of geophysics: 1. Reviews section: comprehensive critical reviews; 2. Recent publications of interest to geophysicists; 3. Contents of English editions of Russian periodicals, namely, Izvestiia (Geophysics series), Geodesy and aerophotography, Oceanography translations, Soviet hydrology, Geomagnetism and aeronomy, Soviet Antarctic Expedition information bulletin; 4. English translations of Russian and other scientific papers and journals; 5. Contents of all the Union's publications.

International dictionary of geophysics; seismology, geomagnetism, aeronomy, oceanography, geodesy, gravity, marine geophysics, meteorology, the earth as a planet and its evolution. Edited by S. K. Runcorn and others. New York: Pergamon, 1967. 2 v.
Encyclopedic treatment. Articles accompanied by bibliographies.

U.S. Geological Survey. Geophysical abstracts. Washington: Government Printing Office, 1929- . Monthly.
Abstracts of current literature pertaining to the physics of the solid earth and to geophysical exploration. Worldwide in scope. Annual indexes. Geophysical abstracts will terminate with the December 1971 issue. Coverage of solid-earth geophysics will continue in the G.S.A.'s Bibliography and index of geology.

"United States National Report, 1963-1967; Fourteenth General Assembly International Union of Geodesy and Geophysics." In: American Geophysical Union. Transactions. v. 48, no. 2, 1967. p. 329-751.
Many useful bibliographies in fields of hydrology, seismology, meteorology, geomagnetism, physical oceanography, and volcanology.

TEXTS AND TREATISES

Advances in geophysics. Edited by H. E. Landsberg and J. Van Mieghem. New York: Academic Press, 1952- . Annual.

Beloussov, Vladimir V. Basic problems in geotectonics. Principal translator: Paul T. Broneer. New York: McGraw-Hill, 1962. 816 p.

Cloos, Hans. Einführung in die Geologie; ein Lehrbuch der inneren Dynamik. Berlin: Borntraeger, 1963, c1936. 211 p.

Coulomb, Jean and G. Jobert. The physical constitution of the earth. Translated by A. E. M. Nairn. Edinburgh: Oliver and Boyd, 1963. 328 p.

Daly, Reginald A. Strength and structure of the earth. New York: Hafner, 1969, c1940. 434 p.

Delany, F. and C. H. Smith, eds. "Deep-seated foundations of geological phenomena." A special issue of Tectonophysics. v. 7, no. 5/6, 1969. p. 359-601.

The earth's crust and upper mantle: structure, dynamic processes, and their relation to deep-seated geological phenomena. Edited by Pembroke J. Hart. Washington: American Geophysical Union, 1969. 735 p. (Monograph no. 13.)

Egyed, László. Physik der festen Erde. Budapest: Akademini Kiadó, 1969. 367 p.

Ewing, William M., W. S. Jardetsky and F. Press. Elastic waves in layered media. New York: McGraw-Hill, 1957. 380 p. (Lamont Geological Observatory. Contribution no. 189.)

Gaskell, Thomas, ed. The earth's mantle. New York: Academic Press, 1967. 509 p.

----------. Physics of the earth. London: Thames and Hudson, 1970. 216 p.
Survey of modern geophysics with many color photographs.

Geophysics. Tulsa: Society of Exploration Geophysicists, 1936- . Bimonthly.

Geotectonics; Translation of Geotektonika. Washington: American Geological Institute, 1967- . Bimonthly.

Gutenberg, Beno. Internal constitution of the earth. 2nd ed. New York: Dover, 1951. 439 p.

----------. Physics of the earth's interior. New York: Academic Press, 1959. 240 p.

Handbuch der Physik. Hrsg. von S. Flugge, Berlin: Springer, 1955- .
V. 47. Geophysik I; v. 48. Geophysik II; v. 49. Geophysik III. 2 v. Encyclopedia of solid earth geophysics, including rotation of the earth, seismology, gravity, tectonic, radioactivity, magnetism, etc.

Howell, Benjamin F. Introduction to geophysics. New York: McGraw-Hill, 1959. 399 p.

Israël, Hans. Einführung in die geophysik. New York: Springer, 1969. 223 p.

Jacobs, John A., R. D. Russell and J. Tuzo Wilson. Physics and geology. New York: McGraw-Hill, 1959. 424 p.

Jeffreys, Sir Harold. The earth: its origin, history, and physical constitution. 5th ed. Cambridge, England: Cambridge University Press, 1970. 525 p.

Jordan, P. The expanding earth. New York: Pergamon, in prep.

Journal of geophysical research. Washington: American Geophysical Union, 1896- . Tri-monthly.
Three sections are published on the 1st, 15th and 20th of each month respectively. They are concerned with:

1. space physics, including solar physics, planetary atmosphere; 2. physics and chemistry of the solid earth, planetology, geodesy; 3. physics and chemistry of the atmosphere.

Liustikh, Evgenii N. Isostasy and isostatic hypotheses. Washington: American Geophysical Union, 1960. 119 p. (Soviet Research in Geophysics in English Translation, v. 2.)

Munk, Walter H. and G. J. F. MacDonald. The rotation of the earth: a geophysical discussion. Cambridge, England: Cambridge University Press, 1960. 323 p.

NATO Advanced Study Institute. Newcastle-upon-Tyne, England. 1967. The application of modern physics to the earth and planetary interiors. Edited by S. K. Runcorn. New York: Wiley, 1969. 692 p.

----------. 1970. Palaeogeophysics. Edited by S. K. Runcorn. New York: Academic Press, 1970. 518 p.

Odishaw, Hugh, ed. Research in geophysics. V. 2. Solid earth and interface phenomena. Cambridge, Mass.: M.I.T. Press, 1964. 595 p.
Includes recent papers on "the nature and structure of the earth, especially its crust and interior; geophysical studies in the oceans; longitude and latitude determinations, earth tides, and gravity; currents and waves of oceans; glaciers and ice sheets; and various aspects of meteorology and the atmosphere... ." Foreward.

Phillips, O. M. The heart of the earth. San Francisco: Freeman, 1968. 236 p.

Phinney, Robert A., ed. The history of the earth's crust. Princeton, N. J.: Princeton University Press, 1968. 244 p.
Papers given at a Goddard Institute for Space Studies concerning the evolution of the earth's crust, 1966, New York City.

Poldervaart, A. W., ed. Crust of the earth. Baltimore, Maryland: Waverly Press, 1955. 762 p. (Geological Society of America. Special Paper no. 62.)

Ringwood, A. E. and D. H. Green, eds. Phase transformations and the earth's interior; Proceedings of a symposium, Canberra, Australia, 1969. Amsterdam: North Holland, 1970. 520 p. (Upper Mantle Project. Scientific Report no. 26.) (Physics of the Earth and Planetary Interiors, v. 3, 1970.)
Review of knowledge about upper and lower mantles of the earth.

Robertson, Eugene C. Interior of the earth. Washington: Government Printing Office, 1966. 10 p. (U.S. Geological Survey. Circular no. 532.) Introductory.

Royal Astronomical Society, London. The earth today, a collection of papers. Edited by A. H. Cook and T. F. Gaskell. London, 1961. 404 p. (Geophysical Journal, v. 4.)

Stacey, F. D. Physics of the earth. New York: Wiley, 2969. 324 p.

Steinhart, John S. and J. T. Smith, eds. The earth beneath the continents; a volume of geophysical studies in honor of Merle A. Tuve. Washington: American Geophysical Union, 1966. 663 p.

Takeuchi, Hitoshi. Theory of the earth's interior. Waltham, Mass.: Blaisdell, 1966. 131 p.

Tozer, D. C., ed. Non-elastic processes in the mantle; proceedings of the International Upper Mantle Committee Symposium, Newcastle-upon-Tyne, 1966. Oxford: Blackwell, 1968. 450 p. (Geophysical Journal, v. 14.)

Tucker, R. H. and others. Global geophysics. New York: Elsevier, 1970. 199 p.

Vening Meinesz, Felix A. The earth's crust and mantle. New York: Elsevier, 1964. 124 p. (Developments in Solid Earth Geophysics, v. 1.)
Review of present knowledge, with emphasis on convection currents.

GRAVITY, GEODESY

Bibliographia geodaetica; International geodaetische Dokumentation. Berlin: Akademie Verlag, 1963- . Monthly

with annual index.
French, English and Russian subtitles. Classified bibliography.

* * *

Bomford, Guy. Geodesy. 3rd ed. Oxford: Clarendon, 1971. 731 p.

Cook, A. H. and V. T. Saunders. Gravity and the earth. London: Wykeham, 1969. 108 p.

Ewing, Clair E. and M. M. Mitchell. Introduction to geodesy. New York: Elsevier, 1970. 304 p.

Garland, G. D. The earth's shape and gravity. Elmsford, N. Y.: Pergamon, 1965. 188 p.

Heiskanen, W. A. and F. A. Vening Meinesz. The earth and its gravity field. New York: McGraw-Hill, 1958. 470 p.

Mueller, Ivan I. and J. D. Rockie. Gravimetric and celestial geodesy; a glossary of terms. New York: Ungar, 1966. 129 p.
Up-dates and supplements Definitions of surveying, mapping, and related terms, American Society of Civil Engineers, Committee on Definitions of Surveying Terms, 1954; and Definitions of terms used in geodetic and other surveys, H. C. Mitchell, 1949.

Ramberg, Hans. Gravity, deformation and the earth's crust; as studied by centrifugal models. New York: Academic Press, 1967. 214 p.

HEAT SOURCES AND TRANSFER

Summers, W. K. Annotated indexed bibliography of geothermal phenomena. Socorro, New Mexico: New Mexico Institute of Mining and Technology, 1971. 670 p.
Worldwide. 14,000 references.

U.S. Geological Survey. Thermal springs of the United States and other countries of the world--a summary. G. A. Waring. Revised by R. R. Blakenship and R. Bentall. Washington: Government Printing Office, 1965. 383 p. (Professional Paper no. 492.)

* * *

"Geothermal problems." A special issue of Tectonophysics. v. 10, nos. 1-3, 1970. p. 1-390.

Kern, Raymond and A. Weisbrod. Thermodynamics for geologists. Translated from the French by Duncan McKie. San Francisco: Freeman, 1967. 304 p.

Lee, William H. K., ed. Terrestrial heat flow. Washington: American Geophysical Union, 1965. 276 p. (Geophysical Monograph no. 8.) (NAS-NRC Publication no. 1288.)

Tozer, D. C. "Heat transfer and convection currents." In: Royal Society of London. Transactions. v. A258, 1965. p. 252-271.

MAGNETISM, PALEOMAGNETISM

Allan, T. D. "A review of marine geomagnetism." In: Earth science reviews. v. 5, no. 3, 1969. p. 217-254.

Chapman, Sydney and J. Bartels. Geomagnetism. 2nd ed. Oxford: Clarendon Press, 1962. 2 v.

Cook, Alan H. and T. F. Gaskell, eds. The earth in space; a collection of papers dedicated to Sydney Chapman on his 80th birthday. London: Blackwell, 1969. 248 p.

Cox, Allan and R. R. Doell. "Review of paleomagnetism." In: Geological Society of America. Bulletin. v. 71, no. 6, 1960. p. 645-768.

Hospers, J. and J. I. Van Andel. "Paleomagnetism and tectonics, a review." In: Earth science reviews. v. 5, 1970. p. 5-44.

Irving, Edward. Paleomagnetism and its application to geological and geophysical problems. New York: Wiley, 1964. 399 p.

Jacobs, John A. The earth's core and geomagnetism. New York: Macmillan, 1963. 137 p.

Journal of geomagnetism and geoelectricity. Tokyo: Japan Publications Trading Co., 1949- . Quarterly.

Matsushita, S. and W. H. Campbell. Physics of geomagnetic phenomena. New York: Academic Press, 1967. 2 v.

NATO Advanced Study Institute. Freising, Germany, 1967. Earth's particles and fields. Edited by Billy M. McCormac. New York: Reinhold, 1968. 464 p.

NATO Advanced Study Institute on Paleomagnetic Methods. Newcastle-upon-Tyne, 1964. Methods in paleomagnetism. Edited by D. W. Collinson, K. M. Creer and S. K. Runcorn. Amsterdam: Elsevier, 1967. 609 p. (Developments in Solid Earth Geophysics, v. 3.)

Nagata, Takeshi. Rock magnetism. 2nd ed. New York: Plenum, 1961. 350 p.

Rikitake, Tsuneji. Electromagnetism and the earth's interior. New York: Elsevier, 1966. 308 p. (Development in Solid Earth Geophysics, v. 2.)

Strangway, David W. The history of the earth's magnetic field. New York: McGraw-Hill, 1970. 168 p.
Includes discussion of rock magnetism, continental drift, and sea floor spreading.

Wienert, K. A. Notes on geomagnetic observatory and survey practice. Paris: UNESCO, 1970. 217 p.
Magnetic measurements.

SEISMOLOGY (See Also Geologic Hazards)

Bibliography of seismology. Ottawa: Dominion Observatory, 1929- . Semi-annual with annual and decennial subject indexes.

Lomnitz, C. Earthquakes--atlas of world seismicity. New York: Pergamon, 1969. 130 p.

* * *

Adams, W. M. Earthquakes; an introduction to observational seismology. Boston: Heath, 1964. 122 p.

Akademiia Nauk SSSR. Institut Fiziki Zemli. Problems in deep seismic sounding. Edited by S. M. Zverev. Translated by T. J. Smith. New York: Consultants Bureau, 1967. 166 p.

----------. Problems of theoretical seismology and physics of the earth's interior. Edited by V. A. Magnitskii. Translated by Z. Lerman. Jerusalem: Israel Program for Scientific Translations, 1963. 148 p. (IPST no. 814. Available from NTIS, Arlington, Va.)

Andrianova, Z. S. and others. Seismic Love waves. Translated by F. M. C. Goodspeed. New York: Consultants Bureau, 1967. 91 p.

Båth, Markus. Mathematical aspects of seismology. New York: Elsevier, 1968. 415 p. (Developments in Solid Earth Geophysics, v. 4.)

Bullen, Keith E. An introduction to the theory of seismology. 3rd ed. Cambridge, England: Cambridge University Press, 1963. 381 p.

Cagniard, Louis. Reflection and refraction of progressive seismic waves. Translated by E. A. Flinn and C. H. Dix. New York: McGraw-Hill, 1962. 282 p.

"Earthquake mechanics." A special issue of Tectonophysics. v. 9, no. 2/3. 1970.
Emphasis on earthquake prediction.

Gutenberg, Beno and C. F. Richter. Seismicity of the earth and associated phenomena. 2nd ed. Princeton, N. J.: Princeton University Press, 1954. 310 p.

Hodgson, John H. Earthquakes and earth structure. Englewood Cliffs, N. J.: Prentice-Hall, 1964. 166 p.

Matuzawa, Takeo. Study of earthquakes. Tokyo: Uno Shoten, 1964. 213 p.

National Research Council. Division of Earth Sciences. Committee on Seismology. Seismology: responsibilities and requirements of a growing science. Washington: National Academy of Science, 1969. 97 p.

Richter, Charles F. Elementary seismology. San Francisco: Freeman, 1958. 768 p.

Rothe, J. P. The seismicity of the earth (1953-1965). Paris: UNESCO, 1969. 336 p. (Earth Sciences Series no. 1.)

A continuation of Gutenberg and Richter. Seismicity of the earth.

Seismic methods for monitoring underground explosions. Stockholm: International Institute for Peace and Conflict Research, 1968. 130 p.
Current techniques of seismic verification to distinguish earthquakes and underground explosions.

Seismic refraction prospecting. Edited by Albert W. Musgrave. Tulsa: Society of Exploration Geophysicists, 1967. 604 p.

Seismological Society of America. Bulletin. Baltimore, 1911- . 6/yr.

White, James E. Seismic waves; radiation, transmission, and attenuation. New York: McGraw-Hill, 1965. 302 p.

APPLIED AND EXPLORATORY GEOPHYSICS

Dobrin, Milton B. Introduction to geophysical prospecting. 2nd ed. New York: McGraw-Hill, 1960. 446 p.

Eve, Arthur S. and D. A. Keys. Applied geophysics in the search for minerals. 4th ed. Cambridge, England: Cambridge University Press, 1954. 382 p.

Exploration geophysics. v. 48- . Edited by M. K. Polshkov. Translation editor, G. V. Keller, New York: Plenum, 1968- .
Translated from Russian. Includes papers on topics as seismic exploration, electrical methods, gravity and magnetic field surveys.

The geophysical directory. 25th ed. Houston: Geophysical Directory, Inc., 1970. 435 p.
Includes contractors, suppliers, oil and mining companies.

Grant, F. S. and G. F. West. Interpretation theory in applied geophysics. New York: McGraw-Hill, 1965. 583 p.

Griffiths, D. H. and R. F. King. Applied geophysics for engineers and geologists. Elmsford, N. Y.: Pergamon, 1965. 223 p.

Guyod, Hubert and L. E. Shane. Geophysical well logging. V. 1. Introduction to geophysical well logging, acoustical logging. Houston: H. Guyod, 1969. 256 p.

Heiland, Carl A. Geophysical exploration. New York: Prentice-Hall, 1940. 11, 013 p. (reprinted by Hafner, 1963).

Jakosky, Jay L. Exploration geophysics. 2nd ed. Los Angeles: Trija Publishing Co., 1950. 1195 p.

Keller, George V. and F. C. Frischknecht. Electrical methods in geophysical prospecting. New York: Pergamon, 1966. 519 p.

Kelley, Dana R. A summary of major geophysical logging methods. Harrisburg: Pennsylvania Geological and Topographical Survey, 1969. 82 p. (Mineral Resources Report no. 61.)

Krinitzsky, E. L. Radiography in the earth sciences and soil mechanics. New York: Plenum, 1970. 180 p.

Lehrbuch der Angewandten Geologie. Edited by A. Bentz and H. J. Martini. Stuttgart: Enke, 1961- . V. 1. Allgemeine Methoden. 1961. 1071 p.; v. 2. Geowissenschaftliche Methoden. Pt. 1. 1968. 1354 p. A comprehensive survey of specialized geologic techniques.

Maliuga, Dimitrii P. Biogeochemical methods of prospecting. New York: Consultants Bureau, 1964. 205 p.

Musgrave, Albert W. Seismic refraction prospecting. Tulsa, Oklahoma: Society of Exploration Geophysicists, 1967. 604 p.

Nettleton, Lewis I. Geophysical prospecting for oil. McGraw-Hill, 1940. 444 p.

Parasnis, D. S. Mining geophysics. New York: Elsevier, 1966. 356 p. (Methods in Geochemistry and Geophysics, v. 3.)
A practical handbook which includes description of instruments and field techniques as related to Swedish ore field exploration.

----------. Principles of applied geophysics. London: Methuen, 1962. 176 p.

Runcorn, S. K., ed. Methods and techniques in geophysics. New York: Interscience, 1960-1966. 2 v.

Society of Exploration Geophysicists. Mining Geophysics Volume Editorial Committee. Society of Exploration Geophysicists' Mining Geophysics. Tulsa, Oklahoma, 1966-67. 2 v.
V. 1. Case histories; v. 2. Theory.

Wyllie, Malcolm R. J. Fundamentals of well log interpretation. 3rd ed. New York: Academic Press, 1963. 238 p.

EXTRA-TERRESTRIAL GEOLOGY

REFERENCE WORKS

Brown, Harrison S. A bibliography on meteorites. Chicago: University of Chicago Press, 1953. 686 p. (An International Catalogue of Meteorites, v. 1.)
Covers literature published from 1941-1950.

Directory of meteorite collections and meteorite research. Paris: UNESCO, 1968. 50 p.

Freeberg, Jacquelyn H. Terrestrial impact structures--a bibliography. Washington: Government Printing Office, 1966. 91 p. (U.S. Geological Survey. Bulletin no. 1220.)
Worldwide in scope.

----------. 1965-1968. Washington: Government Printing Office, 1969. 39 p. (U.S. Geological Survey. Bulletin no. 1320.)

Hey, Max H. Catalogue of meteorites. London: British Museum (Natural History), 1966. 637 p.

Horback, Henry and E. J. Olson. "Catalog of the collection of meteorites in Chicago Natural History Museum." In: Fieldiana: Geology, v. 15, no. 3, 1965. p. 175-319.

U.S. Department of the Air Force. Office of Aerospace Research. Cambridge Research Laboratories. Bibliography of lunar and planetary research 1961- . Edited by John W. Salisbury and others. Springfield, Virginia: National Technical Information Service, 1962- . (AD285 646; AD608 825; AD679 995; AD697 845.)

U.S. Department of the Army. Redstone Scientific Information Center (Alabama). Tektites: a bibliography. C. F. Shenk. Washington: Office of Technical Services, 1964. 122 p. (AD-430 350.)

TEXTS AND TREATISES

French, Beavan M. and N. M. Shorts, eds. Shock metamorphism of natural materials. Baltimore: Mono Book Corp., 1968. 644 p.
Proceedings of a conference on meteorite impacts as a geologic process, held in 1966 at the Goddard Space Flight Center.

Geiss, J. and E. D. Goldberg, eds. Earth science and meteoritics... . Amsterdam: North-Holland, 1963. 312 p.

Haber, Heinz. Our blue planet; the story of the earth's evolution. Translated by E. Stuhlinger. New York: Scribner, 1969. 86 p.

Heide, Fritz. Meteorites. Translated by E. Anders. Chicago: University of Chicago Press, 1964. 144 p.

Kaula, W. M. An introduction to planetary physics: the terrestrial planets. New York: Wiley, 1968. 490 p.

Kelley, Frederic R. "Geologic aspects of space exploration," "Studying the moon's surface," "Recent studies in space geology." In: California. Division of Mines and Geology. Mineral information service. v. 17, nos. 10, 11, and 12, 1964.

King, J. W. and W. S. Newman, eds. Solar terrestrial physics. New York: Academic Press, 1967. 390 p.

Krinov, E. L. Giant meteorites. New York: Pergamon, 1966. 397 p.
"One of the better primers on the subject of meteorites." Geophysical journal. June, 1967.

----------. Principles of meteoritics. Edited by M. Brown. Translated by I. Vidzinuas. New York: Pergamon, 1960. 535 p.

Kuiper, Gerard P. and B. Middlehurst, eds. The solar system. Chicago: University of Chicago Press, 1953-1963. V. 1. The solar system. 1953; v. 2. The earth as a planet. 1954; v. 3. Planets and satellites. 1961; v. 4. The moon, meteorites and comets. 1963.

Lowman, Paul D., Jr. Space panorama. Zurich, Switzerland: Reinhold A. Muller, 1968.
Color photographs made on Gemini flights with geological annotations.

Marsden, B. G. and A. G. W. Cameron, eds. Earth-moon systems. New York: Plenum, 1966. 288 p.
Proceedings of an international conference, sponsored by the Institute for Space Studies of the Goddard Space Flight Center, NASA, in 1964.

Mason, Brian. Meteorites. New York: Wiley, 1962. 274 p.

"Meteorites and tektites." In: Geochimica et cosmochimica acta. v. 31, 1967. p. 1535-2108.
A special issue of 43 papers dedicated to Harvey H. Nininger.

Millman, Peter M., ed. Meteorite research. Dordrecht, the Netherlands: Reidel, 1969. 940 p.
Proceedings of a symposium held in Vienna, Austria, Aug. 1968.

Moore, Carleton B., ed. Researches on meteorites. New York: Wiley, 1962. 227 p.

NATO Advanced Study Institute, Newcastle-upon-Tyne, 1966. Mantles of the earth and terrestrial planets. Edited by S. K. Runcorn. New York: Interscience, 1967. 584 p.
A symposium volume which brings together the new discoveries in the study of planetary interiors.

----------, 1967. The application of modern physics to the earth and planetary interiors. Edited by S. K. Runcorn. New York: Wiley, 1969. 692 p.

Odishaw, Hugh, ed. Earth in space. New York: Basic Books, 1967. 340 p.

Physics of the earth and planetary interiors. Amsterdam: North Holland, 1967- . 6/yr.

Ronca, Luciano. "An introduction to the geology of Mars." In: Geologists Association (London). Proceedings. v. 81, pt. 1, 1970. p. 111-128.

Special Committee for the International Years of the Quiet Sun. *Annals of the IQSY.* Edited by C. M. Minnis. Cambridge: M.I.T. Press, 1968- .
Solar-terrestrial physics: terrestrial aspects, 1969. 460 p.

Wood, John A. *Meteorites and the origin of planets.* New York: McGraw-Hill, 1968. 117 p.

LUNAR STUDIES

Freeberg, Jacquelyn H. *Bibliography of the lunar surface.* Springfield, Va.: National Technical Information Service, 1970. 374 p. (PB 194 206.)

Lunar surface studies; a continuing bibliography. Washington: Government Printing Office, 1964- . (National Aeronautics and Space Administration SP-7003, with 4 supplements to date.)
Annotated bibliography with indexes.

* * *

Alter, Dinsmore, ed. *Lunar atlas.* New York: Dover, 1968.
Outstanding photographs.

Apollo 11 Lunar Science Conference. Edited by Philip H. Abelson. Washington: American Association for the Advancement of Science, 1970. 368 p.
144 reports presented at the Lunar Sample Analysis Conference, Houston, 1970, and originally published in Jan. 30, 1970 issue of *Science.*

Baldwin, Ralph B. *A fundamental survey of the moon.* New York: McGraw-Hill, 1965.

----------. *The measure of the moon.* Chicago: University of Chicago Press, 1963. 488 p.

Fielder, Gilbert, ed. *Geology and physics of the moon; a study of some fundamental problems.* New York: American Elsevier, 1971. In press.

----------. *Lunar geology.* London: Butterworth, 1965. 184 p.

Kopal, Zdenek. The moon. 2nd ed. Dordrecht, the Netherlands: Reidel, 1969. 525 p.

----------, Josef Klepesta, and Thomas W. Rackham. Photographic atlas of the moon. New York: Academic Press, 1965. 277 p.

Kuiper, Gerard P. Photographic lunar atlas. Tucson, Arizona: University of Arizona Press, 1960. 23 p.
Superb photographs. Supplements 1 and 2. Whitaker, E. A. and D. W. G. Arthur, comp. Orthographic atlas of the moon. Tucson: University of Arizona Press, 1961, 1963. (The Lunar and Planetary Laboratory, nos. 2 and 3.)

Levinson, A. A., ed. Proceedings of the Apollo 11 lunar science conference, Houston, Texas, 1970. New York: Pergamon, 1970. 3 v.
V. 1. Mineralogy and petrology; v. 2. Chemical and isotope analyses; v. 3. Physical properties. This was also published as supplement no. 1 of Geochimica et cosmochimica acta. v. 34, 1970.

Lowman, Paul D., Jr. Lunar panorama; a photographic guide to the geology of the moon. Feldmeilen, Switzerland: Reinhold A. Muller, 1969. 136 p.
Photos from Surveyors, Lunar Orbitors and Apollo 8.

Mason, Brian and W. G. Nelson. The lunar rocks. New York: Wiley, 1970. 192 p.

The moon; international journal of lunar studies. Dordrecht, the Netherlands. Reidel, 1969- . Quarterly.

Mutch, Thomas A. Geology of the moon. Princeton, N. J.: Princeton University Press, 1970. 324 p.

"Special papers--Apollo 11 Symposium." In: Journal of geophysical research. v. 75, 1970. p. 6453-6574.

Spurr, J. E. Geology applied to selenology. Lancaster, Pennsylvania: Science Press, 1944, 1945 (v. 1, 2). Concord, N. H.: Rumford Press, 1948, 1949 (v. 3, 4).

Whipple, H. E., ed. Geological problems in lunar research; a symposium. New York: New York Academy of Sciences, 1965. (Annals v. 123, article 2.)

GEOCHEMISTRY
(See Also Mineralogy - Methodology)

REFERENCE WORKS

Chemical abstracts. Washington: American Chemical Society, 1907- . Bi-weekly.
Includes sections on phase and chemical equilibriums, thermodynamics and thermochemistry, crystallization and crystal structure, mineralogical and geological chemistry.

Clark, Sydney P., Jr. Handbook of physical constants. Revised ed. New York: Geological Society of America, 1966. 587 p. (Memoir no. 97.)

Handbook of chemistry and physics. 52nd ed. Cleveland, Ohio: Chemical Rubber Company, 1971-1972.

Helgeson, Harold C., T. H. Brown, and R. H. Leeper. Handbook of theoretical activity diagrams depicting chemical equilibria in geologic systems involving an aqueous phase at one atm and 0° to 300° C. San Francisco: Freeman, 1969. 256 p.

Lange, Norbert, and Gordon M. Forker. Handbook of chemistry. 10th ed., revised, New York: McGraw-Hill, 1967.

Meites, Louis. Handbook of analytical chemistry. New York: McGraw-Hill, 1963.

Robie, Richard A. and D. R. Waldbaum. Thermodynamic properties of minerals and related substances at 298.15° K (25.0° C) and one atmosphere (1.013 bars) pressure and at higher temperatures. Washington: Government Printing Office, 1968. 256 p. (U.S. Geological Survey. Bulletin no. 1259.)

Signeur, Austin V. Guide to gas chromatography literature. New York: Plenum, v. 1, 1964; v. 2, 1967.

U.S. Geological Survey. Data of geochemistry. 6th ed. Edited by M. Fleischer. Washington: Government Printing Office, 1962- . (Professional Paper no. 440.)
Scheduled to have 36 parts.

Wedepohl, Karl H., ed. Handbook of geochemistry. Berlin: Springer-Verlag, 1969- . 2 v.
V. 1. Includes fundamental facts of geochemistry, geophysics, and cosmochemistry; v. 2. Loose-leaf, with yearly additions, arranged according to atomic numbers.

TEXTS AND TREATISES

Abelson, Philip H., ed. Researches in geochemistry. New York: Wiley, 1959-1967. 2 v.

Ahrens, Louis H. Distribution of the elements in our planet. New York: McGraw-Hill, 1965. 110 p.

----------, ed. Origin and distribution of the elements; proceedings of a symposium. New York: Pergamon, 1968. 1200 p.

Akademiia nauk SSSR. Institut geokhimii i analiticheskoi khimii. Chemistry of the earth's crust. Edited by A. P. Vinogradov. Jerusalem: Israel Program for Scientific Translations, 1966-67. 2 v.

Barnes, Hubert L. Geochemistry of hydrothermal ore deposits. New York: Holt, Rinehart and Winston, 1967. 688 p.

Breger, Irving A., ed. Organic geochemistry. New York: Pergamon, 1963. 658 p.

Calvin, Melvin. Chemical evolution. Oxford: Clarendon Press, 1969. 278 p.

Charlot, G. Colorimetric determination of elements, principles and methods. New York: Elsevier, 1964. 449 p.

Chemical geology. Amsterdam: Elsevier, 1966- . 6/yr.

Chemie der Erde. Jena: Gustav Fischer, 1914- . Irregular.

Cotton, F. Albert and G. Wilkinson. Advanced inorganic chemistry, a comprehensive text. 2nd ed. New York: Interscience, 1966. 1136 p.

Day, Frank H. The chemical elements in nature. New York: Reinhold, 1964, c1963. 372 p.
"...practical sources of the chemical materials and the various kinds of transformations they undergo under natural conditions."

Degens, Egon T. Geochemistry of sediments: a brief survey. Englewood Cliffs, N. J.: Prentice-Hall, 1965. 320 p.

---------- and D. A. Ross, eds. Hot brines and recent heavy metal deposits in the Red Sea. Berlin: Springer, 1969. 600 p.
76 contributors. A geochemical and geophysical account.

Eglinton, Geoffrey and Sister Mary T. J. Murphy, eds. Organic geochemistry; methods and results. New York: Springer, 1969. 828 p.
Chemical, biological, and geological aspects of the subject presented by 36 contributors.

Energlyn, W. D. and L. Brealy. Analytical geochemistry. Amsterdam: Elsevier, in press. (Methods in Geochemistry and Geophysics, v. 5.)

Ernst, W. Geochemical facies analysis. Amsterdam: Elsevier, 1970. 152 p. (Methods in Geochemistry and Geophysics, v. 11.)

Fyfe, W. S. Geochemistry of solids; an introduction. New York: McGraw-Hill, 1964. 199 p.

Garrels, Robert M. Mineral equilibria at low temperature and pressure. New York: Harper, 1960. 254 p.

---------- and C. L. Christ. Solutions, minerals, and equilibria. New York: Harper, 1965. 450 p.

Geochemistry and mineralogy of rare elements and genetic types of their deposits. Edited by K. A. Vlasov. Translated by Z. Lerman. Jerusalem: Israel Program for Scientific Translations, 1966-68. 3 v.
V. 1. Geochemistry of rare elements; v. 2. Mineral-

ogy of rare elements; v. 3. Genetic types of rare-element deposits. A comprehensive treatise on the geochemistry, mineralogy, and geology of rare elements collectively written by the staff of the Institute of Mineralogy, Geochemistry and Crystal Chemistry of Rare Elements, Moscow.

Geochemistry international. Washington: The Geochemical Society, 1964- . 6/yr.
Translations of selected articles from the Russian Geokhimiya and other geochemical papers from other languages, primarily Russian, German, French, and Japanese. Succeeds Geochemistry, the cover-to-cover translation of Geokhimiya for the years 1956-1963.

Geochimica et Cosmochimica Acta. New York: Pergamon, 1950- . Monthly.

Goldschmidt, Victor M. Geochemistry. Edited by Alex Muir. Oxford: Clarendon Press, 1958. 730 p.

International Congress on Organic Geochemistry. 2nd, 1964- . Proceedings; advances in organic geochemistry. New York: Pergamon, 1966- . Irregular. Latest is 4th, Amsterdam, 1968. Pergamon, 1969. 617 p.

Keller, Walter D. Chemistry in introductory geology. 4th ed. Columbia, Mo.: Lucas Brothers, 1969. 108 p.

Krauskopf, Konrad B. Introduction to geochemistry. New York: McGraw-Hill, 1967. 721 p.

Manskaya, S. M. and T. V. Drozdova. Geochemistry of organic substances. Translated and edited by L. Shapiro and I. A. Breger. New York: Pergamon, 1968. 345 p.

Mason, Brian H. Principles of geochemistry. 3rd ed. New York: Wiley, 1966. 329 p.

Olphen, Hendrik van. An introduction to clay colloid chemistry, for clay technologists, geologists, and soil scientists. New York: Interscience, 1963. 301 p.

Perelman, Aleksandr I. Geochemistry of epigenesis. Translated from the Russian by N. N. Kohanowski. New York: Plenum, 1967. 266 p.

Physics and chemistry of the earth, a progress series. Edited by L. H. Ahrens, K. Rankama, and S. K. Runcorn. New York: Pergamon, 1956- . Irregular.

Povarennykh, A. S. Crystal chemical classification of minerals. Translated by J. E. S. Bradley. New York: Plenum, 1970. 600 p.

Rankama, Kalervo. Isotope geology. New York: McGraw-Hill, 1954. 535 p.

----------. Progress in isotope geology. New York: Interscience, 1963. 705 p.
A continuation of Isotope geology.

---------- and Th. G. Sahama. Geochemistry. Chicago: University of Chicago Press, 1950. 911 p.

Ritchie, Arthur S. Chromatography in geology. Amsterdam: Elsevier, 1964. 185 p. (Methods in Geochemistry and Geophysics, v. 1.)

Russell, R. D. and R. M. Farquhar. Lead isotopes in geology. New York: Interscience, 1960. 243 p.

Smales, A. A. and L. R. Wagner, ed. Methods in geochemistry. New York: Interscience, 1960. 464 p.

Smith, Frederick G. Physical geochemistry. Reading, Mass.: Addison-Wesley, 1963. 624 p.

Volborth, Alexis. Elemental analysis of geochemistry--A: major elements. Amsterdam: Elsevier, 1969. 373 p. (Methods in Geochemistry and Geophysics, v. 8.)

Wedepohl, Karl H. Geochemistry. Translated by E. Althaus. New York: Holt, Rinehart and Winston, 1970. 224 p.

Wells, Alexander F. Structural inorganic chemistry. 3rd ed. Oxford, England: University Press, 1962. 1055 p.

APPLIED AND EXPLORATORY GEOCHEMISTRY

Boyle, R. W. and R. G. Garrett. "Geochemical prospecting --a review of its status and future." In: Earth-science reviews. v. 6, no. 1, 1970. p. 51-75.

Ginzburg, I. I. Principles of geochemical prospecting; techniques of prospecting for non-ferrous ores and rare metals. Translated by V. P. Sokoloff. New York: Pergamon, 1960. 311 p.

Hawkes, Herbert E. and J. S. Webb. Geochemistry in mineral exploration. New York: Harper, 1962. 415 p.

Hillebrand, William F. and G. E. F. Lundell. Applied inorganic analysis, with special reference to the analysis of metals, minerals, and rocks. 2nd ed. New York: Wiley, 1953. 1034 p.

Nagy, Bartholomew and U. Colombo, eds. Fundamental aspects of petroleum geochemistry. Amsterdam: Elsevier, 1967. 388 p.

ECONOMIC GEOLOGY

REFERENCE WORKS

Annotated bibliography of economic geology. Mount Pleasant, Mich.: Economic Geology Publishing Co., 1929-1971(?).
This annotated bibliography of economic and applied geology will cease with v. 39, as the bibliographies by the U.S. Geological Survey and the Geological Society of America now cover the same topics. Cumulative index was published for 1928-1954.

Cagnacci-Schwicker, Angelo. International dictionary of metallurgy, mineralogy, geology, mining and oil industries: English, French, German, Italian. Milan: Technoprint International, 1968; reprinted by McGraw-Hill, 1970. 1530 p.
A dictionary of equivalents in 4 languages.

Economic geology. Mount Pleasant, Mich.: Economic Geology Publishing Co., 1905- . Semiquarterly.
Since 1961, a regular feature in the "reviews section" of Economic geology has been "Geology of ore deposits," abstracts of Geologiy Rudnykh Mestorozhdeniy published by the Academy of Sciences of the U.S.S.R.

Kaplan, Stuart R., ed. A guide to information sources in mining, minerals, and geosciences. New York: Wiley, 1965. 599 p. (Guides to Information Sources in Science and Technology, v. 2.)
Pt. 1. Organizations; pt. 2. Literature.

Nelson, Archibald and K. D. Nelson. Dictionary of applied geology; mining and civil engineering. London: Newnes, 1967. 421 p.

U.S. Bureau of Mines. List of publications issued by the Bureau of Mines from July 1, 1910 to January 1, 1960 with subject and author index. Washington: Government Printing Office, 1960. 826 p.
Kept up-to-date by annual supplements, which cumulate every 5 years.

----------. Mineral facts and problems. Washington: Government Printing Office, 1970. 1291 p. (Bulletin no. 650.)

----------. Minerals yearbook. Washington: Government Printing Office, 1932/33- . Annual.
Worldwide statistics on metals, non-metals, petroleum, and minerals products. Continues Mineral Resources of the United States, published by the Geological Survey and Bureau of Mines from 1883-1933. Production figures may be supplemented by referring to the February or March issue of Engineering and Mining journal, devoted each year to mineral production, prices, and outlook. Latest Minerals yearbook divided into: v. 1-2. Metals, Minerals and fuels; v. 3. Area reports: domestic; v. 4. Area reports: international.

Wöhbier, Herbert and others. Worldwide directory of mineral industries education and research. Houston: Gulf Publishing Co., 1968. 451 p.
An international directory of geoscience departments in universities and institutes, with an indication of current research.

TEXTS AND TREATISES

Economic geology: fiftieth anniversary volume economic geology 1905-1955. Edited by Alan M. Bateman. Urbana, Illinois: Economic Geology Publishing Co., 1955. 2 v.

Flawn, Peter T. Mineral resources; geology, engineering, economics, politics, law. Chicago: Rand McNally, 1966. 406 p.
Introduction to mineral economics.

Herfindahl, Orris C. Natural resource information for economic development. Baltimore: Johns Hopkins Press, 1969. 212 p.

Lovering, Thomas N. Minerals and world affairs. Englewood Cliffs, N. J.: Prentice-Hall, 1963. 394 p.

National Academy of Sciences/National Research Council. Committee on Resources and man. Resources and man: a study and recommendations. San Francisco: Freeman, 1969. 259 p.
Eight authorities in different scientific fields summarize

a two year study of world resources.

Noble, James A. "Metal provinces of the western United States." In: Geological Society of America. Bulletin. v. 81, 1970. p. 1607-1624.

Park, Charles F., Jr. Affluence in jeopardy: minerals and the political economy. San Francisco: Freeman, 1969. 368 p.
Written by an economic geologist.

Voskuil, Walter H. Selected readings in mineral economics. Reno: Nevada Bureau of Mines, 1966. 18 p. (Report no. 12.)

TECHNOLOGY, EXPLOITATION, MINING

Colorado School of Mines. Mining Engineering Department. Annotated bibliography on selected mining subjects. Golden, Colo.: Colorado School of Mines, 1966. 147 p. (Colorado School of Mines. Quarterly, v. 61, no. 2.)
Topics include drilling and blasting; large hole boring; grouting; and reclamation of mined lands.

Peele, R. Mining engineer's handbook. 3rd ed. New York: Wiley, 1941. 2 v. in 1.

Schenck, George H. K. "Selected bibliography serves as a guide to modern mine valuation methods." In: Engineering and mining journal. v. 167, no. 9, 1966. p. 206-208, 260.

Thrush, Paul W. A dictionary of mining, mineral and related terms. Washington: Government Printing Office, 1968. 1269 p. (U.S. Bureau of Mines, Special Publication.)

* * *

Commonwealth Mining and Metallurgical Congress. 6th, Canada, 1957. Methods and case histories in mining geophysics. Ottawa: Canadian Institute of Mining and Metallurgy, 1957. 357 p.

Griffith, Stanley V. Alluvial prospecting and mining. 2nd ed. New York: Pergamon, 1960. 245 p.

International symposium on operations research and computer applications in the mineral industry. 1961- . Imprint varies.
The 7th, 1968 was published as *Colorado School of Mines. Quarterly.* v. 64, no. 3, 1969. 393 p.

Lewis, Robert S. *Elements of mining.* 3rd ed. revised by George B. Clark. New York: Wiley, 1964. 768 p.

McKinstry, Hugh E. *Mining geology.* New York: Prentice-Hall, 1948. 680 p.

Michelson, Robert A. *Prospecting for uranium.* Golden, Colo.: Colorado School of Mines, 1966. 9 p. (*Colorado School of Mines. Mineral Industries Bulletin,* v. 9, no. 4.)

Parks, R. D. *Examination and valuation of mineral property.* 4th ed. Reading, Massachusetts: Addison-Wesley, 1957. 507 p.

Rudawsky, Oded. *Economic evaluation techniques for mining investment projects.* Golden, Colorado: Colorado School of Mines, 1970-1971. 2 v. (*Colorado School of Mines. Mineral Industries Bulletin,* v. 13, no. 6; v. 14, no. 1.)

Sinclair, John. *Quarrying, opencast and alluvial mining.* New York: Elsevier, 1970. 375 p.

Sloane, Howard N. and L. L. Sloane. *A pictorial history of American mining.* New York: Crown, 1970. 342 p.

Staley, William W. *Introduction to mine surveying.* 2nd ed. Stanford, California: Stanford University Press, 1964. 303 p.

Stoček, Bohuslav. *Introduction to mining.* London: Lange, Maxwell and Springer, 1954. 2 v.

Stout, K. S. *Mining methods and equipment illustrated.* Butte, Montana: Montana Bureau of Mines and Geology, 1967. 97 p. (Bulletin no. 63.)

ENERGY RESOURCES

Averitt, Paul and M. Devereux Carter. *Selected sources of information on United States and world energy resources:*

an annotated bibliography. Washington: Government Printing Office, 1970. 21 p. (U.S. Geological Survey. Circular no. 641.)

Landsberg, Hans H. and S. H. Schurr. Energy in the United States: sources, uses, and policy issues. New York: Random House, 1968. 242 p.

PETROLEUM

REFERENCE WORKS

American Institute of Mining, Metallurgical, and Petroleum Engineers. Petroleum Branch. Production Review Committee. Statistics of oil and gas development and production. New York, 1945-1958.
Beginning 1959 (v. 14), publication is continued as International oil and gas development review of 19 .

American Petroleum Institute. API abstracts of refining literature. New York, 1961- . Weekly.
Petrochemical, geological and refining literature abstracts from U.S., French, German, British and Soviet journals. Patents covered under a companion service, API abstracts of patents, 1961- . These abstracts, together with Tulsa University's Petroleum abstracts, provide the principal access to technical literature for the U.S. petroleum industry.

----------. Petroleum facts and figures. New York, 1928- . Biennial.
The 1959 Centennial edition gives historical statistics of the petroleum industry.

Elsevier's oilfield dictionary in five languages: English/American, French, Spanish, Dutch and German. Edited by J. Moltzer. New York: Elsevier, 1965. 162 p.

Institute of Petroleum. Abstracts quarterly. London, 1969-

Abstracts formerly incorporated in the Journal of the Institute of Petroleum.

International Oil Scouts Association. International oil and gas development yearbook. Austin, Texas, 1931- . Annual.
Currently issued in 2 parts--exploration and production.

International petroleum encyclopedia. Tulsa: Petroleum Publishing Co., 1970. 400 p.
Arranged by area giving summary and forecast, maps, current reserves and production, refining and consumption, historical data, cost indexes, and government agencies.

Koninklijke Nederlandsche Petroleum Maatschappij, N.V. The petroleum handbook. 5th ed. London: Shell International, 1966. 318 p.

Organization of Petroleum Exporting Countries. Sources of petroleum statistical information. Vienna, 1966. Looseleaf.
Identification of main sources of petroleum statistics as they appear in the major trade periodicals.

Seismograph Service Corporation. Bibliography of stratigraphic traps, a compilation of geological and geophysical references, with annotations. Edited by W. E. Pugh and B. G. Preston. Tulsa, 1953, c1951. 195 p.

Special Libraries Association. Petroleum Section. Committee on U.S. Sources of Petroleum and Natural Gas Statistics. U.S. sources of petroleum and natural gas statistics. Margaret M. Rocq. New York: Special Libraries Association, 1961. 94 p.

Swanson, E. G. A century of oil and gas in books, a descriptive bibliography. New York: Appleton-Century, 1960. 214 p.

Tulsa. University. Information Services Department. Petroleum abstracts. Tulsa, 1961- . Weekly.
Approximately 10,000 abstracts a year from world literature. This service, together with the API abstracts to refining literature, provide the U.S. petroleum industry with its principal access to technical information.

U.S.A. oil industry directory. Tulsa: Petroleum Publishing Co., 1970. 352 p.

U.S. Bureau of Mines. Selected list of Bureau of Mines publications on petroleum and natural gas, 1910-1962. V. V. Hutchinson. Washington: Government Printing Office, 1964. 98 p. (Information Circular no. 8240.)

U. S. Geological Survey. Oil and gas investigations map, OM- .
A series of geologic and structure maps. Indexed in U. S. Geological Survey. Publications of the Geological Survey, 1879-1961 and subsequent annual cumulations.

TECHNOLOGY, EXPLOITATION

Canadian Petroleum Association. Surveying and mapping colloquium for the petroleum industry; proceedings of a colloquium, October 1969. Edmonton: University of Alberta, Extension Department, 1969. 175 p.

Dix, Charles H. Seismic prospecting for oil. New York: Harper, 1952. 414 p.

Dunstan, A. E. and others, eds. The science of petroleum; a comprehensive treatise of the principles and practice of the production, refining, transportation, and distribution of the mineral oil. London: Oxford Press, 1938-

Multivolume.

Frick, Thomas C., ed. Petroleum production handbook. New York: McGraw-Hill, 1962. 2 v.
V. 1. Mathematics and production equipment; v. 2. Reservoir engineering.

Haun, John D. and L. W. LeRoy, eds. Subsurface geology in petroleum exploration, a symposium. Golden, Colorado: Colorado School of Mines, 1958. 887 p.
Interpretation of well data and logs.

Heroy, William B., Sr., ed. Unconventional methods in exploration for petroleum and natural gas. Dallas: Institute for the Study of Earth and Man, Southern Methodist University, 1969. 241 p.

Kartsev, A. A. and others. Geochemical methods of prospecting and exploration for petroleum and natural gas. Berkeley: University of California Press, 1959. 349 p.

Keller, George V. Electrical prospecting for oil. Golden, Colo.: Colorado School of Mines, 1968. 268 p. (Colorado School of Mines. Quarterly, v. 63, no. 2.)

Komar, C. A. "Rock mechanics--its role in oil recovery." In: Producers monthly. v. 32, no. 11, 1968. p. 2-7.

LeRoy, Leslie W. and J. W. Low. Graphic problems in petroleum geology. New York: Harper, 1954. 238 p.

Lynch, Edward J. Formation evaluation. New York: Harper, 1962. 422 p.

Merriam, Daniel F., ed. Symposium on computer applications in petroleum exploration. Lawrence: Kansas Geological Survey, 1969. 41 p. (Computer Contribution no. 40.)

Meurs, P. H. van. Petroleum economics and offshore mining legislation, a geological evaluation. New York: Elsevier, in press.

Moody, Graham B., ed. Petroleum exploration handbook; a practical manual summarizing the application of earth sciences to petroleum exploration. New York: McGraw-Hill, 1961. 1 v.

Muskat, Morris. Physical principles of oil production. New York: McGraw-Hill, 1949. 922 p.

Nind, T. E. W. Principles of oil well production. New York: McGraw-Hill, 1964. 364 p.

Pirson, Sylvain J. Handbook of well log analysis for oil and gas formation evaluation. Englewood Cliffs, N. J.: Prentice-Hall, 1963. 326 p.
Complete text and treatment of quantitative log analysis.

----------. Oil reservoir engineering. 2nd ed. New York: McGraw-Hill, 1958. 735 p.

Southwestern Legal Foundation. Exploration and economics of the petroleum industry. Houston: Gulf Publishing Co., 1963-1967. 5 v.

Uren, Lester C. Petroleum production engineering. New York: McGraw-Hill, 1953-1956.
V. 1. Oil field development. 4th ed. 1956. 764 p.;
v. 2. Oil field exploitation. 3rd ed. 1953. 807 p.;
v. 3. Petroleum production economics. 6th ed. 1956. 792 p.

Wheeler, Robert R. and M. Whited. Oil: from prospect to pipeline. 2nd ed. Houston: Gulf Publishing Co., 1971. 136 p. Guidebook for oil company personnel and investors. Expanded to global oil and increasing fuel deficits.

Williamson, Harold and A. R. Daum. American petroleum industry; the age of energy, 1899-1959. Evanston, Illinois: Northwestern Univ., 1964. 928 p.

Zaba, Joseph and W. T. Dohert. Practical petroleum engineer's handbook. 5th ed. Houston: Gulf Publishing Co., in press.

OCCURRENCES, GEOLOGY

American Association of Petroleum Geologists. Bulletin. Tulsa, 1917- . Monthly.
This periodical includes book reviews and lists new publications from the world literature. Noteworthy annual features are "Developments in foreign fields" and "Check list of 19-- field trips in United States, Canada, and selected foreign areas," issued in the July or August issue; and "North American drilling activity in 19--," issued in the June issue.

American Association of Petroleum Geologists. Habitat of oil... . Edited by L. G. Weeks. Tulsa, 1959. 1384 p. Symposium on petroleum occurrence in various parts of the world, and recent research on origin of oil.

----------. Possible future oil provinces of North America. Tulsa, 1951. 361 p. (Bulletin, v. 35, no. 2.)
Good for general information on geologic structures in North America.

----------. Structure of typical American oilfields; a symposium on the relation of oil accumulation to structure. Tulsa, 1929-1948. 3 v.

Andreev, P. F. and others. Transformation of petroleum in nature. New York: Pergamon, 1968. 466 p.

Braunstein, J. and G. D. O'Brien. Diapirs and diapirism. Tulsa: American Association of Petroleum Geologists, 1968. 444 p. (Memoir no. 8.)
Stress is on economic processes of diapirism particularly in search for gas, oil, sulphur and salt. Bibliogra-

phy p. 358-414.

Delta systems in the exploration for oil and gas; a research colloquium. W. L. Fisher and others. Austin: Bureau of Economic Geology, University of Texas, 1969. 210 p.

Dott, Robert H., Sr. and M. J. Reynolds. Sourcebook for petroleum geology. Tulsa: American Association of Petroleum Geologists, 1969. 471 p. (Memoir no. 5.) Part 1. Genesis of petroleum; part 2. Secondary migration and accumulation of petroleum. Semicentennial commemorative volume.

Eremenko, N. A. Petroleum geology. 2nd revised ed. Moscow: "Nedra," 1968. 385 p.
Includes latest scientific achievements in petroleum geology.

Fox, A. F. The world of oil. New York: Pergamon, 1965. 237 p.

Geology of giant petroleum fields; a symposium. Edited by Michel T. Halbouty. Tulsa: American Association of Petroleum Geologists, 1970. 575 p. (Memoir no. 14.)

Hedberg, Hollis D. "Geologic aspects of origin of petroleum." In: American Association of Petroleum Geologists. Bulletin. v. 48, no. 11. Tulsa, 1964. p. 1755-1803.

International Geological Congress, 20th, Mexico, 1956. Symposium sobre yacimientos de petroleo y gas. Edited by Eduardo J. Guzman. Mexico, 1956. 5 v.
Summary of the geology of world petroleum occurrences (except the USSR).

Landes, Kenneth K. Petroleum geology. 2nd ed. New York: Wiley, 1959. 443 p.

----------. Petroleum geology of the United States. New York: Wiley, 1970. 568 p.

Levorsen, Arville I. Geology of petroleum. 2nd ed. San Francisco: Freeman, 1967. 724 p.

----------, ed. Stratigraphic type oil fields. Tulsa: American Association of Petroleum Geologists, 1941. 902 p.

Moore, Carl A. Handbook of subsurface geology. New York: Harper and Row, 1963. 235 p.

National Petroleum Council. Petroleum resources under the ocean floor. Washington, 1969. 108 p.

Organization for Economic Cooperation and Development. The search for and exploitation of crude oil and natural gas in the European area of the OECD. New York: McGraw-Hill, 1962. 153 p.
A good summary of European activity in the field of petroleum.

Pratt, Wallace E. and D. Good, eds. World geography of petroleum. Princeton, N. J.: Princeton University Press, 1950. 464 p. (American Geographical Society. Special Publication no. 31.)

Russell, William L. Principles of petroleum geology. 2nd ed. New York: McGraw-Hill, 1960. 503 p.

----------. Structural geology for petroleum geologists. New York: McGraw-Hill, 1955. 427 p.

"Symposium on petroleum geology of South America." In: Tulsa Geological Society. Digest, v. 31, 1963. p. 75-212.

Trask, Parker D. and H. W. Patnode. Source beds of petroleum. Tulsa: American Association of Petroleum Geologists, 1942. 566 p.

Ver Wiebe, Walter A. North American petroleum, a complete summary of geological information on the occurrence of oil and gas in the United States and Canada, with a concluding chapter on the oil fields of the Middle East. 3rd ed. Wichita, Kansas, 1958. 461 p.

World Petroleum Congress. Proceedings. 1933- . Imprint varies.
1st, London, 1933; 2nd, Paris, 1937; 3rd, The Hague, 1951; 4th, Rome, 1955; 5th, New York, 1959; 6th, Frankfurt, 1963; 7th, Mexico, 1967; 8th, Moscow, 1971.

NATURAL GAS

Beebe, B. Warren and Bruce F. Curtis, eds. Natural gases of North America. Tulsa: American Association of Petroleum Geologists, 1968. 2 v. (Memoir no. 9.)
V. 1. Gases in Mesozoic and Paleozoic rocks; v. 2. Gases in Paleozoic rocks and general papers.

Katz, D. Handbook of natural gas engineering. New York: McGraw-Hill, 1959. 802 p.

Tiratsoo, E. N. Natural gas. New York: Plenum, 1967. 386 p.

OIL SHALES

Oil shale conversion: an annotated bibliography. Salt Lake City: Utah Industrial Services Agency, University of Utah, 1969. 30 p.

Rogers, Marianne P. List of Bureau of Mines publications on oil shale and shale oil, 1917-1968. Washington: U. S. Bureau of Mines, 1969. 61 p. (Information Circular no. 8429.)

* * *

Bell, Harold S. Oil shales and shale oils. New York: Van Nostrand, 1948. 157 p.

"Oil shale." In: Colorado School of Mines. Mineral industries bulletin. 1962-1965.
Pt. 1. Nomenclature, uses, reserves and production. v. 5, no. 2, 1962; pt. 2. Geology and mineralogy of the oil shales of the Green River Formation, Colorado, Utah and Wyoming. v. 5, no. 3, 1962; pt. 3. The natures and origins of kerogens. v. 7, no. 5, 1964; pt. 4. Retorting and other beneficiation of oil shales. v. 7, no. 6, 1964; pt. 5. Oil shale retorts. v. 8, no. 2, 1965.

"Shale oil...the problems and prospects." In: Oil and gas journal. v. 62, no. 10, 1964. p. 65-80.

Symposium on oil shale. 1st, 1964- . In: Colorado School of Mines. Quarterly.
First, v. 59, no. 3, 1964, and supplement; 2nd, v. 60,

no. 3, 1965; 3rd, v. 61, no. 3, 1966; 4th, v. 62, no. 3, 1967; 5th, v. 63, no. 4, 1968.

Welles, Chris. The elusive bonanza; the story of oil shale--America's richest and most neglected natural resource. New York: Dutton, 1970. 256 p.

COAL

Averitt, Paul. Coal resources of the United States, January 1, 1967. Washington: Government Printing Office, 1969. 116 p. (U.S. Geological Survey. Bulletin no. 1275.) Concerns the quantity and distribution of coal in the United States.

International Committee for Coal Petrology. International handbook of coal petrology. Paris: Centre national de la recherche scientifique, 1963. 1 v. (Looseleaf.)

Tomkeieff, Sergei I. Coals and bitumens and related fossil carbonaceous substances; nomenclature and classification. London: Pergamon, 1954. 122 p.

Yingst, Parke O. An annotated bibliography on coal. Golden, Colorado: Colorado School of Mines, 1958. 102 p. (Colorado School of Mines. Quarterly, v. 53, no. 3.)

* * *

American Conference on Coal Science, University Park, Pa., 1964. Coal Science. Washington: American Chemical Society, 1966. 743 p. (Advances in Chemistry Series no. 55.)

Cady, G. H. "Modern concepts of the physical constitution of coal." In: Journal of geology. v. 50, no. 4, 1942. p. 337-356.

Dapples, E. C. and M. E. Hopkins, eds. Environments of coal deposition. Boulder, Colorado: The Geological Society of America, 1969. 204 p. (Special Paper no. 114.)

Francis, Wilfrid. Coal, its formation and composition. 2nd ed. London: Arnold, 1961. 806 p.

Inter-University Geological Congress 13th, University of Newcastle-Upon-Tyne, 1965. Coal and coal-bearing strata. Edited by Duncan Murchison and T. Stanley Westoll. New York: Elsevier, 1968. 418 p.
Coal as a rock, sedimentary environment of coal formation; biological environment of coal formation; geochemistry and metamorphism; and some aspects of coal basins.

Krevelen, Dirk W. van. Coal: typology, chemistry, physics, constitution. Revised ed. Amsterdam: Elsevier, 1961. 514 p.

Moore, Elwood S. Coal... . 2nd ed. New York: Wiley, 1940. 473 p.

Raistrick, A. and C. E. Marshall. The nature and origin of coal seams. London: The English Universities Press, 1939. 282 p.

Stutzer, Otto. Geology of coal. Translated by A. C. Noe. Chicago: University of Chicago Press, 1940. 461 p. (Available from University Microfilms, Ann Arbor, Michigan.)

U.S. Bureau of Mines. The origin of coal. David White and R. Thiessen. Washington: Government Printing Office, 1913. 390 p. (Bulletin no. 38.)

Williamson, Iain A. Coal mining geology. London: Oxford University Press, 1967. 266 p.

ORES AND STRATEGIC MINERALS, ORE DEPOSITS

REFERENCE WORKS

Cooper, Margaret. Bibliography and index of literature on uranium and thorium and radioactive occurrences in the United States. New York: Geological Society of America, 1953- .
Pts. 1 and 2. G.S.A. Bulletin. v. 64; pt. 3. G.S.A. Bulletin. v. 65; pt. 4. G.S.A. Bulletin. v. 66; pt. 5. G.S.A. Special Paper. no. 67, 1958.

Crosby, James W. A descriptive glossary of radioactive minerals. Pullman, Washington, 1955. 148 p. (Washington State College. Institute of Technology. Division of Industrial Research. Bulletin, Series B, no. 230.)

Fischer, R. P. and J. P. Ohl. Bibliography on the geology and resources of vanadium to 1968. Washington: Government Printing Office, 1970. 168 p. (U. S. Geological Survey. Bulletin no. 1316.)

Ridge, John D. Selected bibliographies of hydrothermal and magmatic mineral deposits. New York: Geological Society of America, 1958. 199 p. (Memoir no. 75.)

Special Libraries Association. Metals/Materials Division. Guide to metallurgical information. 2nd ed. Edited by Eleanor B. Gibson and Elizabeth W. Tapia. New York: Special Libraries Association, 1965. 222 p. (SLA Bibliography no. 3.)

U. S. Bureau of Mines. Helium: bibliography of technical and scientific literature from its discovery (1868) to January 1, 1947. Compiled by Henry P. Wheeler, Jr. and L. B. Swenarton. Washington: Government Printing Office, 1952. 76 p. (Bulletin no. 484.)

----------. Helium: bibliography of technical and scientific literature Jan. 1, 1947 to Jan. 1, 1962. Compiled by Harold W. Lipper and C. W. Cherry. Washington: Government Printing Office, 1968. 525 p. (Information Circular no. 8373.)

----------. Helium: bibliography of technical and scientific literature, 1963. Compiled by Philip C. Tully, B. J. King and E. Dowdy. Washington: Government Printing Office, 1970. 831 p. (Information Circular no. 8467.)

U. S. Geological Survey. Glossary of uranium and thorium bearing minerals. J. W. Frondel, M. Fleischer, and R. S. Jones. 4th ed. Washington: Government Printing Office, 1967. (Bulletin no. 1250.)

TEXTS AND TREATISES

American Institute of Mining, Metallurgical and Petroleum Engineers. A decade of digital computing in the mineral industry. Edited by Alfred Weiss. New York, 1969. 952 p.

----------. Ore deposits of the United States, 1933-1967; The Graton-Sales Volume. Edited by John D. Ridge. New York, 1968. 2 v.

Successor to Ore deposits of the western states.

----------. Ore deposits of the western states. Edited by the Committee on the Lindgren Volume, John W. Finch, Chairman. New York, 1933. 797 p. (Available from Ann Arbor, Michigan: University Microfilms, 1967.) Descriptions of ore deposits in the 50 United States discovered since 1933, except those of porphyry copper deposits, which are covered in Titley, S. R. and C. L. Hicks, eds. Geology of porphyry copper deposits--southwestern United States: The Wilson Volume. Tucson: University of Arizona Press, 1966. 287 p.

Amstutz, G. C. Sedimentology and ore genesis. New York: Elsevier, 1964. 184 p.

Bastin, Edson S. Interpretation of ore textures. New York: Geological Society of America, 1950. 101 p. (Memoir no. 45.)

Bateman, Alan M. Economic mineral deposits. 2nd ed. New York: Wiley, 1955, c1950. 916 p.

----------. Formation of mineral deposits. New York: Wiley, 1961, c1951. 371 p.

Blanchard, Roland. Interpretation of leached outcrops. Reno: Nevada Bureau of Mines, 1968. 196 p. (Bulletin no. 66.)

Brobst, Donald A. Barite: world production, reserves, and future prospects. Washington: Government Printing Office, 1970. 46 p. (U.S. Geological Survey. Bulletin no. 1321.)

Brown, J. S., ed. Genesis of stratiform lead-zinc-barite-fluorite deposits (Mississippi Valley type deposits); a symposium. Mount Pleasant, Mich.: Economic Geology Publishing Co., 1968. 443 p. (Monograph no. 3.)

Cameron, Eugene N. Ore microscopy. New York: Wiley, 1961. 293 p.

Economic geology. Mount Pleasant, Mich.: Economic Geology Publishing Co., 1905- . Semiquarterly.

Edwards, Austin B. Textures of the ore minerals and their significance. Melbourne: Australasian Institute of Mining and Metallurgy, 1960, c1954. 242 p.

Freund, Hugo and others. Applied ore microscopy; theory and technique. New York: Macmillan, 1966. 607 p.

Gocht, W. Der metalische Rohstoff Zinn. Berlin: Duncker and Humblot, 1969. 260 p.
Geology and economics of tin is included.

Heinrich, Eberhardt W. Mineralogy and geology of radioactive raw materials. New York: McGraw-Hill, 1958. 654 p.

Helgeson, Harold C. Complexing and hydrothermal ore deposits. New York: Pergamon, 1964. 128 p.

Helium Symposia Proceedings in 1968--a hundred years of helium. Compiled by Harold W. Lipper. Washington: Government Printing Office, 1968. 292 p. (U.S. Bureau of Mines. Information Circular no. 8417.)
Discusses history, development and spectrum of uses.

International Atomic Energy Agency. Geology of uranium and thorium (a bibliography). Vienna, 1962. 134 p. (Bibliographical Series no. 4.) (Distributed by UNIPUB, New York.)

----------. Nuclear techniques and mineral resources. Vienna, 1970. 546 p. (Distributed by UNIPUB, New York.)

International Geological Congress. 20th, Mexico, 1956. Symposium sobre yacimientos de manganeso. Mexico, 1956. 5 v.
V. 1. El manganeso en general; v. 2. Africa; v. 3. Asia y Oceania; v. 5. Europa.

James, C. H. Sedimentary ores: ancient and modern. Revised ed. Leicester: Department of Geology, University of Leicester, 1969. 305 p. (Special Publication no. 1.)

Jenkins, Olaf P. Geology of placer deposits. San Francisco: California Division of Mines and Geology. 1970? 27 p. (Special Publication no. 34--reprint from Mineral Information service, v. 17, nos. 1-9, 1964.)

Jerome, S. E. and D. R. Cook. Relation of some metal mining districts in the western United States to regional tectonic environments and igneous activity. Reno, Nevada: Nevada Bureau of Mines, 1967. 35 p. (Bulletin no. 69.)

Koschmann, A. H. and M. H. Bergendahl. Principal gold-producing districts of the United States. Washington: Government Printing Office, 1968. 283 p. (U.S. Geological Survey. Professional Paper no. 610.)
A description of the geology, mining history, and production of the major gold-mining districts in 21 states. Selected bibliography p. 263-276.

Lamey, Carl A. Metallic and industrial mineral deposits. New York: McGraw-Hill, 1966. 567 p.

Lindgren, Waldemar. Mineral deposits. 4th ed. New York: McGraw-Hill, 1953. 930 p.

Mineralium deposita; international journal for geology, mineralogy, and geochemistry of mineral deposits. Berlin: Springer, 1966- . Quarterly.

National Research Council. Committee on Processes of Ore Deposition. Ore deposits as related to structural features. Edited by W. H. Newhouse. Princeton, N. J.: Princeton University Press, 1942. 280 p.

Park, Charles F., Jr. and R. A. MacDiarmid. Ore Deposits. 2nd ed. San Francisco: Freeman, 1970. 522 p.

Perfil'ev, B. V. and others. Applied capillary microscopy; role of microorganisms in the formation of iron-manganese ores. New York: Consultants Bureau, 1965. 122 p.

Ramdohr, Paul. The ore minerals and their intergrowth. New York: Pergamon, 1970. 1174 p.
Translated from the 3rd German edition with additions and corrections.

Roubault, Marcel. Geologie de l'uranium. Paris: Masson, 1958. 462 p.

Routhier, Pierre. Les gisements metalliferes; géologie et principes de recherche. Paris: Masson, 1963. 2 v.

Sainsbury, C. L. Tin resources of the world. Washington: Government Printing Office, 1969. 55 p. (U. S. Geological Survey. Bulletin no. 1301.)
A description of the types of tin deposits and main tin-producing areas of the world.

Shand, Samuel J. Eruptive rocks; their genesis, composition, classification, and their relation to ore-deposits with a chapter on meteorites. New York: Hafner, 1969, c1947. 488 p.

Survey of world iron ore resources: occurrence and appraisal. Paris: UNESCO, 1970. 479 p.

Titley, Spencer R. and C. L. Hicks, eds. Geology of the porphyry copper deposits, southwestern North America. Tucson: University of Arizona, 1966. 287 p.

U. S. Geological Survey and U. S. Bureau of Mines. Heavy metals program progress report, 1966 and 1967. Washington: Government Printing Office, 1968. (U. S. Geological Survey. Circular no. 560.)

Valeton, I. Bauxites. New York: Elsevier, 1971, in press.

Varentsov, I. M. Sedimentary manganese ores. Translation-editor, D. S. Stewart. New York: Elsevier, 1964. 119 p.

White, Donald E. "Environments of generation of some base-metal ore deposits." In: Economic geology. v. 63, 1968. p. 301-335.

Wilson, H. D. B., ed. Magmatic ore deposits. Mount Pleasant, Mich.: Economic Geology Publishing Co., 1969. 366 p. (Economic Geology Monograph no. 4.)

SALT, EVAPORITES

Cramer, Howard B. "Evaporties--a selected bibliography." In: American Association of Petroleum Geologists. Bulletin. v. 53, no. 4, 1969. p. 982-1011.
Part of a special issue devoted to evaporites.

* * *

Borchert, Hermann and Richard O. Muir. *Salt deposits; the origin, metamorphism, and deformation of evaporites.* Princeton, N. J.: Van Nostrand, 1964. 338 p.

Braitsch, O. *Salt deposits, their origin and composition.* New York: Springer, 1971. 305 p. (Minerals, Rocks and Inorganic Materials, v. 4.)

Halbouty, Michel T. *Salt domes: Gulf Region, United States and Mexico.* Houston: Gulf Publishing Co., 1966. 425 p.

Lefond, Stanley J. *Handbook of world salt resources.* New York: Plenum, 1969. 384 p.
A compendium of world salt deposits and an essential book for work on evaporites.

Lotze, Franz. *Steinsalz und Kalisalze.* 2nd ed. Berlin: Borntraeger, v. 1, 1957- .

Mattox, Richard B., ed. *Saline deposits; a symposium based on papers from the International Conference on Saline Deposits, Houston, Texas, 1962.* Boulder, Colorado: Geological Society of America, 1968. 701 p. (Special Paper no. 88.)

Salt domes in Texas, Louisiana, Alabama, and offshore tidelands. Washington: Government Printing Office, 1966. 78 p. (U.S. Bureau of Mines. Information Circular no. 831-C.)

Symposium on Salt, Cleveland. Proceedings, 1962. *Symposium on salt; geology, mining, evaporated salt, solution mining, underground storage.* Edited by A. C. Bersticker. Cleveland: Northern Ohio Geological Society, 1963. 661 p.

----------, 1965. *Second symposium on salt.* Cleveland: Northern Ohio Geological Society, 1966. 2 v.

----------, 1969. *Third symposium on salt.* Cleveland: Northern Ohio Geological Society, 1970. 2 v.

CONSTRUCTION AND DECORATIVE MATERIALS

American Institute of Mining, Metallurgical and Petroleum Engineers. Industrial minerals and rocks (nonmetallics other than fuels). 3rd ed. Edited by Joseph L. Gillson. New York, 1960. 934 p.

Bates, Robert L. Geology of the industrial rocks and minerals. New York: Harper, 1960. 441 p. (Reprinted by Dover, 1969.)

Forum on geology of industrial minerals. Proceedings. Publications in this series are:

"A symposium on geology of industrial limestone and dolomite." In: Ohio journal of science. v. 66, 1966. p. 97-224.

A symposium on geology of cement raw materials. Bloomington, Indiana: Indiana Geological Survey, 1966. 197 p.

A symposium on industrial mineral exploration and development. Lawrence, Kansas: Kansas Geological Survey, 1967. 183 p. (Special Distribution Publication no. 34.)

Forum on geology of industrial minerals. 4th, Austin, 1968. Proceedings. Austin: Texas Bureau of Economic Geology, 1968. 174 p.

Forum on geology of industrial minerals. 5th, Harrisburg, 1968. Proceedings. Harrisburg: Pennsylvania Topographic and Geologic Survey, 1969. 278 p. (Mineral Resources Report no. M64.)

Grant, M. H. The marbles and granites of the world. Chelsea, London: J. B. Shears and Sons, 1955. 158 p.

Industrial minerals. London: Metal Bulletin Ltd., 1967- . Monthly.

Lamey, Carl A. Metallic and industrial mineral deposits. New York: McGraw-Hill, 1966. 567 p.

GEMOLOGY (See Also Mineralogy)

Bank, H. Edelsteine und mineralien. Friedrichsdorf: Alles und Brilliant, 1969. 126 p.

Bauer, Max. Precious stones; with an appendix on pearls and coral. Translated from German with additions by L. J. Spencer. Rutland: Charles E. Tuttle Company, 1969. 647 p.
New appendices on synthetic gems and the cultured pearl.

MacFall, Russell P. Gem hunter's guide. 4th ed. New York: Thomas Y. Crowell Co., 1969. 279 p.

Pearl, Richard M. American gem trails. New York: McGraw-Hill, 1964. 173 p.

Sinkankas, John. Gemstones of North America. Princeton, N. J.: Van Nostrand, 1959. 675 p.

----------. Prospecting for gemstones and minerals. 2nd ed. New York: Van Nostrand-Reinhold, 1970. 397 p.

----------. Van Nostrand's standard catalog of gems. Princeton, N. J.: Van Nostrand, 1968. 553 p.

Webster, Robert. Gems: their sources, descriptions, and identification. 2nd ed. Hamden, Conn.: Archon Books, 1970. 836 p.

MINERAL RESOURCES OF THE SEA

Firth, F. E., ed. The encyclopedia of marine resources. New York: Van Nostrand-Reinhold, 1969. 740 p.

* * *

Bardach, John. Harvest of the sea. New York: Harper and Row, 1968. 284 p.

"International Symposium on Materials--Key to Effective Use of the Sea. Proceedings." In: Ocean Engineering. v. 1, no. 2-4, 1968-69.
Symposium held in New York City, 1967.

McKelvey, V. E. and others. Subsea mineral resources and problems related to their development. Washington:

Government Printing Office, 1969. 26 p. (U.S. Geological Survey. Circular no. 619.)

Mero, John L. Mineral resources of the sea. New York: Elsevier, 1965. 312 p. (Oceanography Series no. 1.)

Pings, W. B. and D. A. Paist. "Minerals from the oceans." In: Colorado School of Mines. Mineral industries bulletin. v. 13, nos. 2-3, 1970.

ENVIRONMENTAL GEOLOGY

This heading has been used to cover the study of the interaction between man and the geologic environment. As a rapidly expanding interdisciplinary field, there is currently a large amount of literature being generated. A very limited number of selections have been made to indicate the type of work being published and the growing interest in this broad field.

American Geological Institute. Environmental geology; short course (Milwaukee, 1970; Dallas, 1971) lecture notes. Washington, 1970. 429 p.

Cannon, Helen L. and D. F. Davidson, eds. Relation of geology and trace elements to nutrition. Boulder, Colo.: Geological Society of America, 1968. 68 p. (Special Paper no. 90.)
Abstracts and papers presented at a symposium at the G. S. A. annual meeting, New York, 1963.

Cannon, Helen and H. Hopps. Geochemical environment in health and disease. Boulder, Colo.: Geological Society of America, in prep. (Memoir no. 123.)

Ehrlich, Paul R. and A. H. Ehrlich. Population, resources, environment. San Francisco: Freeman, 1970. 383 p. Chap. 4--Limits of the earth.

Engelstad, Orvis, ed. Nutrient mobility in soils; accumulation and losses; papers presented at a 1969 symposium. Madison, Wisconsin: Soil Science Society of America, 1970. 64 p.

Falkie, Thomas V. and Lee W. Saperstein. "Land reclamation for the mining industry." In: Earth and mineral sciences (College of Earth and Mineral Sciences, Pennsylvania State University). v. 40, no. 2, 1970. p. 9-12.

Flawn, Peter T. "The environmental geologist and the body politic." In: Geotimes. v. 13, no. 6, 1968. p. 13-14.

Includes some of the geologically oriented organizations working in this area.

----------. Environmental geology; conservations, land-use, and resource management. New York: Harper and Row, 1970. 313 p.

The Governor's Conference on Environmental Geology. Denver: Colorado Geological Survey, 1969. 78 p. (Special Publication no. 1.)

Hayes, William C. and Jerry D. Vineyard. Environmental geology in towne and country. Rolla, Missouri: Missouri Geological Survey and Water Resources, 1969. 42 p. (Educational Series no. 2.)

Hoult, David P., ed. Oil on the sea; proceedings of a symposium. New York: Plenum, 1969. 114 p.

Illinois. State Geological Survey. Environmental geology notes. 1, 1965- .

Legget, Robert F. Man as a geological agent. Ottawa: National Research Council of Canada, 1969. 248 p.

Leopold, Luna B. Hydrology for urban land planning--a guidebook on the hydrologic effects of urban land use. Washington: Government Printing Office, 1968. (U.S. Geological Survey. Circular no. 554.)

Man...an endangered species? Washington: Government Printing Office, 1968. 100 p. (U.S. Department of the Interior. Conservation Yearbook no. 4.)

McCrone, W. C., R. G. Draftz, and J. G. Delly. The particle atlas: a photomicrographic reference for the identification of particulate substances. Ann Arbor, Michigan: Ann Arbor Science Publishing Co., 1967. 406 p.
A notable contribution to the compilation of inorganic and organic particles information which will be of particular interest to those working in the field of environmental geology.

McGill, J. T. Growing importance of urban geology. Washington: Government Printing Office, 1964. 4 p. (U.S. Geological Survey. Circular no. 487.)

McHarg, Ian L. Design with nature. New York: Natural History Press, 1969. 197 p.

Pestrong, Raymond. "The role of the urban geologist in city planning." In: California. Division of Mines and Geology. Mineral information service. v. 21, no. 10, 1968. p. 151-152.

"A series on urban geology of Greater Waco." In: Baylor Geological Studies. Bulletin. no. 8-14, 1965-1968. Pt. 1. Geology; pt. 2. Soils; pt. 3. Water; pt. 4. Engineering; pt. 5. Socio-economic geology. Example of geological studies applied to a city.

Singer, Fred, ed. Global effects of environmental pollution: an A. A. A. S. symposium. New York: Springer, 1970. 331 p.

U. S. Department of the Interior. Surface mining and our environment. Washington: Government Printing Office, 1967. 124 p.

U. S. Geological Survey. Disposal of liquid wastes by injection underground--neither myth nor millennium. Washington: Government Printing Office, 1969. 15 p. (Circular no. 631.)

----------. Water in the urban environment. Washington: Government Printing Office, 1969-1970. (Circular no. 601.)

Wayne, William J. "Urban geology--a need and a challenge." In: Indiana Academy of Science. Proceedings. v. 78, 1969. p. 49-64.

Wilson, Carroll L. and W. H. Matthews, IV, eds. Man's impact on the global environment: assessment and recommendations for action. Cambridge: M. I. T. Press, 1970. 319 p.
Report of the Study of Critical Environmental Problems.

ENGINEERING GEOLOGY (See Also Rock Mechanics, Economic Geology, Hydrology, and Soil Science)

REFERENCE WORKS

DePuy, G. W. Tunnel geology bibliography. Denver: U. S. Bureau of Reclamation, Office of Engineering Reference,

1969. 50 p. (Bibliography no. 245.)

Engineering index. New York: The Engineering Index, Inc., 1885- . Monthly.
Of interest to geoscientists is its coverage in civil engineering, geology, geophysics, metallography, mining, petroleum technology and soil mechanics.

Happ, Stafford. "Engineering geology reference list." In: Geological Society of America. Bulletin. v. 66, no. 8, 1955. p. 993-1030.

MacFarlane, Ivan C., ed. Muskeg engineering handbook. Toronto: University of Toronto Press, 1969. 297 p.

U.S. Geological Survey. Bibliography and discussion of flood-routing methods and unsteady flow in channels. V. M. Yevdjevich. Washington: Government Printing Office, 1964. 235 p. (Water Supply Paper no. 1690.)

----------. Selected abstracts on engineering geology and related subjects. Washington: Government Printing Office, 1951, 1954. (Circular nos. 75, 259.)

Van Horn, Richard. "Physical property and construction use data sheet for surficial deposits." In: Association of Engineering Geologists. Bulletin. v. 5, no. 1, 1968. p. 18-22.

TEXTS AND TREATISES

American Society of Civil Engineers. Soil Mechanics and Foundation Division. Journal. New York, 1956- . Bimonthly.

Bondarik, G. K. Dynamic and static sounding of soils in engineering geology. Jerusalem: Israel Program for Scientific Translations, 1967. 144 p.

Brahtz, John F. Ocean engineering; goals, environment, technology. New York: Wiley, 1968. 720 p.

Bundred, J. Basic geology for engineers. London: Butterworth, 1969. 256 p.

Butler, Stanley S. Engineering hydrology. Englewood Cliffs, N. J.: Prentice-Hall, 1957. 356 p.

Dapples, Edward C. Basic geology for science and engineering. New York: Wiley, 1959. 609 p.

Duncan, Neil. Engineering geology and rock mechanics. London: Hill, 1969. 2 v.

Eckel, Edwin B., ed. Nevada test site. Boulder, Colo.: Geological Society of America, 1968. 290 p. (Memoir no. 110.)
A summary of the investigations on the geology and hydrology of the Nevada Test Site as presented in papers at the 18th Annual Meeting of the Rocky Mountain Section of GSA, Las Vegas, Nevada, 1966.

Engineering geology. Amsterdam: Elsevier, 1965- . 4/yr.

Engineering geology case histories. Boulder, Colorado: Geological Society of America, 1957- .
No. 7 is composed of case histories involved with Legal aspects of geology in engineering practice.

Farmer, Ian W. Engineering properties of rocks. London: Spon, 1968. 180 p.
An introductory textbook on rock mechanics with an extensive bibliography.

Galley, John E. Subsurface disposal in geologic basins--a study of reservoir strata. Tulsa, Oklahoma: American Association of Petroleum Geologists, 1968. 253 p.
Underground waste and radioactive disposal.

Geological Society of America. Application of geology to engineering practice; the Berkey volume. Papers by John L. Savage and others. New York, 1950. 327 p. (Reprinted by G.S.A., 1955, 1962.)

Gillott, J. E. Clay in engineering geology. New York: Elsevier, 1968. 296 p.

Highway Geology Symposium. Proceedings. 1st, 1950- . Annual. Imprint varies.
The 19th was published by the West Virginia Geological Survey, Morgantown, West Virginia, 1968.

International Conference on Soil Mechanics and Foundation Engineering. Proceedings. 1st, 1963- . Imprint

varies.
The 6th conference available from Ottawa, Ontario: National Research Council, 1968. The 7th conference was held in Mexico City, 1969.

International Research Conference on Marine Geotechnique, 1966. Proceedings; Marine geotechnique. Edited by Adrian R. Richards. Urbana: University of Illinois Press, 1967. 327 p.
Summarizes engineering properties of sediments.

Kiersch, G. A. "Trends in engineering geology in the United States." In: Felsmechanik und Ingenieurgeologie. Supplement no. 1. Principles in the field of geomechanics. New York: Springer, 1964. p. 31-57.

Krynine, Dimitri P. and W. R. Judd. Principles of engineering geology and geotechnics. New York: McGraw-Hill, 1957. 730 p.

Legget, Robert F. Geology and engineering. 2nd ed. New York: McGraw-Hill, 1962. 884 p.

Leonards, G. A., ed. Foundation engineering. New York: McGraw-Hill, 1962. 1136 p.

Linsley, Ray K. and Joseph B. Franzini. Water-resources engineering. 2nd ed. New York: McGraw-Hill, 1964. 410 p.

----------, M. A. Kohler, and J. L. H. Paulhus. Hydrology for engineers. New York: McGraw-Hill, 1958. 340 p.

Myers, John J., C. H. Holm and R. F. McAllister. Handbook of ocean and underwater engineering. New York: McGraw-Hill, 1969. 800 p.

Obert, Leonard and Wilbur I. Duvall. Rock mechanics and the design of structures in rock. New York: Wiley, 1967. 650 p.

Ocean engineering. New York: Pergamon, 1968- . Bimonthly.

Peck, Ralph B., W. E. Hanson and T. H. Thornburn. Foundation engineering. New York: Wiley, 1953. 410 p.

Quarterly journal of engineering geology. London: Geological Society of London, 1969- .

Reviews in engineering geology. Boulder, Colorado: Geological Society of America, 1962- . Irregular. V. 1, 1962; v. 2, 1969.

Schultz, J. R. and A. B. Cleaves. Geology in engineering. New York: Wiley, 1955. 592 p.

Sowers, George B. and G. F. Sowers. Introductory soil mechanics and foundations. 3rd ed. New York: Macmillan, 1970. 556 p.

Stagg, K. G. and O. C. Zienkiewicz. Rock mechanics in engineering practice. New York: Wiley, 1968. 442 p. Chapter 1 deals with purely geologic considerations.

Trefethen, Joseph M. Geology for engineers. 2nd ed. Princeton, N. J.: Van Nostrand, 1959. 632 p.

Walters, Rupert C. S. Dam geology. London: Butterworth, 1962. 335 p.

Wiegel, Robert L. Oceanographical engineering. Englewood Cliffs, N. J.: Prentice-Hall, 1964. 532 p.

GEOLOGIC HAZARDS (See Also Seismology)

Adams, William M. An index to tsunami literature to 1966. Honolulu: Hawaii Institute of Geophysics, University of Hawaii, 1967. (HIG Data Report no. 8; HIG Publication no. 67-21.)

Duke, C. Martin. Bibliography of effects of soil conditions on earthquake damage. San Francisco: Earthquake Engineering Research Institute, 1958. 47 p.

Hollis, Edward P. Bibliography of engineering seismology. 2nd ed. San Francisco: Earthquake Engineering Research Institute, 1958. 144 p.

National Research Council. Highway Research Board. Landslides; a selected and annotated bibliography. Compiled by J. M. Thompkin and S. H. Britt. Washington, 1951. 53 p. (HRB Bibliography no. 10.)

Sansburn, William A. "An earthquake reading list--1." In: California. Division of Mines and Geology. Mineral information service. v. 18, no. 4, 1965. p. 70-71. List is oriented toward earthquakes as a geologic hazard, with emphasis on damage and behavior of buildings during earthquakes.

* * *

American Society for Testing and Materials. Vibration effects of earthquakes on soils and foundations; a 1968 symposium. Philadelphia, 1969. 262 p. (STP no. 450.)

Hayes, Miles O. Hurricanes as geological agents: case studies of hurricanes, Carla, 1961, and Cindy, 1963. Austin: Bureau of Economic Geology, University of Texas, 1967. 56 p. (Report of Investigation no. 61.)

Iacopi, Robert. Earthquake country; how, why and where earthquakes strike in California. Menlo Park, California: Lane Book Company, 1964. 192 p.

Jahns, Richard H. "Geologic hazards--problems for joint solutions." In: Journal of geological education. v. 18, no. 2, 1970. p. 81-84.

National Research Council. Committee on the Alaska Earthquake. The great Alaska earthquake of 1964. Washington: National Academy of Sciences, 1968- . (Publication no. 1603.)

----------. Toward reduction of losses from earthquakes. Washington: National Academy of Sciences, 1969. 34 p.

National Research Council. Highway Research Board. Committee on Landslide Investigations. Landslides and engineering practice. Edited by E. B. Eckel. Washington, 1958. 232 p. (HRB Special Report no. 29.)

Sharpe, Charles F. S. Landslides and related phenomena; a study of mass-movements of soil and rock. Paterson, N. J.: Pageant Books, 1960, c1938. 137 p. (Columbia University Geomorphic Studies no. 2.)

Tsunamis in the Pacific Ocean; Proceedings of the International Symposium on Tsunamis and Tsunamis Research. Honolulu, University of Hawaii, 1969. Edited by William

M. Adams. Honolulu: East-West Center Press, 1970. 513 p.

U.S. Coast and Geodetic Survey. Earthquake history of the United States. Washington: Government Printing Office, 1958-1961. 2 v.

U.S. Executive Office of the President. Emergency Preparedness Office. Geologic hazards and public problems. Edited by Robert A. Olson and M. M. Wallace. Washington: Government Printing Office, 1969. 335 p.
Proceedings of a conference held in San Francisco, May, 1969. Includes information on volcanoes, earthquakes, seismic sea waves, land subsidence and landslides.

U.S. Geological Survey. The Alaska earthquake, March 27, 1964. Washington: Government Printing Office, 1966-1970.
A comprehensive geologic study covered in Professional Papers 541-546.

----------. Mercury in the environment. Washington: Government Printing Office, 1970. 67 p. (Professional Paper no. 713.)

Zaruba, Quido and V. Mencl. Landslides and their control. New York: Elsevier, 1969. 203 p.
Includes case histories, classification of landslides, and methods of geologic investigation.

HYDROLOGY, HYDROGEOLOGY

REFERENCE WORKS

Annotated bibliography on hydrology and sedimentation. For a breakdown of this series, see under "SEDIMENTOLOGY."

Chow, Ven Te, ed. Handbook of applied hydrology; a compendium of water-resources technology. New York: McGraw-Hill, 1964. 1418 p.
Basic theories, principles and required data for the study and management of water and water-resource projects. Wide range of topics from geology, soil physics, ice and glaciers, through water law. Bibliographies accompany each topic.

Dictionary catalogue of the Water Resources Center Archives (Berkeley campus). Boston: G. K. Hall, in process.

International Association of Scientific Hydrology. International bibliography of hydrology. Gentbrugge, Belgium.
Bibliographies prepared by the I. A. S. H. National Committees for their countries. Up-dated periodically. Titles, dates, publishers, and places of publication vary as to country.

International glossary of hydrology. 2nd ed. Vienna: International Atomic Energy Agency, 1968. 214 p.
English, French, Russian, and Spanish equivalents. A 1971 revision is planned.

International Hydrological Decade. Guidebook on nuclear techniques in hydrology. Vienna: International Atomic Energy Agency, 1968. 214 p.

Johnson, A. I. Selected bibliography on laboratory and field methods in ground-water hydrology. Washington: Government Printing Office, 1964. 21 p. (Water Supply Paper no. 1779-Z.)

Margat, Jean. Guide bibliographique d'hydrogéologie. Paris: Bureau de recherches géologiques et minières, 1964. 113 p.

Pfannkuch, Hans-Olaf. Elsevier's Dictionary of hydrogeology. New York: Elsevier, 1969. 168 p.
English, French, and German terms.

U.S. Department of the Interior. Water Resources Scientific Information Center. Selected water resources abstracts. Springfield, Virginia: NTIS, 1968- . Semimonthly.

U.S. Geological Survey. Annotated bibliography on artificial recharge of ground water, through 1954. David K. Todd. Washington: Government Printing Office, 1959. 115 p. (Water Supply Paper no. 1477.)

----------. Annotated bibliography on artificial recharge of ground water 1955-67. D. C. Signor, D. J. Growitz, and William Kam. Washington: Government Printing Office, 1970. 141 p. (Water Supply Paper no. 1990.)

----------. Bibliography of hydrology of the United States, 1963. Compiled by J. R. Randolph and R. G. Deike. Washington: Government Printing Office, 1966. 166 p. (Water Supply Paper no. 1863.)
This is the first of an annual series which includes books, journal articles, and other publications on hydrology published in the United States and in international journals. It supercedes in part the Annotated bibliography on hydrology and sedimentation.

----------. Bibliography of hydrology of the United States and Canada, 1964. Compiled by J. R. Randolph, N. M. Baker, and R. G. Deike. Washington: Government Printing Office, 1969. 232 p. (Water Supply Paper no. 1864.)

----------. Water resources data. Washington: Government Printing Office, 1961- . Annual for each state. Pt. 1. Surface water records; pt. 2. Water quality records.

Waring, Gerald A. and O. E. Meinzer. Bibliography and index of publications relating to ground water. Washington: Government Printing Office, 1947. 412 p. (Water Supply Paper no. 992.)

Williams, John R. Ground water in permafrost regions--an annotated bibliography. Washington: Government Printing Office, 1965. 294 p. (Water Supply Paper no. 1792.)

TEXTS AND TREATISES

Advances in hydroscience. New York: Academic Press, 1964- . Annual.
Latest is v. 6, 1970.

Alabama. Geological Survey. Hydrology of limestone terranes. University, Alabama, 1970- . (Issued as Parts of Bulletin no. 94.)

Beeby-Thompson, A. Exploring for water. London: Villiers, 1969. 373 p.
The story of 40 years of hydrological investigations in many countries.

Biswas, A. K. History of hydrology. New York: Elsevier, 1970. 336 p.

Briggs, Peter. Water: the vital essence. New York: Harper and Row, 1967. 223 p.
Emphasis on conservation.

Bruce, J. P. and R. H. Clark. Introduction to hydrometeorology. New York: Pergamon, 1966. 319 p.

Castany, G. Traité pratique des eaux souterraines. Paris: Dunod, 1963. 657 p.

Chorley, Richard, ed. Water, earth, and man. New York: Barnes and Noble, 1969. 588 p.

Davis, Stanley N. and Roger J. M. De Wiest. Hydrogeology. New York: Wiley, 1966. 463 p.

De Wiest, Roger J. M., ed. Flow through porous media. New York: Academic, 1969. 530 p.

----------. Geohydrology. New York: Wiley, 1965. 366 p.

Deutsche Hydrographische Zeitschrift. Hamburg: Deutsches Hydrographisches Institut, 1948- . 6/yr.

Includes a section entitled "Hydrographische Bibliographie," which is a bibliography to German, French, and English literature.

Eagleson, Peter S. Dynamic hydrology. New York: McGraw-Hill, 1970. 462 p.

Ground water. Urbana, Illinois: Water Well Journal Publishing Co., 1963- . Quarterly.

Guilcher, André. Précis d'hydrologie marine et continentale. 2nd ed. Paris: Masson, 1965. 390 p.

Harr, Milton E. Groundwater and seepage. New York: McGraw-Hill, 1962. 315 p.

Heath, Ralph C. and F. W. Trainer. Introduction to groundwater hydrology. New York: Wiley, 1968. 284 p.

Hem, John D. Study and interpretation of the chemical characteristics of natural water. 2nd ed. Washington: Government Printing Office, 1970. 363 p. (U.S. Geological Survey. Water Supply Paper no. 1473.)

Hubbert, M. King. The theory of ground-water motion and related papers. Darien, Conn.: Hafner, 1970. 311 p.

International Association of Scientific Hydrology. Bulletin. Gentbrugge, Belgium, 1956- . 4/yr.

International hydrographic review. Monte Carlo, Monaco: International Hydrographic Bureau, 1924- . Semiannual.
Includes the section "International hydrographic bibliography."

International Hydrology Symposium, Colorado State University, 1967. Proceedings. Comptes Rendu. Fort Collins, Colorado: Colorado State University, 1967. 2 v.

Journal of hydrology. Amsterdam: North-American Publishing Co., 1963- . Quarterly.

Kazman, Raphael, G. Modern hydrology. New York: Harper and Row, 1965. 301 p.
Broad coverage of the science and its practical applications.

Lamarck, J. B. Hydrogeology. Translated by A. V. Carozzi. Urbana: Univ. of Illinois, 1964. 152 p. Originally published 1802.

Meinzer, Oscar E. Hydrology. New York: Dover, 1955. 723 p.

----------. Outline of ground-water hydrology with definitions. Washington: Government Printing Office, 1923. 71 p. (U. S. Geological Survey. Water Supply Paper no. 494.)

Morley, L. W., ed. Mining and groundwater geophysics, 1967; Proceedings of Canadian Centennial Conference on Mining and Groundwater Geophysics, 1967, Niagra Falls, Ontario. Ottawa: Geological Survey of Canada, 1970. 722 p. (Economic Geology Report no. 26.)

Perrault, Pierre. On the origin of springs. Translated by A. LaRocque. New York: Hafner, 1967. First published in 1674. Sometimes called the "first book on hydrology."

Polubarinova-Kochina, P. IA. Theory of ground water movement. Translated by J. M. R. De Wiest. Princeton, N. J.: Princeton University Press, 1962. 613 p.

Stringfield, V. T. and H. E. LeGrand. "Hydrology of carbonate rock terranes--a review." In: Journal of hydrology. v. 8, 1969. p. 349-417.

Swenson, H. A. and H. L. Baldwin. A primer on water quality. Washington: Government Printing Office, 1965. 27 p. (U. S. Geological Survey. Special Publication.)

Termier, Henri and G. Termier. Erosion and sedimentation. Princeton, N. J.: Van Nostrand, 1963. 433 p.

Todd, David K. Ground water hydrology. New York: Wiley, 1959. 336 p.

Tolman, Cyrus F. Ground water. New York: McGraw-Hill, 1937. 593 p.

Walton, William C. Groundwater resource evaluation. New York: McGraw-Hill, 1970. 664 p.

Ward, R. C. Principles of hydrology. New York: McGraw-Hill, 1968. 402 p.

Williams, J. R. Ground water in the permafrost regions of Alaska. Washington: Government Printing Office, 1970. 83 p. (U.S. Geological Survey. Professional Paper no. 696.)

Wisler, Chester O. and E. F. Brater. Hydrology. 2nd ed. New York: Wiley, 1959. 408 p.

SOIL SCIENCE, WEATHERING

REFERENCE WORKS

Elsevier's dictionary of soil mechanics in four languages: English/American, French, Dutch, and German. Compiled by A. D. Visser. New York: Elsevier, 1965. 380 p.

Geodex soil mechanics information service. Glen Ellen, California: Geodex International, Inc., 1968- . Keyword cards and abstracting service on soil mechanics.

Jacks, Graham V., R. Tavernier, and D. H. Boalch, eds. Multilingual vocabulary of soil science. 2nd ed. Rome: Food and Agriculture Organization of the United Nations, 1960. 430 p.
English, Spanish, German, Portugese, Italian, Dutch, Swedish, and Russian terms.

Mitra, Devi and J. E. Willson. A bibliography on geomechanics. Salt Lake City: University of Utah, 1962. 210 p. (Utah. University. Bulletin, v. 53, no. 22.)
Covers soil and rock mechanics.

National Research Council. Highway Research Board. Glossary of pedologic (soils) and landform terminology for soil engineers. Washington, 1957. 32 p. (Special Report no. 25.)

Soil Science Society of America. Glossary of soil science terms. Madison, Wisconsin, 1970. 27 p.

U.S. Department of Agriculture. Soil survey manual. Washington: Government Printing Office, 1951. 503 p. (Handbook no. 18.)

----------. Library. Bibliography of agriculture. Washington: Government Printing Office, 1942-1969, v. 1-33. New York: CCM, 1970- , v. 34- . Monthly.

Includes fields of soil physics, chemistry, and mineralogy; soil genesis and classification.

----------. Soil Conservation Service. Soil classification, comprehensive, system. 7th approximation. Washington: Government Printing Office, 1960. 265 p.

TEXTS AND TREATISES

Alpan, I. "The geotechnical properties of soil." In: Earth-science reviews. v. 6, no. 1, 1970. p. 5-49.

Baver, Leonard D. Soil physics. 3rd ed. New York: Wiley, 1961, c1956. 489 p.

Bear, Firman E., ed. Chemistry of the soil. 2nd ed. New York: Rinehold, 1964. 515 p. (American Chemical Society. Monograph Series no. 160.)

Brewer, Roy. Fabric and mineral analyses of soils. New York: Wiley, 1964. 470 p.

Bridges, E. M. World soils. New York: Cambridge Univ. Press, 1970. 89 p.

Buckman, Harry O. and N. C. Brady. The nature and properties of soils. 7th ed. New York: Macmillan, 1969. 653 p.

Carroll, Dorothy. Rock weathering. New York: Plenum, 1970. 203 p.
Appendix of 360 important weathering studies classified according to rock type and Koppen' climatic index.

Drew, J. V., ed. Selected papers in soil formation and classification. Madison, Wisconsin: Soil Science of America, 1967. 428 p.

Eyre, S. R. Vegetation and soils, a world picture. London: Arnold, 1963. 324 p.

Geotechnique; international journal of soil mechanics. London: Institute of Civil Engineering, 1948- . Quarterly.

Gibson, J. Sullivan and J. W. Batten. Soils; their nature, classes, distribution, uses and care. University, Ala.:

University of Alabama, 1970. 296 p.

International Congress of Soil Science. Transactions. 1st, 1927- .
Papers apply to soil physics, chemistry, mineralogy, genesis, biology, etc. Congresses are held irregularly at different locations. The transactions of the 9th, held in Australia, 1968, were published in 4 volumes by Elsevier, 1968.

Jenny, Hans. Factors of soil formation; a system of quantitative pedology. New York: McGraw-Hill, 1941. 281 p.

Joffee, Jacob S. Pedology. 2nd ed. New Brunswick, N. J.: Pedology Publications, 1949. 662 p.

Journal of soil science. London: Oxford University Press, 1949- . Semiannual.

Jumikis, Alfred R. Introduction to soil mechanics. Princeton, N. J.: Van Nostrand, 1967. 436 p.

----------. Theoretical soil mechanics, with practical applications to soil mechanics and engineering. New York: Van Nostrand-Reinhold, 1969. 432 p.

Keller, Walter D. The principles of chemical weathering. Revised ed. Columbus, Mo.: Lucas Bros., 1957. 111 p.

Krinitzsky, E. L. Radiography in the earth sciences and soil mechanics. New York: Plenum, 1970. 163 p.

Kunze, G. W. and others, eds. Mineralogy in soil science and engineering. Madison, Wisconsin: Soil Science Society of America, 1968. 106 p. (Special Publication no. 3.)

Lambe, T. W. and R. V. Whitman. Soil mechanics. New York: Wiley, 1969. 553 p.

Lee, I. K., ed. Soil mechanics. New York: Elsevier, 1968. 625 p.

Legget, Robert F. "Soil: its geology and use." In: Geological Society of America. Bulletin. v. 78, no. 12, 1967. p. 1433-1460.

Loughnan, F. C. Chemical weathering of the silicate minerals. New York: Elsevier, 1969. 154 p.
Includes case histories of various rock types.

Maignien, R. Review of research on laterites. Paris: UNESCO, 1966. 148 p. (Natural Resources Research Series no. 4.) (Available from UNIPUB, New York.)

Marshall, Charles E. The physical chemistry and mineralogy of soils. New York: Wiley, 1964. 388 p.

Millar, Charles E., L. M. Turk, and H. D. Foth. Fundamentals of soil science. 4th ed. New York: Wiley, 1965. 491 p.

Mohr, Edward C. J. and F. A. Van Baren. Tropical soils; a critical study of soil genesis as related to climate, rocks, and vegetation. The Hague: W. VanHoeve, 1959, c1954. 498 p.

Morrison, R. B. and H. E. Wright, eds. Quaternary soils. Reno: University of Nevada, Desert Research Institute, 1967. 340 p. (International Association for Quaternary Research. VII Congress. Proceedings, v. 9.)

Ollier, Cliff, ed. Weathering. New York: Elsevier, 1969. 304 p. (Geomorphology Texts, v. 2.)

Papadakis, Juan. Soils of the world. New York: Elsevier, 1969. 220 p.

Persons, Benjamin S. Laterite; genesis, location, use. New York: Plenum, 1970. 103 p.

Reiche, Parry. A survey of weathering processes and products. Revised ed. Albuquerque: University of New Mexico Press, 1950. 95 p. (New Mexico. University. Publications in Geology no. 3.)

Robinson, Gilbert W. Soils; their origin, constitution, and classification. 3rd ed. New York: Wiley, 1967, c1949. 573 p.

Schelling, J. "Soil genesis, soil classification and soil survey." In: Geoderma. v. 4, no. 3, 1970. p. 165-193.

Soil science. Baltimore: Williams and Wilkins, 1916- . Monthly.

Soil Science Society of America. Proceedings. Madison, Wis., 1936- . Bimonthly.

Suklje, L. Rheological aspects of soil mechanics. New York: Interscience, 1969. 571 p.

Terzaghi, Karl. From theory to practice in soil mechanics. New York: Wiley, 1960. 425 p. Selections from the writings of Karl Terzaghi prepared by L. Bjerrum and others.

---------- and R. B. Peck. Soil mechanics in engineering practice. 2nd ed. New York: Wiley, 1967. 729 p.

U.S. Department of Agriculture. Soil. Washington: Government Printing Office, 1957. 784 p. (Yearbook, 1957.)

Vilenskii, D. G. The Russian school of soil cartography and its influence on the soil cartography of the world. Jerusalem: Israel Program for Scientific Translations, 1967. 112 p.

Volney, Constantin, F. C. A view of the soil and climate of the United States of America. Translated by C. B. Brown. New York: Hafner, 1968. A facsimile of the Philadelphia 1804 edition.

PERMAFROST

Black, R. "Permafrost; a review." In: Geological Society of America. Bulletin. v. 65, no. 9, 1954. p. 839-856.

Brown, Roger E. Permafrost in Canada; its influence on northern development. Toronto: Toronto University Press, 1970. 234 p.

Cailleux, A. and G. Taylor. Cryopedologie, étude des sols geles. Paris: Hermann, 1954. 218 p.

Ferrians, Oscar J., Jr., Reuben Kachadoorian and Gordon W. Greene. Permafrost and related engineering problems in Alaska. Washington: Government Printing Office, 1969. 37 p. (U.S. Geological Survey. Professional Paper no. 678.)

International Conference on Permafrost, Lafayette, Ind., 1963. Proceedings. Washington: National Academy of Sciences-National Research Council, 1966. 563 p. (NAS-NRC Publication no. 1287.)

Muller, S. W. Permafrost, or permanently frozen ground and related engineering problems. Ann Arbor, Mich.: Edwards, 1947. 231 p.

Swinzow, G. K. "Certain aspects of engineering geology in permafrost." In: Engineering geology. v. 3, 1969. p. 177-215.

Terzaghi, K. "Permafrost." In: Boston Society of Civil Engineers. Journal. v. 39, 1952. p. 1-50.

Tiutiunov, Ivan A. An introduction to the theory of frozen rocks. Translated by J. O. H. Mulhaus. New York: Macmillan, 1964. 94 p.

REGIONAL SECTION

In this section the authors have attempted to list general geologic bibliographies and other references for each continent and country, including states and provinces for the U.S. and Canada.

In a few instances, the citations cover only a specific aspect of the country's geology (e.g., paleontology). Additional coverage, or coverage for a country not listed in the section, can be found by referring to a more general heading (continent) in the regional section or by consulting the bibliographies and abstracting services cited in the general section of the guide.

While compiling this section, the authors noted that there are bibliographies (single or continuing) which pertain to a specific region within a country (e.g., France--west-central, north, French Alps; Germany--various states). To trace these specific bibliographies, we recommend that the geologist consult the general bibliographies listed in the general section; e.g., Geological Society of America. Bibliography and index of geology.

Also, it appears that certain bibliographies, which for many years could be traced in a given serial bibliography, are no longer compiled, or at least no longer published with the same frequency. Probably the cessation or irregular nature of many specialized bibliographies is the result of the inability of small staffs to cope with the increasing output of geological information, preferring, instead, to defer in favor of the major bibliographical and indexing services.

This section is indexed in the table of contents. An alphabetical geographic index appears at the conclusion of the guide.

WORLD

Collective works on the regional geology and geomorphology of the world.

MONOGRAPHS

Bederke, Erich and H-G. Wunderlich. Atlas zur geologie. Germany: Bibliographisches Institut Mannheim, 1968. 50 p.
Only compilation of its kind--areal geological maps and structural features of the world.

King, Lester C. The morphology of the earth; a study and synthesis of world scenery. 2nd ed. New York: Hafner, 1967. 726 p.

Machatschek, Fritz. Das relief der erde. 2nd ed. Berlin: Borntraeger, 1955. 2 v.

Mazarovich, Aleksandr N. Grundlagen der regionalen Geologie der Erdteile. Berlin: Deutscher Verlag der Wissenschaften, 1958. 530 p.

Suess, Eduard. The face of the earth (Das Antlitz der Erde.) Translated by Hertha Sollas. Oxford: Clarendon Press, 1904-1924. 5 v.

SERIES

INCOMPLETE SERIES

These three classic German works essentially ceased publication with the advent of World War II.

Andrée, Karl, H. A. Brouwer and W. H. Bucher, eds. Regionale Geologie der Erde. Leipzig: Akademie-Verlag, 1938-1941.
1. Die alten Kerne. v. 1b. Fennoskandien im Quatär. M. Sauramo. 1941; v. 3b. Quaternary of North America. E. Antevs. 1941; v. 5. Afrika. E. Hennig. 1938; v. 6. The Indian peninsula and Ceylon. G. de P. Cotter; v. 7. Middle and West Australia. E. de C.

Clarke. 1938; v. 8. Antarctica. G. Taylor. 1940; 2. Paläozoische tafelin und gebirge. v. 2. Northwestern Europe: Caledonides. E. B. Bailey and O. Holtedahl. 1938; v. 3. Mittel- und westeuropa. H. Becker. 1938; 3. Jüngere orogenische zonen. v. 4a. The North American cordillera and the Caribbean region. A. C. Waters and H. D. Hedberg. 1939; v. 4b. Die Kordilleren von Südamerika. H. Gerth. 1939.

Krenkel, Erich, ed. Geologie der Erde. Berlin: Borntraeger, 1925-1941, 1955.
1. Grundzüge der physikalischen Verhältnisse der festen der Erde. E. Tams. 1932-1937; 2. Geologie von Europa. Serge von Bubnoff. v. 1. Einfuhrung. Osteuropa. Baltischer Schild. 1926. v. 2. Das ausseralpine Westeuropa. 1930-1936; 3. Geologie Südamerikas. H. Gerth. v. 1. pts. 1-3. 1932-1941. v. 2. 1955; 4. Geologie von Asien. Kurt Leuchs. v. 1. pts. 1-2. 1935-1937; 5. Geologie von Grönland. Lauge von Koch. 1935; 6. Geologie von Spitzbergen, der Bäreninsel, des Königs-Karl und Franz-Joseph Landes. H. Frebold. 1935; 7. Geologie Afrikas. Erich Krenkel. v. 1-3. 1925-1938. (Extensively revised as Geologie und Bodenschatze Afrikas. Leipzig: Geest and Portig, 1957. 597 p.); 8. Geology of North America. R. Ruedemann and R. Balk, eds. v. 1. 1939.

Steinmann, G. and O. Wilkens, eds. Handbuch der regionalen Geologie. Heidelberg: Winter, 1910-1944. 30 v. (Reprinted by Johnson Reprint Co., New York, 1965.)
1. Danemark. N. V. Ussing. 1910; 2. Island. H. Pjeturss. 1910; 3. Philippine Islands. W. D. Smith. 1910; 4. Die mittelatlantischen Vulkaninseln. C. Gagel. 1910; 5. New Zealand and adjacent islands. P. Marshall. 1911; 6. Madagascar. P. Lemoine. 1911; 7. Péninsule ibérique. A. Espagne. R. Douvillé. 1911; 8. Persien. A. F. Stahl. 1911; 9. Oceania. P. Marshall. 1911; 10. Armenien. F. Oswald. 1912; 11. United States of North America. E. Blackwelder. 1912; 12. Niederlande. G. A. F. Molengraaf. 1913; 13. Norwegen, Schweden, Finnland. A. G. Högbom. 1913; 14. Afrique occidentale. P. Lemoine; and English colonies on west coast of Africa and Liberia. J. Parkinson. 1913; 15. Antarktis. O. Nordenskjöld. 1913; 16. Balkanhalbinseln. A: Die Küstenländer Österreich-Ungarns. R. Schubert. 1914; 17. Syrien, Arabien und Mesopotamien. M. Blanckenhorn. 1914;

18. Die österreichischen und deutschen Alpen bis zur alpinodinarischen Grenze. F. Heritsch. 1915; 19. Zentralasien. K. Leuchs. 1916; 20. British Isles. P. G. H. Boswell; and Channel Islands. J. Parkinson. 1916; 21. Grönland. O. B. Böggild. 1917; 22. Kleinasien. A. Philippson. 1918; 23. Aegypten. M. Blanckenhorn. 1921; 24. Die Nordatlantischen Polarinseln. O. Nordenskjöld. 1921; 25. Kaukasus. A. F. Stahl. 1924; 26. Abessinien und Somalien. E. Krenkel. 1926; 27. Union of South Africa. A. W. Rogers and A. L. Hall. 1929; 28. Die Baltischen Länder. L. Rüger. 1934; 29. Mittelamerika. K. Sapper. 1937; 30. Die Sudetenländer. W. Petrascheck. 1944.

RECENT SERIES

Bentz, Alfred, ed. Beiträge zur Regionalen Geologie der Erde. Berlin: Borntraeger, 1961- .
Issued to date: V. 1. Die Geologie Mittelamerikas. Richard Weyl. 1961; v. 2. Geologie von Paraguay. Hannfrit Putzer. 1962; v. 3. Geologie von Chile. Werner Zeil. 1964; v. 4. Geologie der Antillen. Richard Weyl. 1965; v. 5. Rocky Mountains; der geologische Aufbau des kanadischen Felsengebirges. Dietrich Hans Roeder. 1967; v. 6. Geologie von Syrien und dem Libanon. Reinhard Wolfart. 1967; v. 7. Geologie von Jordanien. Friedrich Bender. 1968; v. 8. Geologie von Ungarn. László Trunkó. 1969; v. 9. Geologie von Brasilien. Karl Beurlen. 1970; v. 10. Geology of the South Atlantic islands. R. C. Mitchell-Thome. 1970; v. 11. Geologie von Ecuador. W. Sauer. In press.

De Sitter, L. U., ed. Regional geology series. New York: Wiley, 1964- . Irregular.
Published to date: Geology of the Himalayas. A. Gansser. 1964. 289 p.; The tectonics of the Appalachians. J. Rodgers. 1970. 288 p. In press: East Greenland Caledonides. J. Haller; Scandinavian Caledonides. T. Strand and O. Kulling.

Lexique stratigraphique international. Publié sur l'initative de la Commission de stratigraphie du Congress géologique international. Paris: Centre national de la recherche scientifique (CNRS), 1956- .
Each part of this fundamental reference work describes the stratigraphic units of a country or group of countries. Coverage extends to every country and region of the

world. Many of the parts are prefaced by a summary of the geology of the region. For the general plan of the Lexique see its breakdown in the general section of this guide.

THE ARCTIC

Arctic Institute of North America. The Arctic basin. Revised ed. Coordinated by John E. Sater. Washington: The Institute, 1969. 337 p.

----------. Arctic bibliography. Edited by Marie Tremaine. Washington: U. S. Government Printing Office, 1953-1965, v. 1-12. Montreal: McGill University Press, 1967, v. 13. Montreal: McGill-Queen's University Press, 1969- , v. 14- .
Abstracts of publications originating from scientific investigations in the arctic and subarctic regions of all countries.

----------. Catalogue of the library of the Arctic Institute of North America, Montreal. Prefaced by a description of the library and its arrangement by Nora T. Corley. Boston: G. K. Hall, 1968. 4 v.
This library has cataloged over 71, 000 items and subscribes to over 1, 000 serials; it is the definitive library on the arctic and subarctic of this continent.

Christie, R. L. Publications of the geology of the Arctic Islands by the Geological Survey of Canada. Ottawa: Geological Survey of Canada, 1971. 21 p. (Paper no. 71-10.)

International Symposium on Arctic Geology. 1st, Calgary, 1960. Geology of the Arctic; proceedings of the First International Symposium on Arctic Geology. Edited by Gilbert O. Raasch. Toronto: University of Toronto Press, 1961. 2 v.

National Research Council. Committee on Polar Research. Polar research; a survey. Washington: National Academy of Sciences, 1970. 204 p. (NAS Publication no. 1753.)

"Recent polar literature." In: Polar record, publication of the Scott Polar Research Institute. 3/yr.

A list of recent accessions to the Scott Polar Research Institute library. In 1969, the Polar record listed over 1800 items, including those in many branches of geology. Coverage includes both arctic and antarctic areas.

SVALBARD (Spitsbergen)

Orvin, Anders. Bibliography of literature about geology, physical geography, useful minerals and mining of Svalbard. Oslo: Norges Svalbard- og Ishavs-Undersøkelser, 1947. 121 p. (Skrifter no. 89.)

Sokolov, V. N. and others. "The main features of the tectonic structure of Spitsbergen." In: Geological magazine. v. 105, no. 2, 1968. p. 95-209.

Winsnes, Thore S. and others. Aspects of the geology of Svalbard. Oslo, 1960. 34 p. (International Geological Congress. 21st, Norway, 1960. Guide to Excursion no. A 16.)

NORTH AMERICA

REFERENCE WORKS

B. B. O. A. G. Boletin bibliografico de geofísica y oceanográfia americanas. Mexico: Instituto Panamericano de Geografia e Historia, 1958- .
Issued in two parts: Geofísica and Oceanográfia. Summaries of work and publications in the countries of the Americas.

Earth Science Curriculum Project. Selected maps and earth science publications for the states and provinces of North America. Compiled by William H. Matthews, III. Englewood Cliffs, N. J.: Prentice Hall, 1965. 42 p. (Reference Series no. RS-4.)
Publications listed are primarily non-technical.

U. S. Geological Survey. Abstracts of North American geology. Washington: Government Printing Office, 1966- . Monthly.
For annotation see citation in the general section.

----------. Bibliography of North American geology. Washington: Government Printing Office, 1919- . Annual.
The annual cumulation of the Abstracts of North American geology.

TEXTS AND TREATISES

Childs, Orlo E. and B. Warren Beebe, eds. Backbone of the Americas, a symposium; tectonic history from pole to pole. Tulsa: American Association of Petroleum Geologists, 1963. 320 p. (Memoir no. 2.)

Clark, Thomas H. and Colin W. Stearn. Geological evolution of North America. 2nd ed. New York: Ronald Press, 1968. 570 p.

Eardley, Armand J. Structural geology of North America. 2nd ed. New York: Harper, 1962. 743 p.

King, Philip B. The evolution of North America. Princeton, N. J.: Princeton University Press, 1959. 189 p. Reprinted by Hafner in 1969 with a new introduction by the author.

----------. The tectonics of North America--a discussion to accompany the tectonic map of North America, scale 1:5, 000, 000. Washington: Government Printing Office, 1969. 95 p. (U.S. Geological Survey. Professional Paper no. 628.)

SPECIFIC REGIONS

APPALACHIANS

Clark, Thomas H. Appalachian tectonics. Toronto: University of Toronto Press, 1967. 99 p. (Royal Society of Canada. Special Publication no. 10.)

King, Philip B. and others. Tectonics of the southern Appalachians. Edited by W. D. Lowry. Blacksburg: Virginia Polytechnic Institute, Department of Geological Sciences, 1964. 114 p. (Memoir no. 1.)

Rodgers, John. The tectonics of the Appalachians. New York: Wiley, 1970. 288 p. (Regional Geology Series.)

Studies in Appalachian geology: central and southern. Edited by George W. Fisher and others. New York: Interscience, 1970. 460 p.

Studies in Appalachian geology: northern and maritime. Edited by E-an Zen and others. New York: Interscience, 1968. 475 p.

U.S. Geological Survey. Mineral resources of the Appalachian region. Washington: Government Printing Office, 1968. 492 p. (Professional Paper no. 580.) Includes a summary of geology.

Zen, E-an. Taconic stratigraphic names: definitions and synonyms. Washington: Government Printing Office, 1964. 95 p. (U.S. Geological Survey. Bulletin no. 1174.)

ATLANTIC AND GULF COASTS

Eensminger, Henry R. and W. T. Morton. Selected bibliography of geology and geophysics for the Gulf of Mexico. Washington, D. C.: Naval Oceanographic Office, 1968. 18 p. (Informal Report no. IR 68-69.)

Murray, Grover E. Geology of the Atlantic and Gulf Coast Province of North America. New York: Harper, 1961. 692 p.

Neale, E. R. W. and H. Williams, eds. Collected papers on geology of the Atlantic region. Toronto: Geological Association of Canada, 1967. 292 p. (Special Paper no. 4.)

U. S. Geological Survey. Bibliography of Gulf Coast geology. Houston: Gulf Coast Association of Geological Societies, 1970. 2 v.
Covers through 1968.

GREAT LAKES

Conference on Great Lakes Research. Proceedings. Ann Arbor, Michigan: International Association for Great Lakes Research. Annual.
Proceedings for 1st-9th conferences published in Michigan. University. Great Lakes Research Division. Publications.

Hough, Jack L. Geology of the Great Lakes. Urbana: University of Illinois Press, 1958. 313 p.

Pincus, Howard J., ed. Great Lakes Basin, a symposium... Washington: American Association for the Advancement of Science, 1959. 308 p. (Publication no. 72.)

ROCKY MOUNTAINS

Haun, John D. and Harry C. Kent. "Geologic history of the Rocky Mountain region." In: American Association of Petroleum Geologists. Bulletin. v. 49, no. 11, 1965. p. 1781-1800.
Lead article in a special issue on the Rocky Mountain sedimentary basins.

Rocky Mountain Association of Petroleum Geologists. Geologic atlas of the Rocky Mountain region. Edited by William W. Mallory. Denver, 1971. In press.

Roeder, Dietrich Hans. Rocky Mountains; der geologische Aufbau des kanadischen Felsengebirges. Berlin: Borntraeger, 1967. 318 p. (Beiträge zur regionalen Geologie der Erde, v. 5.)

Symposium on the Tectonic History and Mineral Deposits of the Western Cordillera in British Columbia and Neighboring Parts of the United States. Vancouver, October 25-28, 1964. Montreal: Canadian Institute of Mining and Metallurgy, 1966. 353 p. (Special Volume no. 8.)
A review of major tectonic features of western North America.

GREENLAND

Andersen, Axel, ed. Meddelelser om Grønland general index, v. 1-150. Copenhagen: C. A. Reitzels Forlag for the Kommissionen for Videnskabelige Undersøgelser i Grønland, 1969. 707 p. (Meddelelser om Grønland)
Considered an outstanding journal for Arctic and Greenland research. About one-third of the Meddelelser deals with geology. A supplement covering volumes 151-200 is in preparation.

Brandt, Inger and W. Stuart Watt. "Litteraturfortegnelse." Published annually in Dansk Geologisk Forening. Meddelelser.
Pref: "list of papers on geology or similar matters which by subject, author or place of publication are related to Denmark and Greenland." Until 1969 these bibliographies included Iceland.

CANADA

REFERENCE WORKS

Canada. Department of Energy, Mines and Resources. Canadian minerals yearbook. Ottawa, 1886- . Annual.
Reviews of the mineral industry issued for the years

1886-1933 under various titles; for the years 1934-1955 as Mines Branch reports; for 1956-1961 as Canadian mineral industry; for 1962-present as Canadian minerals yearbook. Chapters dealing with specific commodities are issued as separate pamphlets in the year prior to publication of the yearbook.

Canada. Geological Survey. Index of publications, Geological Survey of Canada, 1845-1958. A. G. Johnson. Ottawa: Department of Mines and Technical Surveys, 1961. 378 p.

----------. 1959-1964. H. M. A. Rice. 1965. 163 p. Kept up-to-date by annual supplements in the Survey's Paper series.

----------. Index to reports of the Geological Survey of Canada from 1927-1950. W. E. Cockfield and others. Ottawa: Department of Mines and Technical Surveys, 1962. 723 p.
This index is the sixth in the Index to reports series issued by the survey. A brief outline of earlier indexes: 1st covers reports from 1863-1884; 2nd covers reports from 1884-1904; 3rd covers Separate Reports from 1906-1910 and Summary Reports from 1905-1916; 4th covers Palaeontology in reports from 1847-1916; and 5th covers Memoirs from 1910-1926, Bulletins from 1913-1926 and Summary Reports from 1917-1926.

Canadian Centre for Geoscience Data. Canadian index to geoscience data, Edition 70-1. Ottawa, 1970. 11 v.
"Computer-based coordinate index which identifies sources of data on the geology of Canada. The first edition (70-1) includes about 20, 000 document titles, mainly from the Geological Survey of Canada; but it also includes titles from the Mineral Resources Branch, Quebec Department of Natural Resources, and Ontario Department of Energy and Resources Management."--Announcement. Ten volumes of the index cover the provinces; the 11th is a thesaurus of authorized terms used for indexing. Plans are to up-date when sufficient new data are available.

Canadian government publications. Mines Branch and Mineral Resources. Bibliography no. 12. Ottawa: Queen's Printer, 1967. 401 p. (Sectional Catalogue no. 12.) Periodically revised.

Faessler, Carl. Cross-index to the geological illustration of Canada. Quebec: Université Laval, 1956- . V. 1-2. Cross-index to the maps and illustrations of the Geological Survey and Mines Branch (Bureau of Mines), 1843-1946. Rev. ed. 1957. 525 p. (Laval. Université. Géologie et minéralogie. Contr. no. 75.) 1st supplement, 1946-1956. 1956. 193 p. (Contr. no. 118); v. 3. Geological illustrations published by Ontario Department of Mines, 1891-1955. 1956. 289 p. (Laval. Univ. Géologie et minéralogie. Contr. no. 117.); v. 4. Geological illustration published by Quebec Department of Mines, 1898-1957. 1957. 281 p. (Laval. Univ. Géologie et minéralogie. Contr. no. 127.)

Traill, R. J. A catalogue of Canadian minerals. Ottawa: Geological Survey of Canada, 1970. 649 p. (Paper no. 69-45.)

TEXTS AND TREATISES

Canada. Geological Survey. Geology and economic minerals of Canada. 5th ed. Edited by R. J. W. Douglas. Ottawa, 1970. 838 p. (Economic Geology Report no. 1, 5th ed.)

Canadian Institute of Mining and Metallurgy. Structural geology of Canadian ore deposits. Ottawa, 1948, 1957. 2 v.

Neale, E. R. W., ed. The earth sciences in Canada, a centennial appraisal and forecast. Toronto: University of Toronto Press, 1968. 259 p. (Royal Society of Canada. Special Publication no. 11.)

SPECIFIC REGIONS

Alberta Society of Petroleum Geologists. Geological history of western Canada; an atlas and text. Calgary, 1964. 220 p.

----------. Lexicon of geologic names in the Western Canada Sedimentary Basin and Arctic Archipelago. Calgary, 1960. 380 p.

----------. Western Canada sedimentary basin; a symposium... . Tulsa: American Association of Petroleum Geologists, 1954. 521 p.

Bird, John B. The physiography of arctic Canada. Baltimore: Johns Hopkins Press, 1967. 336 p.

Brown, Roger E. Permafrost in Canada; its influence on northern development. Toronto: Toronto University Press, 1970. 234 p.

Earth Science Symposium on Hudson Bay. Ottawa, 1968. Proceedings. Edited by P. J. Hood, under the sponsorship of the National Advisory Committee on Research in the Geological Sciences. Ottawa: Geological Survey of Canada, 1969. 385 p. (C. G. S. Paper no. 68-53.)

Sanford, B. V. and others. Geology of the Hudson Bay Lowlands (Operation Winisk). Norris, A. W. and others. Bibliography on Hudson Bay Lowlands. Ottawa: Geological Survey of Canada, 1968. 118 p. (Paper no. 67-60.)

THE PROVINCES

Canada. Geological Survey. List of geological reports of . . . (province). Ottawa, irregular. (Issued in the Miscellaneous Report, G- series.)
A series of separate lists covering Survey publications on the following provinces: Alberta, Atlantic provinces, British Columbia, Manitoba, Ontario, Quebec, Northwest Territories, Saskatchewan and Yukon Territory. Periodically revised.

ALBERTA

Alberta Society of Petroleum Geologists. Annotated bibliography of geology of the sedimentary basin of Alberta and of adjacent parts of British Columbia and Northwest Territories, 1845-1955. R. G. McCrossan and others. Calgary, 1958. 499 p.

Root, J. D. Index to current geologic maps of Alberta. Edmonton: Research Council of Alberta, 1970. 20 p. (Report no. 70-2.)

BRITISH COLUMBIA

British Columbia. Department of Mines. Index to annual reports of the Minister of Mines of the Province of British Columbia for the years 1874-1936, inclusive. H. T. Nation. Victoria, 1938. 510 p.

----------. Index no. 3 to publications of the British Columbia Department of Mines. Victoria, 1955. 228 p.

----------. Index no. 4 to publications of the British Columbia Department of Mines and Petroleum Resources. Victoria, 1966. 142 p.
These three indexes cover Annual Reports and Bulletins from 1874-1964 and Miscellaneous Reports through 1965.

Holland, Stuart S. Landforms of British Columbia, a physiographic outline. Victoria: British Columbia Department of Mines and Petroleum Resources, 1964. 130 p. (Bulletin no. 48.)

LABRADOR

Butler, J. and G. Bartlett. Bibliography of the geology of Newfoundland and Labrador, 1814-1968. St. Johns, Newfoundland: Newfoundland and Labrador Mineral Resources Division, 1969. 273 p. (Bulletin no. 38.)

Cooke, A. and F. Caron. Bibliography of the Quebec-Labrador peninsula; Bibliographie de la Péninsule du Québec-Labrador. Published for Centre d'Etudes Nordiques, Université Laval, Québec. Boston: G. K. Hall, 1968. 2 v.

MANITOBA

Bannatyne, B. B., S. C. Zoltai and M. J. Tamplin. Annotated bibliography of the Quaternary in Manitoba and the adjacent Lake Agassiz region (including archaeology of Manitoba). Winnipeg: Geological Division, Department of Mines and Natural Resources, 1970. 142 p. (Geological Paper no. 2/70.)

Manitoba. Mines Branch. Bibliography of geology of the Precambrian area of Manitoba to 1950. G. C. Milligan. Winnipeg: Mines Branch, 1951. 67 p. (Publication no. 51-1.)

----------. 1950-1957. G. S. Barry. 1959. 39 p. (Publication no. 57-3.)

----------. Bibliography of geology, palaeontology, industrial minerals, and fuels in the post-Cambrian regions of Manitoba to 1950. Lillian B. Kerr. Winnipeg: Mines Branch,

1951. 38 p. (Publication no. 51-2.)

----------. 1950-1957. B. A. Mills. 1959. 32 p. (Publication no. 57-4.)

----------. 1958-1965. B. B. Bannatyne. 1966. 37 p. (Publication no. 66-1.)

* * *

Davies, J. R. and others. Geology and mineral resources of Manitoba. Winnipeg: Department of Mines and Natural Resources, 1962. 190 p.

NEW BRUNSWICK

Abbott, D., ed. Bibliography of New Brunswick geology. Revised ed. Fredericton: Research and Productivity Council, 1969. 105 p. (Record no. 8.)

Gussow, W. C. Carboniferous stratigraphy and structural geology of New Brunswick, Canada. Fredericton: Department of Lands and Mines, 1952. 196 p.

NEWFOUNDLAND

Butler, J. and Bartlett, G. Bibliography of the geology of Newfoundland and Labrador, 1814-1968. St. Johns, Newfoundland: Newfoundland and Labrador Mineral Resources Division, 1969. 273 p. (Bulletin no. 38.)

NOVA SCOTIA

Nova Scotia. Department of Mines. Geology of Nova Scotia. Halifax, (1950). 24 p.

----------. Nova Scotia, the mineral province of eastern Canada. Mineral and geological guide book. Halifax, (1948). 138 p.

ONTARIO

Ontario. Department of Mines. General index to reports. Toronto, 1921- . The first 6 volumes cover only the Annual Reports of the Department (v. 1, 1891-1916; v. 2, 1917-1927; v. 3, 1927-1940; v. 4, 1941-1950; v. 5, 1951-1955; v. 6, 1956-1960). Volume 7, issued in 1966, covers Annual Reports 1960-1963, and in addition covers all works of the Department (except Preliminary Reports) which were published between 1939 and 1965 inclusive.

Winder, C. G. Lexicon of Paleozoic names in southwestern Ontario. Toronto: University of Toronto Press, 1961. 121 p.

* * *

Chapman, L. J. and D. F. Putnam. The physiography of Southern Ontario. 2nd ed. Toronto: University of Toronto Press, 1966. 386 p.

QUEBEC

Cooke, A. and F. Caron. Bibliography of the Quebec-Labrador peninsula; Bibliographie de la Péninsule du Québec-Labrador. Published for Centre d'Etudes Nordiques, Université Laval, Québec. Boston: G. K. Hall, 1968. 2 v.

Quebec. Department of Natural Resources. Catalogue of publications since 1883. Quebec: Direction of Information, 1968. Looseleaf.

* * *

Dresser, John and T. C. Denis. Geology of Quebec. Quebec: Quebec Department of Mines, 1941-1949. 3 v. (Geological Report no. 20.)

Geology and mineral resources of Quebec. Quebec: Department of Natural Resources, 1969. 17 p.

SASKATCHEWAN

Saskatchewan. Department of Mineral Resources. Annotated bibliography of Saskatchewan geology (1823-1958). W. O. Kupsch. Regina, 1959. 199 p. (Report no. 9, revised.)

----------. 1959-1960. 1961. 55 p. (Report no. 9, supplement.)

UNITED STATES

REFERENCE WORKS

Lexicon of geologic names of the United States.

Wilmarth, M. Grace. Lexicon of geologic names of the United States including Alaska. Washington: Government Printing Office, 1938. 2 v. (U.S. Geological Survey. Bulletin no. 896.)
Contains more than 9,000 stratigraphic units described in the literature before 1936.

Keroher, Grace C. and others. Lexicon of geologic names of the United States for 1936-1960. Washington: Government Printing Office, 1966. 3 v. (U.S. Geological Survey. Bulletin no. 1200.)
Lists 14,625 names including cross-references.

----------. Lexicon of geologic names of the United States for 1961-1967. Washington: Government Printing Office, 1970. 848 p. (U.S. Geological Survey. Bulletin no. 1350.)

Cohee, G. V. and others. Changes in stratigraphic nomenclature by the U.S. Geological Survey. Washington: Government Printing Office, 1965- . Annual.
Supplements the Lexicon by providing stratigraphic names which have been adopted, revised, reinstated or abandoned. Appears annually in the Bulletin series as follows: changes for 1963 (Bulletin no. 1194-A); 1964 (Bulletin no. 1224-A); 1965 (Bulletin no. 1244-A); 1966 (Bulletin no. 1254-A); 1967 (Bulletin no. 1274-A); 1968 (Bulletin no. 1294-A); 1969 (Bulletin no. 1324-A).

Schmeckebier, L. M. Catalogue and index of the publications of the Hayden, King, Powell and Wheeler Surveys. Washington: Government Printing Office, 1904. 208 p. (U.S. Geological Survey. Bulletin no. 222.)
Catalogs and indexes the following surveys: Geological and Geographical Survey of the Territories; Geological Exploration of the 40th Parallel; Geographical and Geological Surveys of the Rocky Mountain Region; and Geographical Surveys West of the 100th Meridian.

U.S. Bureau of Mines. List of publications issued by the Bureau of Mines from July 1, 1910 to January 1, 1960

with subject and author index. Washington: Government Printing Office, 1960. 826 p.
Kept up-to-date by annual supplements, which cumulate every 5 years.

----------. Minerals yearbook. Washington: Government Printing Office, 1933- . Annual in 4 volumes.
Worldwide statistics on metals, non-metals, petroleum and mineral products. Continues Mineral resources of the United States, published by the Geological Survey and Bureau of Mines from 1883-1933. Preprints of individual chapters are available from the Bureau of Mines prior to publication in the bound yearbook.

U. S. Geological Survey. Publications of the Geological Survey, 1879-1961. Washington: Government Printing Office, 1964. 457 p.
Complete catalog of the books, maps and charts published by the Survey. Revision of the catalog is not intended, but it is kept up-to-date by annual supplements (which will cumulate) and monthly notices. The catalog and annual supplements are indexed by subject, area and author. An index to U. S. G. S. publications by latitude and longitude is Albertson, G. H. Geologic index to the publications of the U. S. Geological Survey. Denver: Geological Publishing Co., 1963. Kept up-to-date by monthly, quarterly and annual supplements.

U. S. Superintendent of Documents. Monthly catalog of United States Government publications. Washington: Government Printing Office, 1895- . Monthly.
A current bibliography of publications issued by all branches of the federal government, including department and bureau publications. Each monthly issue contains an author-subject index. Annual index with the December issue.

TEXTS AND TREATISES

GENERAL GEOLOGY

For books about the general geology of the United States, see the texts and treatises listed under North America.

SPECIFIC SUBJECTS

American Institute of Mining, Metallurgical and Petroleum Engineers. Ore deposits of the United States, 1933-1967; the Graton-Sales Volume. Edited by John D. Ridge. New York: A.I.M.E., 1968. 2 v. Successor to: American Institute of Mining and Metallurgical Engineers. Ore deposits of the western states. Edited by the Committee on the Lindgren Volume, John W. Finch, Chairman...the Lindgren Volume. New York: A.I.M.E., 1933. 797 p. The present volume includes descriptions of ore deposits in the 50 states discovered since 1933, except those of porphyry copper deposits, which are covered in: Titley, S. R. and C. L. Hicks, eds. Geology of porphyry copper deposits--southwestern United States. The Wilson Volume. Tucson: University of Arizona Press, 1966. 287 p.

"A coast-to-coast tectonic study of the United States." In: University of Missouri, Rolla. Journal. no. 1, 1968. 156 p.

Koschmann, A. H. and M. H. Bergendahl. Principal gold-producing districts of the United States. Washington: Government Printing Office, 1968. 283 p. (U.S. Geological Survey. Professional Paper no. 610.)

Landes, Kenneth K. Petroleum geology of the United States. New York: Wiley, 1970. 571 p.

Thornbury, William D. Regional geomorphology of the United States. New York: Wiley, 1965. 609 p.

STATE GEOLOGICAL SURVEYS

Some state surveys have embarked on systematic publication programs, which, when complete, will insure geologic coverage of each county or area in the state. Reports resulting from such programs are not always so marked, but publication lists from the various state geological surveys generally provide a good indication of these and other activities.

Publication lists are of two types: a complete bibliography, usually indexed, of all publications and maps published by the survey, with out-of-print materials generally indicated with an asterisk; or a "list of available publications," listing only those publications still available for purchase.

Addresses of state geological surveys are given in Earth Science Curriculum Project. Selected maps and earth science publications for the states and provinces of North America. Englewood Cliffs, N. J.: Prentice-Hall, 1965. p. 32-33. (Reference Series no. RS-4.) (Available from American Geological Institute, Washington, D.C.)

THE STATES

Corbin, John B. An index of state geological surveys issued in series. New York: Scarecrow Press, 1965. 667 p.
A compilation of bulletins and other serial reports of each state geological survey published through 1962. Certain periodical type materials and map series are omitted. Arrangement by state, with author and subject indexes.

Long, Harriet K. A bibliography of bibliographies of geology of the states of the United States. Washington: American Geological Institute, 1965. (Geoscience abstracts. Special supplement, v. 7, no. 7. p. 111-125.)
Up-date in progress.

U.S. Congress. Senate. Committee on Interior and Insular Affairs. Mineral and water resources of... (state). Washington: Government Printing Office, 1964- .
A series of reports on the western states of the U.S. prepared for the Committee on Interior and Insular Affairs by the U.S. Geological Survey and cooperating local agencies. Each contains a summary of the geology of the state.

U.S. Geological Survey. Geologic and water-supply reports and maps of... (state). Washington, irregular.
A series of separate lists, one for each state, covering U.S.G.S. publications on that state. Periodically revised.

ALABAMA

Alabama. Geological Survey. Bibliography of Alabama geology. R. M. Harper. University, Alabama, 1935. (Bulletin no. 42, p. 59-108.)

----------. Bibliography of Alabama geology, 1935-1958. Earl L. Hastings. University, Alabama, 1960. 138 p. (Bulletin no. 67.)

ALASKA

Maher, J. C. and W. M. Trollman. Geological literature on the North Slope of Alaska. Tulsa: American Association of Petroleum Geologists, 1970. 133 p.

Orth, Donald J. Dictionary of Alaska place names. Washington: Government Printing Office, 1967. 1084 p. (U.S. Geological Survey. Professional Paper no. 567.)

* * *

Herreid, Gordon. "Geology and ore deposits of Alaska." In: Mining engineering. v. 13, 1961. p. 1316-1325.

U.S. Geological Survey. Landscapes of Alaska; their geologic evolution. Edited by Howell Williams. Berkeley: University of California Press, 1958. 148 p.

Wahrhaftig, Clyde. Physiographic divisions of Alaska. Washington: Government Printing Office, 1966. 51 p. (U.S. Geological Survey. Professional Paper no. 482.)

---------- and G. O. Gates. "Geology (of Alaska)." In: U.S. Congress. Senate. Committee on Interior and Insular Affairs. Mineral and water resources of Alaska. Washington: Government Printing Office, 1964. p. 27-41.

ARIZONA

Moore, Richard T. and Eldred D. Wilson. Bibliography of the geology and mineral resources of Arizona, 1848-1964. Tucson: Arizona Bureau of Mines, 1965. 321 p. (Bulletin no. 173.)

* * *

Galbraith, Frederic W. and Daniel J. Brennan. Minerals of Arizona. Tucson: Arizona Bureau of Mines, 1970. 116 p. (Bulletin no. 181.)

Hayes, Philip T. "Geology (of Arizona)." In: Mineral and water resources of Arizona. Tucson: Arizona Bureau of Mines, 1969. p. 35-58. (Bulletin no. 180.)

Wilson, Eldred D. A resume of the geology of Arizona. Tucson: Arizona Bureau of Mines, 1962. 140 p. (Bulletin no. 171.)

GRAND CANYON

Four Corners Geological Society. Geology and natural history of the Grand Canyon region. n.p., 1969. 212 p. (5th Field Conference Guidebook.)

Hamblin, W. Kenneth and J. Keith Rigby. Guidebook to the Colorado River. Provo, Utah: Brigham Young University, Department of Geology, 1968-1969. (Geology Studies. v. 15, pt. 5 and v. 16, pt. 2.)
Pt. 1. Lee's Ferry to Phantom Ranch in Grand Canyon National Park; pt. 2. Phantom Ranch to Lake Mead, Arizona-Nevada.

McKee, Edwin D. Stratigraphy and history of the Moenkopi Formation of Triassic age. New York: Geological Society of America, 1954. 133 p. (Memoir no. 61.)

---------- and Raymond C. Gutschick. History of the Redwall Limestone of northern Arizona. Boulder, Colorado: Geological Society of America, 1969. 726 p. (Memoir no. 114.)
Latest in a series of monographs describing the principal sedimentary rocks in the Grand Canyon. Former works are: McKee, Edwin D. The environment and history of the Toroweap and Kaibab formations of northern Arizona and southern Utah. Washington, 1938. 268 p. (Carnegie Institution of Washington. Publication no. 492); McKee, Edwin D. and Charles E. Resser. Cambrian history of the Grand Canyon region. Washington, 1945. 232 p. (Carnegie Institution of Washington. Publication no. 563.)

U.S. Geological Survey. The Colorado River region and John Wesley Powell. Washington: Government Printing Office, 1969. 145 p. (Professional Paper no. 669.)

ARKANSAS

Branner, John C. "Bibliography of the geology of Arkansas, 2nd ed." In: The slates of Arkansas by A. H. Purdue. Little Rock: Arkansas Geological Survey, 1909. p. 97-164.

* * *

Banner, George C. Geology of Arkansas. Little Rock: Arkansas Geological and Conservation Commission, 1959. Folded sheet.

Sigma Gamma Epsilon. Summary of Arkansas geology and field trip itineraries. Hot Springs, Ark.: S.G.E. Biennial Convention, 1951.
Includes a bibliography of roughly 700 entries.

CALIFORNIA

California. Division of Mines and Geology. Bibliography of the geology and mineral resources of California to December 31, 1930. Solon Shedd. San Francisco, 1932. 376 p. (Bulletin no. 104.)

----------. Bibliography of the geology and mineral resources of California for the years 1931 to 1936, inclusive. Solon Shedd. San Francisco, 1937. 125 p. (Bulletin no. 115.)
Bibliography for the year 1937 by Solon Shedd published in the California journal of mines and geology. v. 35, no. 3, 1940. p. 275-307.

Egenhoff, Elizabeth L. and Mary H. Rice. Publications of the California Division of Mines to July 1, 1961. San Francisco: California Division of Mines and Geology, 1962. 206 p. (Supplement to the 58th Report of the State Mineralogist.)
A recent and select listing of publications available from the Division can be found in Oakeshott, Gordon B. "A guide to information on the geology of California." In: California Divison of Mines and Geology. Mineral information service. v. 23, no. 10, 1970. p. 195-199.

Terry, Richard D. Bibliography of marine geology and oceanography, California coast. San Francisco: California Division of Mines, 1955. 131 p. (Special Report no. 44.)

* * *

California. Division of Mines and Geology. Geology of northern California. Edited by Edgar H. Bailey. San Francisco, 1966. 508 p. (Bulletin no. 190.)

----------. Geology of southern California. Edited by Richard H. Jahns. San Francisco, 1954. 878 p. (Bulletin no. 170.)

King, P. B. and others. "Geologic history of California and geology of the geomorphic provinces." In: U.S. Congress. Senate. Committee on Interior and Insular Affairs. Mineral and water resources of California. Washington: Government Printing Office, 1966. p. 27-76. (Also as Bulletin 191 of the California Division of Mines and Geology.)

Murdoch, Joseph and R. W. Webb. Minerals of California; centennial volume (1866-1966). San Francisco: California Division of Mines and Geology, 1966. 559 p. (Bulletin no. 189.)

Oakeshott, Gordon B. California's changing landscape--a guide to the geology of the state. New York: McGraw-Hill, 1971. 416 p.

Reed, Ralph D. Geology of California. Tulsa: American Association of Petroleum Geologists, 1933. 355 p.

COLORADO

Johnson, Jesse H. Bibliography of the geology of southwestern Colorado. Golden: Colorado School of Mines, 1924. 7 p. (Circular of Information.)

----------. Southeastern Colorado. 1925. 11 p. (Colorado School of Mines. Circular of Information.)

----------. Northeastern Colorado. 1925. 12 p. (Colorado School of Mines. Circular of Information.)

----------. Northwestern Colorado. 1926. 52 p. (Colorado School of Mines. Quarterly. v. 21, no. 3.)

----------. North-central Colorado. 1927. 38 p. (Colorado School of Mines. Quarterly. v. 22, no. 4.)

----------. South-central Colorado. 1928. 30 p. (Colorado School of Mines. Quarterly. v. 23, no. 4.)

Jones, Olive M. Bibliography of Colorado geology and mining, with subject index from the earliest explorations to

1912. Denver: Colorado Geological Survey, 1914. 493 p. (Bulletin no. 7.)

* * *

Chronic, John and Halka Chronic. Prairie, peak and plateau; a layman's guide to the geology of Colorado. Denver: Colorado Geological Survey, 1971, in press.

Eckel, Edwin B. Minerals of Colorado, a 100-year record. Washington: Government Printing Office, 1961. 399 p. (U.S. Geological Survey. Bulletin no. 1114.)

Guide to the geology of central Colorado. Golden: Colorado School of Mines, 1948. 176 p. (Colorado School of Mines. Quarterly. v. 43, no. 2.)
A field guide to the 33rd annual meeting of the American Association of Petroleum Geologists, 1948.

Guide to the geology of Colorado. Edited by Robert J. Weimer and John D. Haun. Denver: Rocky Mountain Association of Geologists, 1960. 310 p.
A field conference guidebook prepared for the 1960 meeting of the Geological Society of America. The R.M.A.G. guidebooks constitute one of the best sources of material on Colorado geology. There is no definitive work for the state's geology.

Mineral resources of Colorado. Prepared under the supervision of John W. Vanderwilt. Denver: Colorado Mineral Resources Board, 1947. 547 p.

----------, First Sequel. Prepared under the supervision of S. M. Del Rio. Denver: Colorado Mineral Resources Board, 1960. 764 p.

Tweto, Ogden. "Geology (of Colorado)." In: U.S. Congress. Senate. Committee on Interior and Insular Affairs. Mineral and water resources of Colorado. Washington: Government Printing Office, 1964. p. 11-27.

CONNECTICUT

Gregory, Herbert E. Bibliography of the geology of Connecticut. Hartford: Connecticut Geological and Natural History Survey, 1907. 123 p. (Bulletin no. 8.)

Rodgers, John and others. Explanatory text for preliminary geology map of Connecticut, 1956. Stoors: Connecticut Geological and Natural History Survey, 1959. 64 p. (Bulletin no. 84.)

DELAWARE

Marine, Ira W. and W. C. Rasmussen. Preliminary report on the geology and ground-water of Delaware. Newark: Delaware Geological Survey, 1955. 336 p. (Bulletin no. 4.)

DISTRICT OF COLUMBIA

Johnston, Paul M. Geology and ground-water resources of Washington, D.C. and vicinity. Washington: Government Printing Office, 1964. 97 p. (U.S. Geological Survey. Water Supply Paper no. 1776.)

FLORIDA

California Company. Bibliography of geology--Florida. New Orleans: The California Company, 1953. 29 p.

Cooke, Charles W. Geology of Florida. Tallahassee: Florida Geological Survey, 1945. 342 p. (Bulletin no. 29.)

Puri, Harbans S. and Robert O. Vernon. Summary of the geology of Florida and a guidebook to the classic exposures. Tallahassee: Florida Geological Survey, 1964. 312 p. (Special Publication no. 5, revised.)

White, William A. The geomorphology of the Florida peninsula. Tallahassee: Florida Geological Survey, 1970. 164 p. (Bulletin no. 51.)

GEORGIA

Cramer, Howard R. and others. Annotated bibliography of Georgia geology through 1959. Atlanta: Georgia Geological Survey, 1967. 368 p. (Bulletin no. 79.)

HAWAII

Macdonald, Gordon A. Bibliography of the geology and ground water resources of the Island of Hawaii. Honolulu: Hawaii Division of Water and Land Development,

1947. 191 p. (Bulletin no. 10.)

* * *

Hawaii. Division of Water and Land Development. Geology of the Hawaiian Islands. Harold T. Stearns. Honolulu, 1946. 106 p. (Bulletin no. 8.)
The following islands of the Hawaiian Island group are covered in the Bulletin series: Oahu, Bulletin no. 2, 1939; Lanai, Bulletin no. 6, 1940; Kahoolawe, Bulletin no. 6, 1940; Maui, Bulletin no. 7, 1942; Hawaii, Bulletin no. 9, 1946; Molokai, Bulletin no. 11, 1947; Niihau, Bulletin no. 12, 1947; and Kauai, Bulletin no. 13, 1960.

Macdonald, Gordon A. and Agatin T. Abbott. Volcanoes in the sea; the geology of Hawaii. Honolulu: University of Hawaii Press, 1970. 452 p.

Stearns, Harold T. Geology of the state of Hawaii. Palo Alto, California: Pacific Books, 1966. 266 p.

IDAHO

Ross, Clyde P. and Martha S. Carr. An annotated bibliography on the geology of the mining districts of Idaho, 1872-1941. Moscow: Idaho Bureau of Mines and Geology, 1941. (Pamphlet no. 57, p. 111-263.)

----------. Annotated bibliography of papers related to the geology of Idaho, 1941-1957. Moscow: Idaho Bureau of Mines and Geology, 1959. 219 p. (Pamphlet no. 119.)

Ross, Sylvia H. Brief abstracts of publications of the Idaho Bureau of Mines and Geology. Moscow: Idaho Bureau of Mines and Geology, 1968. 67 p. (Information Circular no. 19.)

Savage, C. N. Lexicon of Idaho geologic names. Moscow: Idaho Bureau of Mines and Geology, 1968. 78 p. (Information Circular no. 20.)

* * *

Ross, Clyde P. and J. Donald Forrester. Outline of the geology of Idaho. Moscow: Idaho Bureau of Mines and Geology, 1958. 74 p. (Bulletin no. 15.)

---------- and others. "Geology (of Idaho)." In: U.S. Congress. Senate. Committee on Interior and Insular Affairs. Mineral and water resources of Idaho. Washington: Government Printing Office, 1964. p. 23-40. (Also as Special Report no. 1 of the Idaho Bureau of Mines and Geology.)

ILLINOIS

Willman, H. B. and others. Bibliography and index of Illinois geology through 1965. Urbana: Illinois Geological Survey, 1968. 373 p. (Bulletin no. 92.)

* * *

Illinois. Geological Survey. Geological science field trip guide leaflets. Urbana, 1960- .
Over 100 short studies to areas within the state, periodically revised.

----------. Guide to the geologic map of Illinois. Urbana, 1961. 24 p. (Educational Series no. 7.)

Willman, H. B. and John C. Frye. Pleistocene stratigraphy of Illinois. Urbana: Illinois Geological Survey, 1970. 204 p. (Bulletin no. 94.)

INDIANA

Nevers, George M. and Richard D. Walker. Annotated bibliography of Indiana geology through 1955. Bloomington: Indiana Geological Survey, 1962. 485 p. (Bulletin no. 24.)
Covers from 1776 through 1955.

Shaver, Robert H. and others. Compendium of rock-unit stratigraphy in Indiana. Bloomington: Indiana Geological Survey, 1970. 229 p. (Bulletin no. 43.)

Sigma Gamma Epsilon. Rho Chapter. A survey of Indiana geology, with road logs for two field trips. 2nd ed. Bloomington, Ind.: Indiana University, Dept. of Geology, 1966. 112 p.

IOWA

Iowa. Geological Survey. Annotated bibliography of Iowa geology and mining. Charles R. Keyes. Iowa City, 1913. (Annual Report no. 22, p. 157-908.)

----------. Bibliography of the geology of Iowa, 1960-1964. Paul J. Horlick and others. Iowa City, 1967. 49 p.

Tuttle, Sherwood D. and Stanley J. Prior, Jr. "Geomorphology in Iowa--an annotated bibliography of the literature." In: Iowa Academy of Sciences. Proceedings. v. 75, 1968. p. 253-267.

* * *

Kay, G. F. and J. H. Lees. Sketch of the geology of Iowa. Iowa City: Iowa Geological Survey, n.d. 16 p.

Rose, J. N. Fossils and rocks of eastern Iowa. Iowa City: Iowa Geological Survey, 1967. 147 p. (Educational Series no. 1.)

KANSAS

Merriam, Daniel F. Annotated bibliography of the Kansas Precambrian. Lawrence: Kansas Geological Survey, 1964. 18 p. (Special Distribution Publication no. 8.)

Roberts, Robert S. and Warren G. Hodson. Ground water in Kansas--bibliography and subject index. Lawrence: Geological Survey of Kansas, 1966. 41 p. (Bulletin no. 182.)

VerWiebe, Walter A. and Byron N. Cooper. Bibliography of Kansas geology 1823-1938. Municipal University of Wichita, 1938. 76 p. (University Studies Bulletin no. 5.) (Municipal University of Wichita. Bulletin. v. 13, no. 11.)

* * *

Merriam, Daniel F. The geologic history of Kansas. Lawrence: Kansas Geological Survey, 1963. 317 p. (Bulletin no. 162.)

Zeller, Doris E., ed. The stratigraphic succession in Kansas. Lawrence: Kansas Geological Survey, 1968. 81 p. (Bulletin no. 189.)

KENTUCKY

Jillson, Willard R. A bibliography of county geological reports in Kentucky during the past century, 1856-1956. Frankfort, Kentucky: Roberts Printing Co., 1957. 18 p.

----------. A bibliography of the mineral resources of Kentucky, 1818-1965. Frankfort, Ky.: Roberts Printing Co., 1966. 66 p.

----------. Geological research in Kentucky: a summary account of the several geological surveys of Kentucky, including a complete list of their publications and a general bibliography of 806 titles pertaining to Kentucky geology. Frankfort: Kentucky Geological Survey, 1923. 228 p. (Series 6, volume 15.)
A biography of the directors and a bibliography of the work of the Survey for the years 1838-1923. The following work, now rare, supplements Jillson's book: Sulzer, Elmer G. "Geological research in Kentucky--addenda." In: Kentucky State Historical Society. Register. v. 30, no. 93, 1932. p. 322-334.

McFarlan, Arthur C. Geology of Kentucky. Lexington: University of Kentucky, 1943. 531 p.

LOUISIANA

Louisiana. Geological Survey. Publications of the Louisiana Geological Survey. Baton Rouge, 1969. 20 p.

MAINE

Maine. Geological Survey. Bibliography on Maine geology, 1836-1957. Augusta, 1958. 143 p.

----------. Revised supplement to the Bibliography of Maine geology, 1836-January 1, 1967. Augusta, 1967. 46 p.

MARYLAND

Donohue, Mildred D. and Norma S. Gordon. Fossil finds in Maryland; a retrospective bibliography. College Park, Maryland: University of Maryland Libraries, 1967. 233 p.

Mathews, Edward B. "Bibliography and cartography of Maryland, including publications relating to the physiography, geology and mineral resources." In: Maryland Geological Survey. Report. no. 1, 1897. p. 229-332.

Vokes, Harold E. Geography and geology of Maryland. Baltimore, Md.: Department of Geology, Mines and Water Resources, 1957. 243 p. (Bulletin no. 19.)

MASSACHUSETTS

Cassidy, Martin. A partial bibliography of the geology of Massachusetts through 1958. Cambridge: Harvard University, Department of Geological Sciences, 1962. 90 p.

Johansson, W. I. A bibliography of the paleontology of Massachusetts, 1821-1949. Amherst: Massachusetts University, Department of Geology and Mineralogy, 1952. 28 p. (Special Publication no. 3.)

Strahler, Arthur N. A geologist's view of Cape Cod. New York: Natural History Press, 1966. 115 p.

MICHIGAN

Kelley, R. W. and E. A. Kirkby. Sources of geological information in Michigan. Lansing: Michigan Geological Survey, 1967. 24 p. (Circular no. 5.)

Michigan. Geological Survey. An index of the geology of Michigan, 1823-1955. Helen M. Martin and Muriel T. Straight. Lansing, 1956. 461 p. (Publication no. 50.)

----------. Bibliography of Michigan geology, 1956-1960. Edward A. Kirkby. Lansing, 1961. 66 p.

* * *

Dorr, John A., Jr. and Donald F. Eschman. The geology of Michigan. Ann Arbor: University of Michigan Press,

1970. 476 p.

MINNESOTA

Minnesota. Geological Survey. Bibliography of Minnesota geology. Theodora G. Melone and Leonard W. Weis. Minneapolis, 1951. 124 p. (Bulletin no. 34.)

----------. Bibliography of Minnesota geology, 1951-1961. John F. Splettstoesser and Sally A. Sloan. Minneapolis, 1962. 72 p. (Bulletin no. 34, Supplement.)

Schwartz, George M. and George A. Thiel. Minnesota's rocks and waters, a geological story. Revised ed. Minneapolis: Minnesota Geological Survey, 1963. 365 p. (Bulletin no. 37, revised.)

MISSISSIPPI

McGlothlin, Tom. "General geology of Mississippi." In: American Association of Petroleum Geologists. Bulletin. v. 28, no. 1. Tulsa, 1944. p. 29-62.

MISSOURI

Missouri. Geological Survey and Water Resources. Bibliography of the geology of Missouri. Darling K. Greger. Rolla, 1945. 294 p. (v. 31, 2nd series.)

----------. 1945-1955. Edward L. Clark and others. Rolla, 1956. 146 p. (v. 38, 2nd series.)

----------. 1955-1965. Jerry D. Vineyard and others. Rolla, 1967. 229 p. (v. 42, 2nd series.)

----------. 1966. Jerry D. Vineyard. Rolla, 1967. 57 p. (Information Circular no. 18.)

----------. 1967. Jerry D. Vineyard. Rolla, 1968. 57 p. (Information Circular no. 19.)

----------. 1968. Jerry D. Vineyard. Rolla, 1969. 48 p. (Information Circular no. 21.)

----------. 1969. Jerry D. Vineyard. Rolla, 1970. 44 p. (Information Circular no. 23.)

* * *

Branson, Edwin B. The geology of Missouri. Columbia: University of Missouri, 1944. 535 p. (University of Missouri Studies no. 3.)

Howe, W. B. and J. W. Koenig. The stratigraphic succession in Missouri. Rolla: Missouri Geological Survey and Water Resources, 1961. 185 p. (v. 40, 2nd series.)

Vineyard, Jerry D. and others. "Geology (of Missouri)." In: Mineral and water resources of Missouri. Rolla: Missouri Division of Geological Survey and Water Resources, 1967. p. 13-32. (v. 43, 2nd series.) (Also as U.S. 90th Congress. Senate Document no. 19.)

MONTANA

Billings Geological Society. Selected bibliography of stratigraphy in Montana and adjacent areas. Butte: Montana Bureau of Mines and Geology, 1966. 61 p. (Special Publication no. 38.)

Chelini, J. M. Index of unpublished geologic studies in Montana. Butte: Montana Bureau of Mines and Geology, 1965. 88 p. (Special Publication no. 34.)

Montana Bureau of Mines and Geology. Publications of the Montana Bureau of Mines and Geology, 1919-1968. Butte, 1968. 21 p.

Waldron, Cecil R. and Roy H. Earhard. Bibliography of the geology and mineral resources of Montana. Butte: Montana Bureau of Mines and Geology, 1942. 356 p. (Memoir no. 21.)

* * *

Perry, Eugene S. Montana in the geologic past. Butte: Montana Bureau of Mines and Geology, 1962. 78 p. (Bulletin no. 26.)

Sahinen, U. M. and others. "Geology (of Montana)." In: U.S. Congress. Senate. Committee on Interior and Insular Affairs. Mineral and water resources of Montana. Washington: Government Printing Office, 1963. p. 13-21. (Also as Special Publication no. 28 of the Montana

Bureau of Mines and Geology.)

NEBRASKA

Barbour, Erwin H. and C. A. Fisher. The geological bibliography of Nebraska. Nebraska State Board of Agriculture, 1902. (Annual Report for 1901, p. 248-266.)

Condra, G. E. and E. C. Reed. The geological section of Nebraska. Lincoln: Nebraska Geological Survey, 1959. 82 p. (Bulletin no. 14a, revised ed.)

NEVADA

Gianella, Vincent P. Bibliography of geologic literature of Nevada; and Robert W. Prince. Bibliography of geologic maps of Nevada areas. Reno: Nevada Bureau of Mines, 1945. 205 p. (University of Nevada Bulletin, v. 39, no. 6.) (Geology and Mining Series no. 43.)

Nevada. Bureau of Mines. Geology and mineral resources of... county. Reno, 1957- .
A series of county reports, an example of the type mentioned in the introduction to this section, issued in the Bulletin series of the Bureau. Issued thus far: Elko County, Bulletin no. 54; Mineral County, Bulletin no. 58; Humboldt County, Bulletin no. 59; Clark County, Bulletin no. 62; and Eureka County, Bulletin no. 64.

Wallace, R. E. and others. "Geology (of Nevada)." In: Mineral and water resources of Nevada. Reno: Nevada Bureau of Mines, 1964. p. 11-48. (Bulletin no. 65.) (Also as U.S. 88th Congress. Senate Document no. 87.)

NEW HAMPSHIRE

Goldthwait, James W. and others. The geology of New Hampshire. Concord: New Hampshire Planning and Development Commission, 1951-1956, reprinted 1964. 3 v.
V. 1, Surficial geology; v. 2, Bedrock geology; v. 3, Minerals and mines.

NEW JERSEY

Grametbaur, Agnes B. Bibliography and index of the geology of New Jersey. Trenton: Geologic and Topographic Survey, 1946. 142 p. (Bulletin no. 59.)

Kummel, Henry B. The geology of New Jersey. Revised ed. Trenton: Department of Conservation and Development, 1940. 203 p. (Bulletin no. 50, Geologic Series.)

Widmer, Kemble. The geology and geography of New Jersey. Princeton, N. J.: Van Nostrand, 1964. 193 p. (New Jersey History Series, v. 19.)

NEW MEXICO

Jicha, Henry L., Jr. and Christina Lochman-Balk. Lexicon of New Mexico geologic names: Precambrian through Paleozoic. Socorro: New Mexico Bureau of Mines and Mineral Resources, 1958. 137 p. (Bulletin no. 61.)

New Mexico. Bureau of Mines and Mineral Resources. Bibliography of New Mexico geology and mineral technology through 1950. Marian R. Burks and John H. Schilling. Socorro, 1955. 198 p. (Bulletin no. 43.)

----------. 1951-1955. Constance F. Schilling and John H. Schilling. Socorro, 1956. 136 p. (Bulletin no. 52.)

----------. 1956-1960. Constance F. Schilling and John H. Schilling. Socorro, 1961. 124 p. (Bulletin no. 74.)

----------. 1961-1965. Teri Ray. Socorro, 1966. 124 p. (Bulletin no. 90.)

* * *

Dane, C. H. and G. O. Bachman. "Geology (of New Mexico)." In: U.S. Congress. Senate. Committee on Interior and Insular Affairs. Mineral and water resources of New Mexico. Washington: Government Printing Office, 1965. p. 19-39. (Also as Bulletin 87 of the New Mexico Bureau of Mines and Mineral Resources.)

NEW YORK

Muller, Ernest H. Bibliography of New York Quaternary geology. Albany: New York State Museum and Science Service, 1965. 116 p. (Bulletin no. 398.)

* * *

Broughton, John G. and others. The geology of New York State. Albany: New York Geological Survey, 1962. 42 p. (Map and Chart Series no. 5.)
Accompanies the geologic map of New York, which is issued in a folio of 5 sheets. The Survey's Educational Leaflet no. 20, The geology of New York State, (1966, 49 p.) was adapted from this work.

Schuberth, Christopher J. The geology of New York City and environs. Garden City, N. Y.: Natural History Press for the American Museum of Natural History, 1968. 304 p.

NORTH CAROLINA

Laney, Francis B. and K. H. Wood. Bibliography of North Carolina geology, mineralogy and geography, with a list of maps. Raleigh: North Carolina Geological and Economic Survey, 1909. 428 p. (Bulletin no. 18.)

Stuckey, Jasper L. North Carolina: its geology and mineral resources. Raleigh: North Carolina Department of Conservation and Development, 1965. 550 p.

---------- and Stephen G. Conrad. Explanatory text for geologic map of North Carolina. Raleigh: North Carolina Division of Mineral Resources, 1958. 51 p. (Bulletin no. 71.)

U.S. Geological Survey. Geology of the Great Smoky Mountains, Tennessee and North Carolina. Washington: Government Printing Office, 1961-1965. 4 v. (U.S. Geological Survey. Professional Paper no. 349, A-D.)

NORTH DAKOTA

Budge, Chrissie E. Bibliography of the geology and natural resources of North Dakota, 1814-1944, with supplements 1 (1948) and 2 (1951). Grand Forks: North Dakota Geological Survey, 1959.
Supplements cover through 1951. A new bibliography by Mary W. Scott, superceding the above work, is in preparation by the North Dakota Geological Survey.

Hainer, John L. The geology of North Dakota. Grand Forks: North Dakota Geological Survey, 1958, c1956. 46 p. (Bulletin no. 31.)

OHIO

Ohio. Geological Survey. Bibliography of Ohio geology, 1819-1950. Dorothy G. Watkins. Columbus, 1953. 103 p. (Bulletin no. 52.)

----------. 1951-1960. Pauline Smyth. Columbus, 1963. 63 p. (Information Circular no. 32.)

----------. 1961-1965. Pauline Smyth. Columbus, 1969. 46 p. (Information Circular no. 36.)

OKLAHOMA

Jordan, Louise. Subsurface stratigraphic names of Oklahoma. Norman: Oklahoma Geological Survey, 1957. 220 p. (Guidebook no. 6.)

Oklahoma Geological Survey. Bibliography of Oklahoma geology with subject index. L. E. Trout and G. H. Myers. Norman, 1915. 105 p. (Bulletin no. 25.)

----------. "Bibliography and index of Oklahoma geology." Published annually in Oklahoma geology notes since 1955.

* * *

Barrett, E. "The geologic history of Oklahoma--an outline." In: Oil and gas fields of Oklahoma, v. 1. Oklahoma City Geological Society, 1963. p. 1-32.

Ham, W. E. and others. Basement rocks and structural evolution of southern Oklahoma. Norman: Oklahoma Geological Survey, 1964. 302 p. (Bulletin no. 95.)

Johnson, Kenneth S. "Geology and mineral resources of Oklahoma." In: The compass of Sigma Gamma Epsilon. v. 48, no. 1, 1970. p. 5-18.

Jordan, Louise. "Geology of Oklahoma--a summary." In: Oklahoma geology notes. Norman: Oklahoma Geological Survey, 1967. p. 215-228.
Reprinted with slight modification from Geophysical Society of Tulsa. Proceedings. v. 8, 1961-1964. p. 27-41.

OREGON

Bibliography of the geology and mineral resources of Oregon.

Oregon. State Planning Board. Bibliography of the geology and mineral resources of Oregon, with digests and index to July 1, 1936. Ray C. Treasher and Edwin T. Hodge. Portland, 1936. 224 p.

Oregon. Department of Geology and Mineral Industries. Bibliography of the geology and mineral resources of Oregon. Supplement, July 1, 1936 to December 31, 1945. John E. Allen. Portland, 1947. 108 p. (Bulletin no. 33.)

----------. Second supplement, January 1, 1946 to December 31, 1950. Margaret L. Steere. Portland, 1953. 61 p. (Bulletin no. 44.)

----------. Third supplement, January 1, 1951 to December 31, 1955. Margaret L. Steere and Lillian F. Owen. Portland, 1962. 97 p. (Bulletin no. 53.)

----------. Fourth supplement, 1970. (Bulletin no. 67.)

Youngquist, Walter. Annotated lexicon of names applied to Tertiary stratigraphic units in Oregon and Washington west of the Cascade Mountains, with bibliography. University Station, Oregon: The author, 1961. 92 p.

* * *

Baldwin, Ewart M. Geology of Oregon. Eugene: University of Oregon Cooperative Book Store, 1959. 136 p.

Corcoran, R. E. and others. "Geology (of Oregon)." In: Mineral and water resources of Oregon. Portland: Oregon Department of Geology and Mineral Resources, 1969. p. 23-88. (Bulletin no. 64.)

PENNSYLVANIA

Cramer, Howard R. Annotated bibliography of Pennsylvania geology to 1949. Harrisburg: Pennsylvania Topographic and Geologic Survey, 1961. 453 p. (Bulletin no. G-34.)

----------. Annotated bibliography of Pennsylvania geology--Supplement to 1959. Harrisburg: Pennsylvania Topographic and Geologic Survey, 1965. 237 p. (Bulletin no. G-42.)

Montgomery, Arthur. The mineralogy of Pennsylvania, 1922-1965; supplementing and updating Gordon's The mineralogy of Pennsylvania, 1922. Philadelphia: Academy of Natural Sciences of Philadelphia, 1969. 104 p. (Special Publication no. 9.)

Willard, Bradford. Pennsylvania geology summarized. Harrisburg: Pennsylvania Topographic and Geologic Survey, 1962. 17 p. (Educational Series no. 4.)

RHODE ISLAND

Quinn, Alonzo and David H. Swann. Bibliography of the geology of Rhode Island. 2nd ed. Providence: Rhode Island Port and Industrial Development Commission, 1950. 26 p.

* * *

Quinn, Alonzo W. Bedrock geology of Rhode Island. Washington: Government Printing Office, 1971. 68 p. (U. S. Geological Survey. Bulletin no. 1295.)

----------. "Brief view of Rhode Island geology." In: New England Intercollegiate Geological Conference. Annual meeting guidebook. 55th, Providence, 1963. p. 1-6.

SOUTH CAROLINA

Petty, Julian J. Bibliography of the geology of the State of South Carolina. Columbia: University of South Carolina, 1950. 86 p. (University Publication Series II, Bulletin no. 1.)

----------. Supplement to the bibliography of the geology of the State of South Carolina. Columbia: University of South Carolina, 1957. 9 p. (University Publication Series II, Bulletin no. 1, supplement.)

SOUTH DAKOTA

Parmelee, Gertrude. *A South Dakota bibliography--science and applied science.* South Dakota Library Association, 1960. 24 p.

Tipton, Merlin J. and others. *Bibliography of reports containing maps on South Dakota geology, published before January 1, 1959.* Vermillion: South Dakota Geological Survey, 1966. 71 p. (Circular no. 33.)

* * *

Denson, N. M. "Geology (of South Dakota)." In: *Mineral and water resources of South Dakota.* Vermillion: South Dakota Geological Survey, 1964. p. 17-38. (Bulletin no. 16.)

Rothrock, E. P. *A geology of South Dakota. Pt. 1--The surface. Pt. 3--Mineral resources.* Vermillion: South Dakota Geological Survey, 1943-1944. (Bulletins nos. 13 and 15.)
Intended as a three volume work. V. 2, about the bedrock was never published. However, a work intended to partially fill the gap is: Agnew, Allen F. and Paul C. Tychsen. *A guide to the stratigraphy of South Dakota.* Vermillion: South Dakota Geological Survey, 1965. 195 p. (Bulletin no. 14.)

TENNESSEE

Wilson, Charles W., Jr. *Annotated bibliography of the geology of Tennessee through December 1950.* Nashville: Tennessee Division of Geology, 1953. 308 p. (Bulletin no. 59.)

----------. *Annotated bibliography of the geology of Tennessee January 1951 through December 1960.* Nashville: Tennessee Division of Geology, 1965. 99 p. (Bulletin no. 67.)

* * *

Floyd, Robert J. *Tennessee rock and mineral resources.* Nashville: Tennessee Division of Geology, 1965. 119 p. (Bulletin no. 66.)
Physiography and general geology of the state, p. 3-15.

U.S. Geological Survey. Geology of the Great Smoky Mountains, Tennessee and North Carolina. Washington: Government Printing Office, 1961-1965. 4 v. (U.S. Geological Survey. Professional Paper no. 349, A-D.)

TEXAS

Texas. University. Bureau of Economic Geology. "Bibliography and subject index of Texas geology." E. H. Sellards. In: The geology of Texas, v. 1. Austin, 1932. (Texas. University. Bulletin no. 3232, p. 819-996.)

----------. Bibliography and index of Texas geology, 1933-1950. Roselle M. Girard. Austin, 1959. 238 p. (Texas. University. Publication no. 5910.)

----------. Bibliography and index of Texas geology, 1951-1960. Elizabeth T. Moore and M. D. Brown. Austin, in press.

West Texas Geological Society. Bibliographic Committee. Bibliography of Permian Basin geology, West Texas and Southeastern New Mexico. Midland, 1967. 163 p. (West Texas Geological Society. Publication no. 67-54.)

----------. Lexicon Committee. Lexicon of pre-Pennsylvanian stratigraphic names of West Texas and Southeastern New Mexico. Midland, 1958. 153 p.

* * *

Texas. University. Bureau of Economic Geology. The geology of Texas. Austin, 1932-1937. 3 v. (Bulletin nos. 3232, 3401, and 3701.)

UTAH

Buss, Walter R. Bibliography of Utah geology to December 31, 1950. Salt Lake City: Utah Geological and Mineralogical Survey, 1951. 219 p. (Bulletin no. 40.)

Childers, Barbara S. Abstracts of theses concerning the geology of Utah to 1966. Salt Lake City: Utah Geological and Mineralogical Survey, 1970. 233 p. (Bulletin no. 86.)

Rich, Edith and Bernice Y. Smith. "Utah geology in print." In: Utah Geological and Mineralogical Survey. Quarterly review. Salt Lake City, 1965- . Annual.
Issued thus far: Bibliography for 1964 in Quarterly review, v. 1, no. 4; 1965 in v. 1, no. 8; 1966 in v. 1, no. 12; 1967 in v. 2, no. 2 plus supplement; 1968 in v. 3, no. 2; 1969 in v. 4, no. 2.

* * *

Bullock, Kenneth C. Minerals of Utah. Salt Lake City: Utah Geological and Mineralogical Survey, 1967. 237 p. (Bulletin no. 76.)

Crawford, Arthur L., ed. Surface, structure, and stratigraphy of Utah. Salt Lake City: Utah Geological and Mineralogical Survey, 1963. 170 p. (Bulletin no. 54a.)

Hilpert, L. S. and others. "Geology (of Utah)." In: U.S. Congress. Senate. Committee on Interior and Insular Affairs. Mineral and water resources of Utah. Washington: Government Printing Office, 1964. p. 17-37. (Also as Bulletin 73 of the Utah Geological and Mineralogical Survey.)

Jensen, Mead LeRoy, ed. Guidebook of northern Utah. Salt Lake City: Utah Geological and Mineralogical Survey, 1969. 266 p. (Bulletin no. 82.)

Stokes, William L. and E. B. Heylmun. Outline of the geologic history and stratigraphy of Utah. Salt Lake City: Utah Geological and Mineralogical Survey, n.d. 36 p.

Utah Geological Society. Guidebook to the geology of Utah. Salt Lake City, 1946- . Irregular.
Guidebooks are distributed by the Utah Geological and Mineralogical Survey, University of Utah, Salt Lake City, Utah.

GREAT SALT LAKE

Bissell, Harold J. Bonneville--an Ice-Age lake. Salt Lake City: Department of Geology, Brigham Young University, 1968. 66 p. (Geology Studies. v. 15, pt. 4.) (Studies for Students no. 3.)

Stokes, William L., ed. The Great Salt Lake. Salt Lake City: Utah Geological Society, 1966. 173 p. (Guidebook to the Geology of Utah, no. 20.) (Distributed by the Utah Geological and Mineralogical Survey, University of Utah, Salt Lake City.)

VERMONT

Vermont. Geological Survey. Publications of the Vermont Geological Survey. Montpelier: available from the Vermont State Library.

* * *

Stewart, David P. Glacial geology of Vermont. Montpelier: Vermont Development Department, 1961. 124 p. (Vermont Geological Survey. Bulletin no. 19.)

VIRGINIA

Bibliography of Virginia geology.

Roberts, Joseph K. Annotated geological bibliography of Virginia. Charlottesville: The Alderman Library, 1942. 726 p. (University of Virginia Bibliographical Series no. 2.)

Hoffer, F. B. Bibliography of Virginia geology and mineral resources, 1941-1949. Charlottesville: Virginia Division of Mineral Resources, 1968. 58 p. (Information Circular no. 14.)

Ruhle, James L. Geologic literature of the coastal plain of Virginia, 1783-1962. Charlottesville: Virginia Division of Mineral Resources, 1965. 95 p. (Information Circular no. 9.)

Dietrich, Richard V. Geology and Virginia. Charlottesville: University Press of Virginia, 1970. 213 p.

WASHINGTON

Washington. Division of Mines and Geology. Bibliography and index of the geology and mineral resources of Washington, 1814-1936. W. A. G. Bennett. Olympia, 1939. 140 p. (Bulletin no. 35.)

----------. 1937-1956. William H. Reichert. Olympia, 1960. 721 p. (Bulletin no. 46.)

----------. 1957-1962. William H. Reichert. Olympia, 1969. 375 p. (Bulletin no. 59.)

Youngquist, Walter. Annotated lexicon of names applied to Tertiary stratigraphic units in Oregon and Washington west of the Cascade Mountains, with bibliography. University Station, Oregon: The author, 1961. 92 p.

* * *

Campbell, Charles D. Introduction to Washington geology and resources. Olympia: Washington Division of Mines and Geology, 1962. 44 p. (Information Circular no. 22R.)

Livingston, Vaughn E., Jr. Geologic history and rocks and minerals of Washington. Olympia: Division of Mines and Geology, 1969. 42 p. (Information Circular no. 45.)

Weissenborn, A. E. and others. "Geology (of Washington)." In: U.S. Congress. Senate. Committee on Interior and Insular Affairs. Mineral and water resources of Washington. Washington, D.C.: Government Printing Office, 1966. p. 13-46. (Also as Reprint 9 of the Washington Division of Mines and Geology.)

WEST VIRGINIA

Gillespie, William H. and John A. Clendening. West Virginia geology, archeology and pedology: A bibliography and index. Morgantown: West Virginia University Library, 1964. 241 p.

Lucke, John B. Bibliography and index of West Virginia geology and natural resources to July 1, 1937. Morgantown: West Virginia Geological Survey, 1937. 84 p. (Bulletin no. 4.)

WISCONSIN

Hotchkiss, W. O. and E. F. Bean. A brief outline of the geology, physical geography and industries of Wisconsin. Madison: Wisconsin Geological and Natural History Survey, 1925. 60 p. (Bulletin no. 67, Educational Series no. 9.)

WYOMING

Bovee, Gladys G. "Bibliography and index of Wyoming geology, 1823-1916." In: Wyoming Geologist's Office Bulletin no. 17. Cheyenne, 1918. p. 317-446.

Troyer, Max L. Bibliography of Wyoming geology, 1917-1945. Laramie, Wyoming: Geological Survey of Wyoming, 1969. 73 p. (Bulletin no. 53.)

* * *

Hager, Michael W. Fossils of Wyoming. Laramie: Wyoming Geological Survey, 1970. 51 p. (Bulletin no. 54.)

Thomas, Horace D. Geologic history and structure of Wyoming. Laramie: Wyoming Geological Survey, 1957. 11 p. (Contribution of the Geological Survey of Wyoming. Reprint no. 18.)
Reprinted from Wyoming oil and gas fields. Wyoming Geological Association, 1957. 484 p.

MIDDLE AMERICA

REFERENCE WORKS

B. B. O. A. G. Boletin bibliografico de geofísica y oceanográfia americanas. Mexico: Instituto Panamericano de Geografia e Historia, 1958- .
Issued in two parts: Geofísica and Oceanográfia. Summaries of work and publications in the countries of the Americas.

Lexique stratigraphique international. Volume V: Latin America. Publie sur l'initative de la Commission de stratigraphie du Congress géologique international. Paris: Centre national de la recherche scientifique (CNRS), 1956- .
For a breakdown of the Lexique see entry in general section of this guide.

Maldonado-Koerdell, Manuel. Bibliografia geológica y paleontológica de América Central. Geological and paleontological bibliography of Central America. Mexico City: Instituto Panamericano de Geografía e Historia, 1958. 288 p. (Publicación no. 204.)
Includes an introduction on the history of geological exploration and research in Mexico.

Pan American Union. Department of Economic Affairs. Natural Resources Unit. Annotated index of aerial photographic coverage and mapping of topography and natural resources undertaken in the Latin American member countries of the OAS. Washington, 1964- .
Each of the 19 volumes provides the following information: aerial photography, topographic and planimetric maps, geologic maps, various resource map information.

Smith, Harriet W. Sources of geological information in Latin America. Washington, D.C.: Pan American Union, 1969. 22 p. (Seminar on the Acquisition of Latin American Library Materials. 14th, San Juan, Puerto Rico, 1969. Working Paper no. 4.)

U.S. Geological Survey. Abstracts of North American geology. Washington: Government Printing Office, 1966- . Monthly.
Covers Middle America. For annotation see entry in general section of this guide.

----------. Bibliography of North American geology. Washington: Government Printing Office, 1919- . Annual. The annual cumulation of the Abstracts of North American geology.

TEXTS AND TREATISES

Childs, Orlo E. and B. Warren Beebe, eds. Backbone of the Americas, a symposium; tectonic history from pole to pole. Tulsa: American Association of Petroleum Geologists, 1963. 320 p. (Memoir no. 2.)

Dengo, Gabriel O. Estructura geológica, historia tectónica y morfología de América Central. Mexico: Centro Regional de Ayuda Técnica, 1968. 50 p.

----------. "Geological structure of Central America." In: International Conference on Tropical Oceanography. Miami Beach, Florida, 1965. Proceedings. University of Miami, Institute of Marine Sciences, 1967. (Studies in Tropical Oceanography, no. 5, p. 56-73.)

---------- and Otto Bohnenberger. "Structural development of northern Central America." In: Tectonic relations of northern Central America and the western Caribbean--the Bonacca Expedition. Tulsa: American Association of Petroleum Geologists, 1969. (Memoir no. 11, p. 203-220.)

Eardley, A. J. Structural geology of North America. 2nd ed. New York: Harper, 1962. 743 p.

Murray, Grover E. Geology of the Atlantic and Gulf Coast Province of North America. New York: Harper, 1961. 692 p.

Weyl, Richard. Die Geologie Mittelamerikas. Berlin: Borntraeger, 1961. 226 p. (Beiträge zur Regionalen Geologie der Erde, v. 1.)

----------. "Supplement." In: Zentralblatt fur Geologie und Paläontologie. pt. 1, nos. 7/8, 1970. p. 1003-1051.

ANTILLES AND THE CARIBBEAN

Comitas, L. Caribbeana, 1900-1965. Seattle: University of Washington Press, 1968.
"Geology and land forms" p. 505-525; "Mining and economic geology" p. 689-697.

Paterson, D. S. Preliminary annotated bibliography on the geology, mineral and water resources of the Turks and Caicos Islands, West Indies, 1884-1968. London: Institute of Geological Sciences, Mineral Resources Division, 1969. 8 p. (Sources of Information. Preliminary Notes, Series B, no. 264.)

Puerto Rico. University. Institute of Caribbean Science. Status of geological research in the Caribbean. Mayagüez, 1959- . Annual.

Rutten, Louis M. R. Bibliography of West Indian geology. Utrecht: Oosthoek, 1938. 103 p. (Geographiche en Geologische Mededeelingen. Geologische Physiographische Reeks, no. 16.)

* * *

Barr, K. W. "The structural framework of the Caribbean region." In: Tulsa geological digest. v. 31, 1963. p. 75-102.

Butterlin, Jacques. La constitution géologique et la structure des Antilles. Paris: Centre national de la recherche scientifique (CNRS), 1956. 453 p.

Caribbean Geological Conference. Transactions. 1st, 1955-

1st, Antigua, B.W.I., 1955; 2nd, Mayaguez, 1959; 3rd Kingston, 1962; 4th, Port-of-Spain, 1965.

Gulf Coast Association of Geological Societies. Geology of the American Mediterranean; transactions of the 19th annual meeting. Tallahassee, 1969. 665 p. (Transactions, v. 19.)
Particularly for the Caribbean, the G.C.A.G.S. Collo-

quium, p. 269-321.

Hess, H. H., ed. Caribbean geological investigations. New York: Geological Society of America, 1966. 310 p. (Memoir no. 98.)

Martin-Kaye, P. H. A. "A summary of the geology of the Lesser Antilles." In: Overseas geology and mineral resources. v. 10, no. 2, 1969. p. 172-206.

Schuchert, Charles. Historical geology of the Antillean-Caribbean region. New York: Wiley, 1935. 811 p.

Weyl, Richard. Geologie der Antillen. Berlin: Borntraeger, 1966. 410 p. (Beiträge zur Regionalen Geologie der Erde, v. 4.)

COSTA RICA

Dengo, Gabriel. Bibliografía de la geología de Costa Rica. Cuidad Universitaria, 1962. 67 p. (Costa Rica. Universidad. Publicación. Ser. Ciencias Naturales, no. 3.)

CUBA

Bermúdez, Pedro. Las formaciones geológicas de Cuba. Havana: Instituto Cubano de Recursos Minerales, 1961. 177 p. (Geología Cubana no. 1.)

Calvache, A. Bosquejo histórico del conocimiento de la geología de Cuba. Havana: Academia Ciencias, Dept. Geología, 1965. 107 p.

Furrazola-Bermudez, Gustav and others. Geología de Cuba. Havana: Editora del Consejo Nacional de Universidade, 1964. 2 v.
Translated into English as: U.S. Joint Publications Research Service (JPRS); Geology of Cuba. March, 1968 (JPRS no. 32325). (Microprint no. 4005.) Most figures are available only on the original work in Spanish.

GUATEMALA

Anon. Bibliografía geológica de Guatemala, América Central (hasta 1965). Guatemala: La Edición, 1966. 81 p.

Bonis, Samuel B., ed. *Excursion guidebook for Guatemala.* Guatemala: Geological Society of America, 1967. 71 p. (Instituto Geográfico Nacional, Boletín Geología no. 4.) (Geological Society of America, Guidebook, 1967 Annual Meeting.)

----------. "A summary of the geology of Guatemala." In: *Instituto Centro-americano de Investigación y Techología Industrial (ICAITI). Publ. Geol.* no. 2, 1969. p. 76-80.

HAITI

Butterlin, Jacques. *Géologie générale et régionale de la République d'Haïti.* Paris: Institute des hautes études de l'Amérique latine, 1960. 194 p. (Travaux et Mémoires, v. 6.)

JAMAICA

Hose, H. R. "The geology and mineral resources of Jamaica." In: *Colonial geology and mineral resources (London).* v. 1, no. 1, 1950. p. 11-36.

Zans, Verner A. *Synopsis of the geology of Jamaica.* Kingston: Geological Survey of Jamaica, 1963. 72 p. (Bulletin no. 4.)

MEXICO

Aguilar Y. Santillan, Rafael. *Bibliografía geológica y minera de la República Mexicana.* Mexico City: Secretaría de Fomento, 1898. 158 p. (México. Instituto Geológico. Boletín no. 10.)

----------. *Completada hasta el año de 1904.* Mexico City: Secretaría de Fomento, 1908. 330 p. (México. Instituto Geológico. Boletín no. 17.)

----------. *1905-1918.* Mexico City: Departmento de Minas, 1918. 97 p. (México. Departmento de Minas. Boletín Minero, 1918, Marzo y Abril.)

----------. *1919-1930.* Mexico City: Talleres Gráficos de la Nacion, 1936. 83 p.

Blasquez L., Luis. Bibliografía hidrogeológica de la República Mexicana. Mexico City: Universidad Nacional Autónoma de México, Instituto de Geología, 1961. 101 p.

Maldonado-Koerdell, Manuel. "Nomenclatura, bibliografía y correlación de las formaciones arqueozóicas y paleozoicas de México." In: Asociación Mexicana de Geólogos Petroleros. Boletín. v. 6, nos. 3-4, 1954. p. 113-137.

----------. "Nomenclatura, bibliografía y correlación de las formaciones continentales (y alyunes marinas) del Mesozoico de México." In: Asociación Mexicana de Geólogos Petroleros. Boletín. v. 10, nos. 5-6, 1958. p. 287-308.

* * *

Barrera, Tomas. Guía geológica de Oaxaca. Mexico City: Universidad Nacional Autónoma de Mexico, Instituto de Geología, 1946. 101 p.

Beal, Carl H. Reconnaissance of the geology and oil possibilities of Baja California, Mexico. New York: Geological Society of America, 1948. 138 p. (Memoir no. 31.)

Blasquez L., Luis. Hidrogeología de las regiones desérticas de México. Mexico City: Universidad Nacional Autónoma de México, Instituto de Geología, 1959. 175 p. (Anales, v. 15.)

Esperanza Yarza, Luz. Los volcanes de México. Mexico City: Ediciones de la Sociedad Mexicana de Geografía y Estadística, 1948. 187 p.

Garfias, V. R. and T. C. Chapin. Geología de México. Mexico City: Editorial Jus, 1949. 202 p.

Gonzalez Reyna, Jenaro. Reseña Geológica del Estado de Oaxaca. Mexico City: Consejo de Recursos Naturales no Renovables, 1962. 509 p.

International Geological Congress. 20th, Mexico, 1956. Excursiones. Edited by Manuel Maldonado-Koerdell. Mexico City, 1956.

----------. Riqueza minera y yacimientos minerales de México. 3rd ed. Jenáro González Reyna. Mexico City, 1956. 497 p.

Tamayo, Jorge L. Geografía general de México. 2nd ed. Mexico: Instituto Mexicano de Investigaciones Económicas, 1962. 5 v.

PANAMA

Terry, R. A. A geological reconnaissance of Panama. San Francisco: California Academy of Sciences, 1956. 91 p. (Occasional Paper no. 23.)

Woodring, W. P. Geology and paleontology of Canal Zone and adjoining parts of Panama. Washington: Government Printing Office, 1957. (U.S. Geological Survey. Professional Paper no. 306.)
Geology covered in Part A, p. 10-61.

PUERTO RICO

Hooker, Marjorie. Bibliography and index of the geology of Puerto Rico and vicinity, 1866-1968. San Juan: Geological Society of Puerto Rico, 1969. 53 p.

TRINIDAD

Great Britain. Overseas Geological Surveys. Mineral Resource Division. The general and economic geology of Trinidad, B.W.I. 2nd ed. with revisionary appendix by G. E. Higgins. London: Her Majesty's Stationery Office, 1960. 145 p.

Liddle, R. A. The geology of Venezuela and Trinidad. 2nd ed. Ithaca, New York: Paleontological Research Institution, 1946. 890 p.

SOUTH AMERICA

REFERENCE WORKS

B.B.O.A.G. Boletin bibliografico de geofísica y oceanográfia americanas. Mexico: Instituto Panamericano de Geografia e Historia, 1958- .
Issued in two parts: Geofísica and Oceanográfia. Summaries of work and publications in the countries of the Americas.

Cramer, Howard R. "Selected bibliography of South American geology." In: Tulsa Geological Society. Digest. v. 31, 1963. p. 213-239.

Lexique stratigraphique international. Volume V: Latin America. Publie sur l'initative de la Commission de stratigraphie du Congress géologique international. Paris: Centre national de la recherche scientifique (CNRS), 1956- .
For a breakdown of the Lexique see entry in general section of this guide.

Menendez, Carlos A. Bibliografía paleobotánica de America del Sur. Buenos Aires: Coni, 1968. 229 p. (Museo Argentino de Ciencias Naturales "Bernardino Rivadavia." Instituto Nacional de Investigacion de las Ciencias Naturales. Revista. Paleontología. v. 1, no. 6.)

Pan American Union. Department of Economic Affairs. Natural Resources Unit. Annotated index of aerial photographic coverage and mapping of topography and natural resources undertaken in the Latin American member countries of the OAS. Washington, 1964- .
Each of the 19 volumes provides the following information: aerial photography, topographic and planimetric maps, geologic maps, various resource map information.

Psuty, Norman P. A bibliography of Latin American coastal geomorphology. Mexico City: Pan American Institute of

Geography and History, 1970. 12 p. (Special Publication no. 1.)

Singewald, Joseph. Bibliography of economic geology of South America. New York: Geological Society of America, 1943. 159 p. (Special Paper no. 50.)

Smith, Harriet W. Sources of geological information in Latin America. Washington, D. C.: Pan American Union, 1969. 22 p. (Seminar on the Acquisition of Latin American Library Materials. 14th, San Juan, Puerto Rico, 1969. Working Paper no. 4.)

TEXTS AND TREATISES

Childs, Orlo E. and B. Warren Beebe, eds. Backbone of the Americas, a symposium; tectonic history from pole to pole. Tulsa: American Association of Petroleum Geologists, 1963. 320 p. (Memoir no. 2.)

Gerth, Heinrich. Geologie Südamerikas. Berlin: Borntraeger, 1932-1955.
V. 1, pt. 1. Die geologische Geschichte Südamerikas dis zum Ende des Paläozoicums. 1932; v. 1, pt. 2. Die geologische Geschichte Südamerikas wahrend des Mesozoicums. 1935; v. 1, pt. 3. Die Ablagerungen des Tertiärs. 1941; v. 2. Bau der Südamerikanischen Kordillere. 1955.

Jenks, William F., ed. Handbook of South American geology; an explanation of the geologic map of South America. New York: Geological Society of America, 1956. 378 p. (Memoir no. 65.)

Morrison, R. P. A resumé of the geology of South America. Toronto: Toronto University, Institute of Earth Sciences, 1962. 146 p. Processed. (Contract no. AF 19(628)-22.) (Science Report no. 1.)

Petersen, Ulrich. "Metallogenic provinces in South America." In: Geologie. v. 59, no. 3, 1970. p. 834-897. Entire issue devoted to aspects of the geology of South America.

Weeks, L. G. "Paleogeography of South America." In: Geological Society of America. Bulletin. v. 59, 1948. p. 249-282.

Zeil, Von Werner. "Zur Geologie der Anden." In: Geologische Rundschau. v. 59, no. 3, 1970. p. 827-833. Entire issue devoted to aspects of the geology of South America.

ARGENTINA

Borrello, Angel V., ed. Indice bibliográfico de estratigrafía Argentina. La Plata: Provinca de Buenos Aires, Comisión de Investigación Cientifica, 1965. 638 p.

Buenos Aires (Province) Argentine Republic. Comisión de Investigación Cientifica. Bibliografía geológica y cartográfica de la Provincia de Buenos Aires. La Plata, 1961. 145 p.

----------. 1st Supplemento (1961-1963). La Plata, 1964. 26 p.

Sparn, Enrique. Bibliografía de la geología, mineralogía y paleontología de la República Argentina. Cordoba: Academia Nacional de Ciencias, 1920- . Until 1928, this bibliography was published as Miscelánea of the Academia Nacional de Ciencias, Cordoba (nos. 2, 3, 5, 11 and 17). From 1928-1949, the bibliography appeared in the Boletín of the academy (v. 32, no. 4; v. 33, no. 3; v. 34, no. 3; v. 35, no. 4; v. 38, no. 1-2; and v. 39, no. 1-2.)

* * *

Borrello, Angel V. Geology of Argentina. Buenos Aires: distributed by Librart, in press.

Lambert, Luis R. and others. Explicación del mapa geológico de la República Argentina, escala 1:2, 500, 000. Buenos Aires: Dirección General de Industria y Minera, 1950. 28 p. (Boletín no. 73.)

Windhausen, Anselm. Geología Argentina. Buenos Aires: Casa Jacobo Peuser, 1931. 645 p.

BOLIVIA

Muñoz Reyes, Jorge and others. Bibliografía geológia, mineralógica y paleontológica de Bolivia. La Paz: Ministerio de Minas y Petróleo. Dept. Nacional de

Geología, 1962. 186 p. (Boletín no. 4.)

* * *

Ahlfeld, Federico and Leonardo Branisa. Geología de Bolivia. La Paz: Editorial Don Bosco, 1960. 245 p.

Lohmann, H. H. "Outline of tectonic history of Bolivian Andes." In: American Association of Petroleum Geologists. Bulletin. v. 54, no. 5, 1970. p. 735-757.

BRAZIL

Cruz, P. R. and others. Annotated bibliography and index of the geology of Bahia. Brazil. Divisão de Geologia e Mineralogia, 1969. 175 p. (Boletim no. 242.)

Iglesias, Dolores and Maria de Lourdes Meneghezzi. Bibliografia e índice da geologia do Brazil, 1641-1940. Rio de Janerio: Divisão de Geologia e Mineralogia, 1959. 385 p. (Boletim no. 204.)

----------. 1941-1950. 1960. 167 p. (Boletim no. 206.)

----------. 1951-1960. 1967. 203 p. (Boletim no. 238.)

----------. 1962-1963. 1969. 134 p. (Boletim no. 244.)

Loewenstein, Pedro and others. Bibliografia e índice da geologia da Amazônia legal brasileira, 1641-1964. Belem: Museu Paraense Emílio Goeldi, 1969. 291 p. (Publicações Avulsas no. 11.)

* * *

Azevedo Martins, Emmanoel. Sinopse de geologia do Brazil. Rio de Janerio: Museo Nacional, 1959. 49 p. (Publicações Avulsas, v. 30.)

Beurlen, Karl. Geologie von Brasilien. Berlin: Borntraeger, 1970. 444 p. (Beiträge zur Regionalen Geologie der Erde, v. 9.)

Guimarães, Djalma. Geologia do Brazil. Rio de Janeiro: Divisão de Fomento da Producão Mineral, 1964. 674 p. (Memoir no. 1.)

International Field Institute. Guidebook: Brazil. Washington: American Geological Institute, 1966. n.p.

Olivera, A. I. de and O. H. Leonardos. Geologia do Brazil. 2nd ed. Rio de Janerio: Ministerio da Agricultura. Serviço de Informação Agrícola, 1943. 813 p.

CHILE

Brüggen, Juan. "Bibliografía minera y geológica de Chile." In: Sociedad Nacional Minería. Boletín. Ser. 2, v. 31. Santiago, 1919. p. 441-513, 539-607.

----------. "Bibliografía minera y geológica de Chile." In: Sociedad Nacional Minería. Boletín. Ser. 2, v. 38. Santiago, 1926. p. 870-889, 989-1007.

Muñoz Cristi, Jorge and Juan K. Kokot. Bibliografía geológica de Chile (1927-1953). Santiago: Chile Universidad. Instituto de Geología, 1955. 121 p. (Publicación no. 5.)

* * *

Brüggen, Juan. Fundamentos de la geología de Chile. Santiago: Instituto Geográfico Militar, 1950. 374 p.

Fuller, Carlos R. Geología y yacimientos metalíferos de Chile. Santiago: Instituto de Investigaciones Geológicas, 1965. 386 p.

Zeil, Werner. Geologie von Chile. Berlin: Borntraeger, 1964. 246 p. (Beiträge zur Regionalen Geologie der Erde, v. 3.)

COLOMBIA

Ramírez, Jesús E. Bibliografía de la Biblioteca del Instituto Geofísico de los Andes Colombianos sobre geología y geofísica de Colombia. 2nd ed. Bogota, 1957. 521 p. (Ser. C, Geología, Boletín no. 6.)

Royo y Gómez, José. Bibliografía geológica, geográfica y minera de Colombia. Bogota: Instituto Geológico Nacional, 1945. 123 p. (Compilación de los estudios geológicos oficiales en Colombia, v. 6, Anexo 2.)

----------. "Adicidnes y correccidnes a la bibliografía geológica, geográfica y minera de Colombia." In: Colombia. Servicio Geológico Nacional. Complicación de los estudios geológicos oficiales en Colombia. v. 8. Santiago, 1950. p. 313-344.

Venegas Leyva, Alicia. Bibliografía de los informes del Instituto Geológico Nacional. 1960. 587 p. (Compilación de los estudios geológicos oficiales en Colombia, v. 9.)

----------. Suplemento no. 1 (1960-1964). 1965. 93 p.

* * *

Bürgl, Hans. "Historia geológica de Colombia." In: Academia Colombiana de Ciencias Exactas, Físicas y Naturales, Bogotá. Revista. v. 11, no. 43, 1961. p. 137-194.
There is a mimeographed translation of this work by N. R. Rowlinson (81 leaves, 1962).

Del Río, Alejandro M. Mapa geológico general de la República de Colombia a escala 1:2, 000, 000, breve explicación. Bogota: Ministerio de Minas y Petróleos, Servicio Geológico Nacional, 1945. 34 p. (Complicación de los estudios geológicos oficiales en Colombia, v. 6, anexo 1.)

Radelli, Luigi. Géologie des Andes colombiennes. Grenoble: Grenoble. Université. Laboratoire de géologie de la Faculté des Sciences, 1967. 422 p. (Travaux. Mémoires, no. 6.)

ECUADOR

Larrea, Carlos M. Bibliografía científica del Ecuador. v. 2: Geología, petrografía, mineralogía, paleontología, vulcanología, botánica y zoología. Quito: Casa de la Cultura Ecuatoriana, 1952. 319 p.

Oppenheim, Victor. "The structure of Ecuador." In: American journal of science. v. 248, 1950. p. 527-539.

Sauer, Walther. Geología del Ecuador. Quito: Ministerio de Educación, 1965. 383 p.

----------. Geologie von Ecuador. Berlin: Borntraeger, 1971. 316 p. (Beiträge zur Regionalen Geologie der Erde, v. 11.)

FALKLAND ISLANDS

Sparn, Enrique. Bibliografía de la geología y del clima de las islas Malvinas. Cordoba: Universidad, 1954. 29 p.

* * *

Borrello, Angel V. Sobre la geología de las Islas Malvinas. Buenos Aires: Ediciones Culturales Argentinas, 1963. 70 p. (Biblioteca del Sesquicentenario. Serie Investigaciones.)

FRENCH GUIANA

Abonnenc, Émile, J. Hurault and R. Saban. Bibliographie de la Guyane francaise. Tome 1: Ouvrages et articles de langue française concernant la Guyane et les territoires avoisinants. Paris: Larose, 1957. 278 p.

Choubert, Boris. Géologie et pétrographie de la Guyane francaise. Paris: Office de la recherche scientifique outre-Mer, 1949. 120 p.

GUYANA

Dixon, C. G. and H. K. George. Bibliography of the geology and mining of British Guiana. Georgetown: British Guiana Geological Survey, 1964. 87 p.

Macdonald, J. R. A guide to the mineral exploration in Guyana. Georgetown: Geological Survey of Guyana, 1968. 91 p. (Bulletin no. 38.)

Stockwell, G. M. The geology of British Guiana. Georgetown: British Guiana Geological Survey, 1955. 102 p. (Bulletin no. 25.)

PARAGUAY

Eckel, Edwin B. Geology and mineral resources of Paraguay--a reconnaissance. Washington: Government Printing Office, 1959. 110 p. (U.S. Geological Survey. Professional Paper no. 327.)

Putzer, Hannfrit. *Geologie von Paraguay.* Berlin: Borntraeger, 1962. 182 p. (Beiträge zur Regionalen Geologie der Erde, v. 2.)

PERU

Castro Bastos, Leonidas. *Bibliografía geológica del Peru.* Lima: The author, 1960. 317 p.

The Paracas papers. v. 1, no. 1 (master bibliography). 1000 selected references to the geography, oceanology, geology, ecology and archaeology of coastal Peru and adjacent areas. Norbert P. Psuty and others. Springfield, Virginia: NTIS, 1968. 61 p. (AD 671 870.)

* * *

Bellido, B. E. and F. S. Simons. *Memoria explicativa del mapa geológico (1:2, 000, 000) del Peru.* Lima: Sociedad Geológica del Peru, 1957. 88 p. (Boletín no. 31.)

Steinmann, G. *Geologie von Peru.* Heidelberg: Winter, 1929. 448 p.

SURINAM

Steenhuis, Jakob F. *De geologische literatuur over of van belang voor Nederlandsch-Guyana (Suriname) en de Nederlandsche Westindische Eilandern (Antillen).* The Hague: Moulton, 1934. 80 p.

----------. *1934-1950.* The Hague: Moulton, 1951. 108 p.

* * *

Bosma, W. and W. Groeneweg, eds. *Contributions to the geology of Surinam.* Paramaribo: Surinam Geologisch Mijnbouwkundige Dienst, 1969. 191 p. (Mededeling no. 20.)

Ijzerman, Robert. *Outline of the geology and petrology of Surinam (Dutch Guiana).* The Hague: Nijhoff, 1931. 519 p.

URUGUAY

Lambert, Roger. Bibliographie géologique de la république orientale de l'Uruguay. Montevideo, Imprenta nacional, 1939. 79 p. (Instituto geológico del Uruguay. Boletín no. 26.)

* * *

Bossi, Jorge. Geología del Uruguay. Montevideo: Universidad de la República, 1966. 464 p. (Colección ciencias no. 2.)

VENEZUELA

Hedberg, H. D. and Hedberg, F. "Bibliografía e índice de geología, minería y petróleo de Venezuela." In: Venezuela. Ministerio de Fomento. Revista de Fomento. Ano 8, no. 58-59, 1945. p. 43-123.
Covers 1813-1944.

Korol, B. and C. J. Forjonel. "Bibliografía e índice de geología, minería y petróleo de Venezuela, prima parte, 1950-1958." In: Venezuela. Ministerio de Minas e Hidrocarburos. Dirección de Geología. Boletín de Geología. v. 5, no. 10, 1959. p. 121-211.

Stainforth, R. M. "Internal publications on the geology of Venezuela, 1958 to mid-1965." In: American Association of Petroleum Geologists. Bulletin. v. 49, no. 12, 1965. p. 2289-2294.

Venezuela. Ministerio de Mines e Hidrocarburos. Dirección de Geología. Memoria-Tercer Congreso Geológico Venezolano. Caracas, 1960-1961. 4 v. (Boletín de Geología. Publicación Especial no. 3.)

----------. Stratigraphical lexicon of Venezuela (English Edition). 2nd ed. Caracas, 1970.

* * *

Bucher, W. H. Structure and orogenic history of Venezuela. New York: Geological Society of America, 1952. 113 p. (Memoir no. 49.)

González de Juana, Clemente. "Introducción al estudio de la geología de Venezuela." In: Venezuela. Dirección de Geología. Boletín de Geología. 1951-1952. V. 1, no. 1, p. 117-139; v. 1, nos. 2, 3, p. 195-216, 265-287; v. 2, nos. 5, 6, p. 311-330, 407-416.

Liddle, R. A. The geology of Venezuela and Trinidad. 2nd ed. Ithaca, New York: Paleontological Research Institution, 1946. 890 p.

Martínez, Anibal R. Chronology of Venezuelan oil. London: Allen and Unwin, 1969. 207 p.

Mencher, E. and others. "Geology of Venezuela and its oil fields." In: American Association of Petroleum Geologists. Bulletin. v. 37, no. 4, 1953. p. 690-777.

THE ATLANTIC

Emery, K. O. and others. "Marine geology of the Atlantic continental margin of the southern United States." In: Southeastern geology. v. 9, no. 4, 1968. p. 173-271.

Heezen, Bruce C., Marie Tharp and Maurice Ewing. Floors of the Ocean, I: North Atlantic. New York: Geological Society of America, 1959. 2 parts, text and map. (Special Paper no. 65.)

----------. Physiographic diagram of the South Atlantic Ocean, the Caribbean Sea, the Scotia Sea, and the eastern margin of the South Pacific Ocean. New York: Geological Society of America, 1962.

International Conference on Stratigraphy and Structure Bearing on the Origin of the North Atlantic Ocean. Gander, Newfoundland, 1967. North Atlantic--geology and continental drift. Edited by Marshall Kay. Tulsa: American Association of Petroleum Geologists, 1969. 1082 p. (Memoir no. 12.)

Mitchell-Thome, Raoul. Geology of the South Atlantic islands. Berlin: Borntraeger, 1970. 367 p. (Beiträge zur Regionalen der Erde, v. 10.)

U.S. Geological Survey. Atlantic continental shelf and slope of the United States. Washington: Government Printing Office, 1966- . (Professional Paper no. 529.)

EUROPE

Lexique stratigraphique international. Volume I: Europe. Publie sur l'initative de la Commission de stratigraphie du Congress géologique international. Paris: Centre national de la recherche scientifique (CNRS), 1956- .
For a breakdown of the Lexique see entry in general section of this guide.

* * *

Bogdanov, A. A. and others, eds. Tectonique de l'Europe. Notice explicative pour la carte tectonique internationale de l'Europe au 1:2, 500, 000. Moscow: Nauka, 1964. 360 p.

Brinkmann, Roland. Geologic evolution of Europe. 2nd ed. Transl. by John E. Sanders. Stuttgart: Enke, 1969. 161 p.
Translated from selected parts of the 8th edition of Brinkmann's Abriss der Geologie, v. 2.

Bubnoff, Serge von. Geologie von Europa. Berlin: Borntraeger, 1926-1936. 4 v. (Geologie der Erde.)
V. 1. Einfuhrung. Osteuropa. Baltischer Schild. 1926; v. 2. Das ausseralpine Westeuropa. 3 v. 1930-1936.

Dorn, Paul. Geologie von Mitteleuropa. 3rd ed. Stuttgart: E. Schweizerbart'sche, 1966. 488 p.

Gignoux, Maurice. Stratigraphic geology. San Francisco: Freeman, 1955. 682 p.
Translation of Gignoux's 4th edition of Géologie stratigraphique, 1950. A regional stratigraphy, principally of Europe.

Rutten, Martin G. The geology of western Europe. New York: Elsevier, 1969. 520 p.
Important modern work. Does not include Great Britain and Iceland. Includes section, "Sources for geological

information in western European countries."

Shackleton, M. R. Europe, a regional geography. 7th ed. New York: Praeger, 1964. 572 p.
Includes geologic structure of the regions of Europe.

SPECIFIC REGIONS

ALPS

Cadish, Joos. Geologie der Schweizer Alpen. 2nd ed. Basel: Wepf, 1953. 480 p.

Debelmas, J. and M. Lemoine. "The Western Alps: palaeogeography and structure." In: Earth science reviews. v. 6, no. 4, 1970. p. 221-256.

Gwinner, M. P. Geologie der Alpen; Stratigraphie, Paläogeographie, Tektonik. Stuttgart: E. Schweizerbart'sche, 1971. 477 p.

International Field Institute. The Alps, 1962. Augustin Lombard and others. Washington: American Geological Institute, 1962. 130 p. (International Field Institute. Guidebook, 2nd.)

Kraus, Ernst. Die Baugeschichte der Alpen. Berlin: Akademie-Verlag, 1951. 2 v.

Liniger, Hans. Vom bau der Alpen; gemeinverständliche historische Geologie von Mitteleuropa. Munich: Ott Verlag Thun, 1958. 236 p.
Outlines the geologic history of central Europe, with special emphasis on the alpine zone.

Oxburgh, E. R. "An outline of the geology of the Central Eastern Alps" and "The Eastern Alps--a geological excursion guide." In: Geologists' Association (London). Proceedings. v. 79, pt. 1, 1968. p. 1-124.

BALKANS

"Geološka bibliografija Balkanskog poluostrova. Bibliographie géologique de la péninsule balkanique." Issued irregularly in Geoloski Anali Balkanskoga Poluostrova. Annales géologiques de la péninsule balkanique. Belgrade: Universitet. Geoloski Zavod, 1889-1951.

NORTH AND BALTIC SEAS

Inter-University Geological Congress, 14th, Hull, England, 1967. Geology of shelf seas. Edited by D. T. Donovan. Edinburgh: Oliver and Boyd, 1969. 160 p.
Largely concerned with the North Sea.

Model, Fritz. Geophysikalische Bibliographie von Nord--und Ostsee. Published for the Deutsches Hydrographisches Institut. Berlin: Borntraeger, 1966. 2 v. in 3. (Deutsche Hydrographische Zietschrift. Ergänzungsheft. Reihe A, no. 8.)

SCANDINAVIA

"The earth sciences in Scandinavia." In: Earth science reviews. v. 3, no. 2, 1967. Atlas section, p. A. 65-A. 156.
Includes the countries Denmark, Finland, Iceland, Norway and Sweden. Gives summaries of the general geology, the earth science activities and the organizations and institutes of each country.

"Outline of the geology and physiography of Finland, Norway and Sweden." In: Geochemical prospecting in Fennoscandia. Edited by Aslak Kvalheim. New York: Interscience, 1967. p. 3-95.

ALBANIA

Magnani, Mario. Bibliografia geologica e geografico-fisica della regione Albanese. 2nd ed. Rome: Instituto poligrafico dello stato liberia, 1941. 106 p. (Reale Ufficio Geologico d'Italia. Bollettino. v. 64, Supplement.)

AUSTRIA

"Geologische Literatur Österreichs." In: Austria. Geologische Bundesanstalt. Verhandlungen. Vienna, annual.

* * *

Beck-Mannagatta, P. Erläuterungen zur geologischen und zur Lagerstätten-Karte, 1:1, 000, 000 von Österreich. Vienna: Geologischen Bundesanstalt, 1966. 94 p.
In German, French and English.

Schaffer, F. X., ed. Geologie von Österreich. 2nd ed. Vienna: Deuticke, 1951. 810 p.

BELGIUM

Abstracts of Belgian geology and physical geography. Edited by L. Walschot. Ghent: Geological Institute, University of Ghent, 1969- . Annual.
Abstracts for 1967 published in v. 1, 1969.

Société Géologique de Belgique. Prodrome d'une description géologique de la Belgique. Edited by Paul Fourmarier. Liege: H. Vaillant-Carmanne, 1954. 826 p.

BRITISH ISLES

Clayton, Keith M., ed. A bibliography of British geomorphology. London: George Philip, 1964. 200 p. (British Geomorphological Research Group. Occasional Publication no. 1.)
Covers principally 1945-1962, with some earlier articles. British geomorphology from 1962-1966 covered in Geomorphological abstracts, and from 1966- in Geographical abstracts, A: Geomorphology. For annotation, see under "Geomorphology" in the subject section.

Eastwood, T. Stanford's geological atlas of Great Britain. London: Edward Stanford, 1964. 288 p.
Ireland and Channel Islands are omitted.

Smith, R. A. A bibliography of Lake District geology and geomorphology. Kells, Whitehaven, Cumberland: Cumberland Geological Society, 1965. 44 p.

Willis, Leonard J. A palaeogeographical atlas of the British Isles and adjacent parts of Europe. London: Blackie and Son, 1952. 64 p.

* * *

Bennison, George M. and Alan E. Wright. The geological history of the British Isles. London: Arnold, 1969. 406 p.

Evans, J. W. and C. J. Stubblefield. Handbook of the geology of Great Britain. London: Murby, 1929. 556 p.

Great Britain. Geological Survey. British regional geology. London: Her Majesty's Stationery Office, 1946- .
A set of handbooks covering the geology of Great Britain, continually revised.

Rayner, Dorothy H. The stratigraphy of the British Isles. Cambridge: Cambridge University Press, 1967. 453 p.

Stamp, L. Dudley. Britain's structure and scenery. 5th ed. London: Collins, 1960. 255 p.

----------. Introduction to stratigraphy, British Isles. 3rd ed. London: Murby, 1957. 381 p.

BRITISH COMMONWEALTH

Reed, F. R. C. The geology of the British Empire. 2nd ed. London: Arnold, 1949. 764 p.

"Select bibliography on geology and the mineral industry." In: Overseas geology and mineral resources. London: Her Majesty's Stationery Office, 1950- .
Latest published is no. 37 (Overseas geology and mineral resources, v. 10, no. 1, 1966). Preface--"select bibliography on geology, mineral resources and the mineral industry relates principally to Commonwealth, Colonial and Dependent Territories, but also includes items from the world literature likely to be of interest to workers overseas."

IRELAND

Eager, Alan R. A guide to Irish bibliographical material. London: Library Association, 1964. 392 p.
"Geology" p. 111-116; "Paleontology" p. 117-119.

* * *

Charlesworth, J. K. The geology of Ireland. Edinburgh: Oliver and Boyd, 1953. 276 p.

----------. Historical geology of Ireland. Edinburgh: Oliver and Boyd, 1963. 264 p.

Cole, Grenville A. J. and T. Hallissy. Handbook of the geology of Ireland. London: Murby, 1924. 82 p.

Nevill, W. E. Geology and Ireland, with physical geography and its geological background. Dublin: Figgis, 1963. 264 p.

SCOTLAND

Craig, Gordon Y., ed. The geology of Scotland. Edinburgh: Oliver and Boyd, 1965. 534 p.

Sissons, John Brian. The evolution of Scotland's scenery. Hamden, Conn.: Archon Books, 1967. 259 p.

WALES

Bassett, D. A. Bibliography and index of geology and allied services for Wales and the Welsh borders, 1536-1896. Cardiff: National Museum of Wales, 1963. 246 p.

----------. 1897-1958. 1961. 376 p.

Brown, Eric H. The relief and drainage of Wales, a study in geomorphological development. Cardiff: University of Wales Press, 1960. 186 p.

BULGARIA

Bulgarska Akademiia na Naukite, Sofia. Scientific Information Center for Natural, Mathematical and Social Sciences. Abstracts of Bulgarian scientific literature: geology and geography (English edition). Sofia, 1957- . Semi-annual.

CYPRUS

Avnimelech, Moshe A. Bibliography of Levant geology, including Cyprus, Hatay, Israel, Jordania, Lebanon, Sinai and Syria. Jerusalem: Israel Program for Scientific Translations, 1965- . v. 1, 1965. 192 p.; v. 2, 1969. 200 p.

Pantazis, Th. M. "A revised bibliography of Cyprus geology." In: Cyprus Geological Survey Department. Bulletin. no. 2, 1969. p. 57-81.

* * *

Henson, F. R. S., R. V. Browne and J. McGinty. "A synopsis of the stratigraphy and geological history of Cyprus." In: Geological Society of London. Quarterly journal. v. 105, 1949. p. 1-41.

CZECHOSLOVAKIA

Czechoslovak Republic. Ústřední Ústav Geologický. Mineralogicko-geologická bibliografie ČSSR za rok (yr.). Prague, 1930- . Annual.

* * *

Andrusov, Dimitrij. Geológia československych Karpát. Bratislava: Vydavatel'stvo Slovenskej akadémie vied, 1958-1965. 3 v.
The first 2 volumes have been translated into German.

Bernard, J. H., ed. Mineralogie Ceskoslovenska. Prague: Academia, 1969. 396 p.

Czechoslovakia. Geological Survey. Regional geology of Czechoslovakia. Prague, 1966-1968. 2 v. plus maps. (Distributed by E. Schweizerbart'sche, Stuttgart.)
V. 1--The Bohemian Massif; v. 2--The West Carpathians.

"The earth sciences in Czechoslovakia." In: Earth science reviews. v. 4, no. 1, 1968. Atlas section, p. A 17-A 68.

International Geological Congress. 23rd, Prague, 1968. Guide to excursions. Prague, 1968. 50 nos. in 51.
Covers Czechoslovakia and parts of Austria, Germany, Hungary, Poland and Romania.

DENMARK

Brandt, Inger and W. Stuart Watt. "Litteraturfortegnelse." Published annually in Dansk Geologisk Forening. Meddelelser.
Pref: "list of papers on geology and similar matters which by subject, author or place of publication are related to Denmark and Greenland." Until 1969 these bibliographies included Iceland.

Garboe, Axel. Geologiens historie i Danmark. Copenhagen: C. A. Reitzels Forlag, 1959-1961. 2 v.

Nordmann, V. J. H., ed. Summary of the geology of Denmark. Copenhagen: C. A. Reitzel, 1928. 219 p. (Danmark Geologiske Undersøgelse. Ser. V, no. 4.)

FAEROE ISLANDS

Rasmussen, Jóannes and Arne Noe-Nygaard. Geology of the Faeroe Islands (Pre-Quaternary). Copenhagen: Danmarks Geologiske Undersøgelse, 1970. 142 p. (Series 1, no. 25.)

FINLAND

Laitakari, Aarne. Geologische Bibliographie Finnlands, 1555-1933. Helsinki: Commission géologique de Finlande, 1934. 224 p. (Bulletin no. 108.)

----------. 1555-1933, supplement. Otaniemi: Geologinen Tutkimuslaitos, 1967. 24 p. (Bulletin de la Commission géologique de Finlande, no. 231.)

----------. Index of Finnish minerals with bibliography. Otaniemi: Geologinen Tutkimuslaitos, 1967. 842 p. (Bulletin de la Commission géologique de Finlande, no. 230.)

Okko, Marjatta and Marjetta Hannikainen. Guide to the publications of the Geological Survey of Finland, 1897-1960. Helsinki: Geologinen Tutkimuslaitos, 1960. 104 p.

* * *

Hausen, Hans. The history of geology and mineralogy in Finland, 1828-1918. Helsinki: Societas Scientiarum Fennica, 1968. 147 p. (The History of Learning and Science in Finland, 1828-1918, no. 7a.)

Simonen, Ahti. Pre-Quaternary rocks in Finland. Helsinki: Commission géologique de Finlande, 1960. 49 p. (Bulletin no. 191.)

FRANCE

Bibliographie des sciences de la terre. Orleans, France: Bureau de recherches géologique et minières, 1968- . Monthly.
Computer prepared bibliography of geosciences issued in 9 sections, each with author, KWIC (Key word in context), and geographic indexes. Continuation of index card service previously published by the Bureau. Worldwide.

France. Centre national de la recherche scientifique. Bulletin signalétique. Paris, 1956- . Monthly.
A monthly abstracting service of worldwide scientific literature, issued in sections since 1961. Since 1969, the earth science sections have been: 210. Sciences de la terre I -Minéralogie, géochimie, géologie extraterrestre, pétrographie; 214. Sciences de la terre II--Géologie appliquée; 216. Sciences de la terre III--Géologie, paléontologie.

Société géologique de France. Bibliographie des sciences géologique. Paris, 1923-1955. Annual.
1948-1955, published as Extrait du Bulletin analytique du Centre national de la recherche scientifique (CNRS). Superceded in 1956 by France. Centre national de la recherche scientifique. Bulletin signalétique.

* * *

Abrard, René. Géologie de la France. Paris: Payot, 1948. 607 p.

Bertrand, Léon. Histoire géologique du Sol francais. Paris: Flammarion, 1946. 2 v.

Goguel, Jean. Géologie de la France. 4th ed. Paris: Presses universitaires de France, 1965. 126 p. ("Que sais-je?" Le point des connaisances actuelles, no. 443.)

Guides géologiques regionaux. Paris: Masson, 1968- .
Published so far: Bassin de Paris. C. Pomerol and L. Feugueur. 1968. 174 p.; Alpes (Savoie et Dauphiné). J. Debelmas. 1970. 216 p.

FRENCH TERRITORIES

Blondel, Fernand. Bibliographie géologique et minière de la France d'Outre-mer. Paris: Bureau d'etudes géologiques et minières coloniales, 1941-1952. 2 v.

---------- and G. Daumain. v. 1, 2nd ed. 1952. 568 p.

GERMANY

Montanwissenschaftliche Literaturberichte: Abteilung A: Bergbau; Abteilung C: Geowissenschaften. Leipzig, East Germany: VEB Deutscher Verlag fuer Grundstoffindistrie, 1955- . Monthly.
Worldwide literature in the geosciences and mining.

Neues Jahrbuch für Mineralogie, Geologie und Paläontologie. Stuttgart: E. Schweizerbart'sche, 1830-1949.
Important bibliographical serial with long and complicated history. Neues Jahrbuch included both original articles and bibliographies, until, in 1943, the "Referate" or abstracting component of this journal became the Zentralblatt für Geologie und Paläontologie (1943-1949). In 1950, the Zentralblatt split into the Zentralblatt für Geologie und Paläontologie and the Zentralblatt für Mineralogie.

Zentralblatt für Geologie und Paläontologie. Stuttgart: E. Schweizerbart'sche, 1950- . Irregular.
Teil I. Allgemeine und angewandt Geologie. Regionale Geologie; Teil II. Historische Geologie. Paläontologie. Important abstracting serial with long and complicated history. In 1943, the Zentralblatt superceded the "Referate" component of the Neues Jahrbuch für Mineralogie, Geologie und Paläontologie. Worldwide in scope.

Zentralblatt für Mineralogie. Stuttgart: E. Schweizerbart'sche, 1950- . Irregular.
Teil I. Kristallographie und Mineralogie; Teil II. Gesteinskunde. Technische Mineralogie. Geochemie und Lagerstätenkunde.

* * *

Gellert, Johannes F. Grundzüge der physikalischen Geographie von Deutschland. Leipzig: Teubner, 1958-
V. 1--Geologische Struktur und Overflächengestaltung.

1958. 386 p.

Grundriss der Geologie der Deutschen Demokratischen Republik. Hrsg. vom Zentralen Geologischen Institut im Auftrag des Staatssekretariats für Geologie beim Ministerrat der Deutschen Demokratischen Republik. Berlin: Akademie-Verlag, 1968- .
V. 1. pt. 1. Geologische Entwicklung des Gesamtgebietes; v. 1. pt. 2. Anlagenmappe.

Knetsch, Georg. Geologie von Deutschland und einigen Randgebeiten. Stuttgart: Enke, 1963. 386 p.

GREECE

Haralambous, D. Geological and physiogeographical bibliography of Greece. Athens: Institute for Geology and Subsurface Research, 1961. 236 p.
Covers Greece and Cyprus. Greek entries have English translation.

Petroleum Exploration Society of Libya. Guide to the geology and culture of Greece. Tripoli, 1965. 146 p. (7th Field Conference.)

Philippson, Alfred. Die griechischen Landschaften; eine Landeskunde. Frankfurt am Main: V. Klostermann, 1950-1959. 4 v.

HUNGARY

"A magyar földtani irodalom jegyzéke (yr.). Repetoire bibliographie des publications du domaine des sciences géologiques en Hongrie (yr.)." In: Földtani Közlöny (Bulletin of the Hungarian Geological Society). Budapest, annual.

Trunkó, László. Geologie von Ungarn. Berlin: Borntraeger, 1969. 257 p. (Beiträge zur Regionalen Geologie der Erde, v. 8.)

ICELAND

Brandt, Inger and W. Stuart Watt. "Litteraturfortegnelse." Published annually in Dansk Geologisk Forening. Meddelelser.
Pref: "list of papers on geology and similar matters

which by subject, author or place of publication are related to Denmark, Greenland, or Iceland." Iceland does not appear in the 1969 bibliography.

* * *

Áskelsson, J. and others. On the geology and geophysics of Iceland. Reykjavik, 1960. 74 p. (International Geological Congress, 21st, Norway, 1960. Guide to Excursion no. A2.)

ITALY

Italy. Consiglio Nazionale delle Richerche. Comitato per la Geografia, Geologia e Mineralogia. Bibliografia geologica d'Italia. Naples, 1956- .
In progress. Volumes cover the compartimenti and provinces of Italy.

* * *

"Development of the Northern Appennines geosyncline." A special issue of Sedimentary geology. v. 4, nos. 3/4, 1970. p. 205-647.

International Field Institute. Italy, 1964. Edited by John C. Maxwell. Washington: American Geological Institute, 1964. 400 p. (International Field Institute. Guidebook, 3rd.)
About the Appennines.

MALTESE ISLANDS

Hyde, Herbert P. J. Geology of the Maltese Islands. Malta, 1955. 135 p.

NETHERLANDS

"Geological bibliography of the Netherlands and of adjacent countries of importance to the geology of the Netherlands, period 1949-1964.") In: Netherlands. Rijks Geologische Dienst (formerly Geologische Stichting). Mededelingen. n.s., no. 19, 1968. p. 9-61.
A list of the former bibliographies in this series is given on p. 9. Bibliographies for succeeding years are in the Mededelingen, nos. 18, 19 and 20, and are expected to continue annually in this journal.

* * *

Faber, F. J. Geologie van Nederland. Gorinchen: J. Noorduijn, 1947-1965. 4 v.
V. 1-2. Algemene en historische geologie. 1947; v. 3. Nederlandse landschappen. 1948; v. 4. Aanvullende hoofstukken. 1965.

Nederlandsch Geologisch Mijnvouwkundig Genootschap. Geological history of the Netherlands; explanation to the general geological map of the Netherlands on the scale of 1:200, 000. Edited by A. J. Pannekoek. Hague: Staatsdrukkerij-en Uitgeverijbedrijf, 1956. 147 p.

NORWAY

Hoel, Adolph and Johannes Norvik. Glaciological bibliography of Norway. Oslo: Norsk Polarinstutt, 1962. 242 p. (Skrifter no. 126.)
Covers Norway and its territories.

Rosendahl, Halvor. "Littérature géologique concernant la Norvège avec Svalbarde et la territoire des recherches de la mer glaciale et la mer norvégienne et groenlandienne." In: Norsk geologisk tidsskrift. v. 13, 1933. p. 129-301.
Covers 1916-1932 and refers to earlier compilations.

* * *

Holtedahl, Olaf, ed. Geology of Norway. Oslo: Norges Geologiske Undersøkelse, 1960. 540 p. (Skrifter no. 208.)

POLAND

Bibliografia geologiczna Polski. Warsaw: Wydawnictwa Geologiczne, 1914- . Annual.
The latest bibliography is for 1968, published as Bibliografia geologiczna Polski no. 41 in 1970.

Fleszarowa, Regina. Retrospektywna bibliografia geologiczna Polski. Warsaw: Wydawnictwa Geologiczne, 1957- .
Pt. 1 (covering 1900-1950). 1957-1958. 2 v. plus supplement (1964, 10 p.); pt. 1 (covering 1900-1950, subject arrangement). 1960-1963. 2 v.; pt. 2 (covering 1750-1900). v. 1, 1966. 573 p.

Ksiazkiewicz, Marian, Jan Samsonowicz and Edward Ruhle. An outline of the geology of Poland. Translated by Jan Burchart. Warsaw: Scientific Publications Foreign Cooperation Center, Central Institute for Scientific, Technical and Economic Information, 1968. 414 p. (U. S. National Technical Information Service, TT 60-21279.) Translation of Zarys Geologii Polski, published in 1965 by Wydawnictwa Geologiczne. 380 p.

PORTUGAL

Beato, Maria de Fátima. Catálogo das publicações dos Serviços Geológicos de Portugal, 1865-1968. Lisbon: Serviços de Portugal, 1969. 73 p.

Portugal. Direcção Geral de Minas e Serviços. Geologia de Portugal; ensaio bibliográfico. 2nd ed. Luis de Menezes Correa Acciaiuoli. Lisbon, 1957-58. 2 v.

Thadeu, Décio. Carta geológica de Portugal na escala de 1/1. 000. 000. Noticia explicativa. Lisbon: Serviços Geológicos de Portugal, 1958. 15 p.

PORTUGUESE TERRITORIES

Gonçalves, Francisco and Jaime Caseiro. Bibliográfia geológica do ultramar português. Lisbon: Junta Investigações Ultramar, 1959. 273 p.

Portugal. Junta de Investigações do Ultramar. Curso de geologia do Ultramar. Lisbon, v. 1, 1968- .

ROMANIA

Academia Republicii Socialiste Romania. Centrul de Documentare Stiintifica. Buletin de informare stiintifica: Geologie, geografie. Bucharest, 1964- . Monthly.

Antonovici, Suzana. Bibliografia geologică a Republicii Socialiste România, al IV-lea supliment. Bucharest: Rom. Inst. Geol., 1969. 394 p.
The 4th supplement. The original bibliography was published in 1926, with supplements in 1929, 1939 and 1962.

Oncescu, N. Geologia Republicii Populare Romîne. Bucharest: Editura Technică, 1957. 438 p.

SICILY

Alvarez, Walter and Klaus H. A. Gohrbandt, eds. Geology and history of Sicily. Tripoli: Petroleum Exploration Society of Libya, 1970. 291 p. (12th Field Conference.)

Marchetti, Marco P. "Summary introduction to the geology of Sicily." In: Petroleum Exploration Society of Libya. Excursion to Sicily, May 27-30, 1960. Tripoli, 1960. p. 47-60.

SPAIN

Bibliografía geológica española." In: Acta geológica hispanica. Irregular.
Issued thus far: Acta geológica hispanica. Año 2, no. 2, 1967. p. 25-52. (Covers 1964-1965.); Año 3, no. 6, 1968. p. 145-168. (Covers 1966.)

López de Azcona, Juan M. and A. de Andrés Galindo. Catálogo de publicaciones, 1849-1963. Madrid: Instituto Geológica y Minero de España, 1964. 189 p.
A catalog of the publications of the Spanish Geological Survey.

San Miguel de la Cámara, M. "Bibliografía vulcanológica española." In: Bulletin volcanologique. Ser. 2, v. 12, 1952. p. 203-214.
Covers years 1820-1951.

SWEDEN

Larsson, Walter, ed. Swedish geological literature, 1958-1963. Stockholm: Sveriges Geologiska Undersökning, 1968. 546 p. (Ser. C, no. 630, Avhandlingar och Uppsater.) (Årsbok 62, no. 1.)

----------. "Swedish geological literature, 1956-1957." In: Geologiska Föreningens i Stockholm. Förhandlingar. v. 82, no. 4, 1960. p. 433-604.
Continuation of the bibliography "Revue annuelle de la litterature géologique suedoise" by R. Sandegren published annually in Geologiska Föreningens i Stockholm. Förhandlingar. From 1927-1955.

Sweden. Geologiska Undersökning. Register over Sveriges Geologiska Undersökning publikationer 1858-1958. Stockholm, 1959. 330 p. (Årsbok 53, no. 1.) (Ser. C, no. 563.)

* * *

Koark, Hans J. "Geologie von Schweden." In: Geologische Rundschau. v. 59, no. 2, 1970. p. 763-791.

Lundqvist, G. Description to accompany the Map of Quaternary Deposits of Sweden (English edition). Stockholm: Sveriges Geologiska Undersökning, 1959. (Series Ba, no. 17.)

Magnusson, Nils H. and others. Description to accompany the Map of pre-Quaternary Rocks in Sweden (English edition). Stockholm: Sveriges Geologiska Undersökning, 1960. 177 p. (Series Ba, no. 16.)

----------. Sveriges geologi. 4th ed. Stockholm: Svenska bokförlaget, 1963. 698 p.

SWITZERLAND

Jeannet, Alphonse. Bibliographie geólogique de la Suisse pour les anneés 1910 à 1920. Berne, 1927. 2 v. (Beiträge zur Geologichen Karte der Schweiz. N.S. no. 56.)

----------. 1921 à 1930. Berne, 1936. 496 p. (Beiträge zur Geologischen Karte der Schweiz. N.S. no. 73.)

* * *

Carte géologique generale de la Suisse. Geologische Generalkarte der Schweiz. Carta geologica generale della Svizzera. Bern: Kummerly and Frey, 1956-1965. Eight map sheets, 1:200, 000 plus explanatory brochures to accompany each sheet.

Heim, Albert. Geologie der Schweiz. Leipzig: Tauchnitz, Führer der Schweiz. 2nd ed. Basel: Wepf, 1967. 9 v.

TURKEY

Campbell, Angus S., ed. Geology and history of Turkey. Tripoli: Petroleum Exploration Society of Libya, 1971.

511 p. (Annual Field Conference Guidebook, 13th.)

"Türkiye jeolojik bibliyografyasi. Bibliographie géologique de la Turquie." In: Türkiye Jeoloji Kurumu. Bulteni. v. 1, no. 2, 1948. p. 96-135.

"Türkiye jeolojik bibliyografyasi, 1949." In: Türkiye Jeoloji Kurumu. Bulteni. v. 3, 1951. p. 165-168.

* * *

Erentöz, Cahit. "A general review of the geology of Turkey." In: Turkey. Mineral Research and Exploration Institute. Bulletin, foreign edition. no. 48, 1956. p. 40-58.

Furon, Raymond. "Introduction a la géologie et a l'hydrogéologie de la Turquie." In: Paris. Muséum national d'Histoire naturelle. Mémoires. Ser. C, v. 3, no. 1, 1963. p. 1-127.

YUGOSLAVIA

Bilten dokumentacije: Rudarstvo i geologija. Belgrade: Jugoslovenski Centar za Technicku i Naucnu Dokumentaciju, 1950- . Monthly.
Worldwide literature abstracts on geology and the mineral industry.

Milojević, Sreten. Geološka bibliografija Jugoslavije od XIV veka do 1944 godine. Bibliographie géologique de la Yougoslavie du XIVe siècle a 1944. Belgrad: Institut bibliographique de la Yougoslavie, 1954. 479 p.

Petroleum Exploration Society of Libya. Excursion to Yugoslavia. Tripoli, 1962. 83 p. (4th Field Conference.)
Summarizes the geology and tectonics of the country.

THE SOVIET UNION

Lexique stratigraphique international. Volume II: U. S. S. R. Publie sur l'initative de la Commission de stratigraphie du Congress géologique international. Paris: Centre national de la recherche scientifique (CNRS), 1956. 4 v.

Referativnyi zhurnal. Moscow: Akademiia Nauk SSSR, 1956- . Monthly.
This service abstracts the greatest total number of geological articles. Worldwide. Following a classed arrangement (in Russian), the citations are given in the language of the publication, with a Russian abstract. Annual indexes. Issued in sections; of interest to geologists are "Geologiya," "Geofizika" and "Gornoe Delo (Mining)."

Russia. Vsesoyuznaya Geologicheskaya Biblioteka. Geologicheskaya literatura SSSR; bibliograficheskiy ezhegodnik za (yr.) gody. Leningrad: Izdatel'stvo Nedra, 1930- . Annual.

* * *

Bogdanov, A. A. "The tectonic history of the U. S. S. R. and adjacent countries." In: International Geological Congress. 23rd, Prague, 1968. Report. v. 3. Prague, 1968. p. 25-34.

International geology review. Washington: American Geological Institute, 1959- . Monthly.
Contains a large number of Russian papers in English translation. Prepublication announcement of geologically important Soviet literature is a regular feature.

Kuznetsov, Sergey S. Geologiya SSSR. Moscow: Izd. Vyssh. Shk., 1968. 439 p.

Leningrad. Vsesoiuznyi Geologicheskii Institut. Structure géologique de l'U. R. S. S. Paris: Centre national de la recherche scientifique (CNRS), 1959-1961. 3 v.

French translation of a fundamental Russian work. V. 1, Stratigraphie; v. 2, Magnatismo; v. 3, Tectonique.

Nalivkin, D. V. Geologiia SSSR. Moscow: Akademiia nauk SSSR, 1962. 813 p.

----------. Geology of the U. S. S. R. Translated by N. Rast, Hitchin, England: Wheldon and Wesley, in preparation.

----------. The geology of the U. S. S. R., a short outline. New York: Pergamon, 1960. 170 p.

ASIA

Lexique stratigraphique international. Volume III: Asia. Publie sur l'initative de la Commission de stratigraphie du Congress géologique international. Paris: Centre national de la recherche scientifique (CNRS), 1956- .
For a breakdown of the Lexique see entry in general section of this guide.

* * *

Kobayashi, T. and R. Toriyama, eds. Geology and palaeontology of southeast Asia. Tokyo: University of Tokyo Press, 1964- .
Latest is V. 7, 1969. Mainly paleontological papers dealing with Thailand, Malaysia and the Philippines.

Leuchs, Kurt. Geologie von Asien. Berlin: Borntraeger, 1935-1937. 2 v. (Geologie der Erde.)
V. 1. pt. 1. Überblick uber Asien, Nordasien; v. 1. pt. 2. Zentralasien.

Ogura, Tsutomu, ed. Geology and mineral resources of the Far East. Tokyo: University of Tokyo Press, 1967- .
Issued thus far: v. 1, 1967 and v. 2, 1969. Covers China, Manchuria and Korea. Designed to complement the series, Geology and palaeontology of southeast Asia (see above) published also by the University of Tokyo Press.

UNESCO. Geological map of Asia and the Far East, explanatory brochure. 1961. 74 p.
Short account of the geological history of Asia accompanying the geological map of 6 sheets (scale, 1:5, 000, 000).

SPECIFIC REGIONS

ARABIAN PENINSULA

Brown, Glen F. and Roy O. Jackson. "The Arabian Shield."

In: International Geological Congress. 21st, Copenhagen, 1960. Report. Pt. 9. Copenhagen, 1960. p. 69-77.

Karpoff, Roman. "Esquisse géologique de l'Arabie seoudite." In: Société géologique de France. Bulletin. Ser. 6, v. 7, 1957. p. 653-697.

U. S. Geological Survey. Geology of the Arabian Peninsula. Washington: Government Printing Office, 1966- . (Professional Paper no. 560.)
Accompanied by Geologic map of the Arabian Peninsula, issued as U. S. G. S. Miscellaneous Geologic Investigations, Map no. I-270 (1963, 1:2,000,000). Parts are: A--General (in prep.); B--Yemen; C--Aden Protectorate; D--Sedimentary geology of Saudi Arabia; E--Bahrain; F--Kuwait; G--Southwestern Iraq; H--Eastern Aden Protectorate and part of Dhufar; I--Qatar (in prep.); J--Oman, Trucial Coast and Dhufar (in prep.); K--Jordan (in prep.).

HIMALAYAS

Gansser, Augusto. Geology of the Himalayas. New York: Wiley, 1964. 289 p. (Regional Geology Series.)

Wadia, D. N. and W. D. West. Structure of the Himalayas. New Delhi, 1964. 10 p. (International Geological Congress. 22nd, India, 1964.)

INDIAN OCEAN

Heezen, Bruce and Marie Tharp. Physiographic diagram of the Indian Ocean, the Red Sea, the South China Sea, the Sulu Sea and the Celebes Sea. New York: Geological Society of America, 1964.
Accompanied by a separate descriptive text.

Pepper, J. F. and G. M. Everhart. The Indian Ocean, the geology of its bordering lands and the configuration of its floor. Washington: U. S. Geological Survey, 1963. (Miscellaneous Geologic Investigations, Map I-380.)

Yentsch, Anne E. A partial bibliography of the Indian Ocean. Woods Hole, Massachusetts: Woods Hole Oceanographic Institution, 1962. 395 p.
Includes works relating to submarine geology, geophysics, geochemistry and coral reefs.

INDONESIAN ARCHIPELAGO

Bemmelen, R. W. van. The geology of Indonesia. For the Indonesian Bureau of Mines. The Hague: Government Printing Office, 1949. 2 v. in 3.
Covers Indonesia, Philippines, the Andaman and Nicobar Islands, Christmas Island, New Guinea and Timor. Reprinted by Nijhoff in 1970 with additional bibliographies.

Jennings, J. N. and J. A. Mabbutt, eds. Landform studies from Australia and New Guinea. Canberra: Australian National University Press, 1967. 434 p.

Umbgrove, J. H. F. Structural history of the East Indies. Cambridge: Cambridge University Press, 1949. 62 p.

LEVANT

Avnimelech, Moshe A. Bibliography of Levant geology, including Cyprus, Hatay, Israel, Jordania, Lebanon, Sinai, and Syria. Jerusalem: Israel Program for Scientific Translations, 1965- .
V. 1, 1965. 192 p.; v. 2, 1969. 200 p.

ADEN PROTECTORATE

Greenwood, J. E. G. W. and D. Bleackley. Geology of the Arabian Peninsula--Aden Protectorate. Washington: Government Printing Office, 1967. 96 p. (U.S. Geological Survey. Professional Paper no. 560-C.)

AFGHANISTAN

Kaever, Matthais von. "Literaturbericht zur Geologie und Paläontologie Afghanistans." pt. 1, 1967- . Published irregularly to date in: Zentralblatt für Geologie und Paläontologie, Teil I, 1967-date.

BURMA

Krishnan, M. S. Geology of India and Burma. 3rd ed. Madras: Higginbothams, 1956. 555 p.

Mehdiratta, R. C. Geology of India, Pakistan, and Burma. 2nd ed. Delhi: Atma Ram, 1962. 203 p.

CEYLON

Wadia, D. N. "Bibliography of Ceylon geology." In: Ceylon Department of Mineralogy. Professional Paper. no. 1, 1943. p. 33-38.

* * *

Cooray, Percival G. An introduction to the geology of Ceylon. Colombo: National Museum of Ceylon, 1967. 324 p. (Spolia Zeylanica. v. 31, pt. 1.)

Fernando, L. J. D. "The geology and mineral resources of Ceylon." In: Imperial Institute (London). Bulletin. v. 46, 1948. p. 305-325.

CHINA

Bibliography of Chinese geology.

Yang, T. I. Bibliography of Chinese geology up to 1934. Peiping: National Academy, 1935. 241 p.

Chi, Y. S. Science bibliography of China: V, Section of geology, 1934-1935. Nanking: National Research Council, 1936. 57 p.

----------. Bibliography of Chinese geology for the years 1936-1940. Nanking: Geological Survey of China, 1942. 147 p.

Li, Hsiao-Fung. Bibliography of geology and geography of Sinkiang. Nanking, 1947. 213 p.

Tseng, T. C. Bibliography of Chinese geology. Bibliography of geology and allied sciences of Tibet and regions to the west of Chinshachiang. Nanking: China Geological Survey, 1946. 114 p.

* * *

Ch'ang, Ta. The geology of China. Washington: Office of Technical Services and Joint Publications Research Service, 1963. 623 leaves. (OTS: 63-21820.) (JPRS: 19-209.)
Translation of the Chinese version published in 1959.

Grabau, Amadeus W. The stratigraphy of China. Peking: Geological Survey of China, 1923-1928. 2 v.

International geology review. Washington: American Geological Institute, 1959- . Monthly.
Chiefly translations of Russian articles, but some Chinese, mostly from Acta geologica sinica.

Lee, J. S. The geology of China. London: Murby, 1939. 528 p.

Weller, J. M. "Outline of Chinese geology." In: American Association of Petroleum Geologists. Bulletin. v. 28, 1944. p. 1417-1429.

EASTERN ADEN PROTECTORATE

Beydoun, Z. R. The stratigraphy and structure of the Eastern Aden Protectorate. London: Her Majesty's Stationery Office, 1964. 107 p. (Overseas Geology and Mineral Resources. Supplement Series no. 5.)

----------. Geology of the Arabian Peninsula--Eastern Aden Protectorate and part of Dhufar. Washington: Government Printing Office, 1966. 49 p. (U.S. Geological Survey. Professional Paper no. 560-H.)

INDIA

La Touche, T. H. D. A bibliography of Indian geology and physical geography. Calcutta: Geological Survey of India, 1917-1926. 5 v.
Continued by annual bibliographies in the Records of the Geological Survey of India.

* * *

Krishnan, M. S. Geology of India and Burma. 3rd ed. Madras: Higginbothams, 1956. 555 p.

Mehdiratta, R. C. Geology of India, Pakistan, and Burma. 2nd ed. Delhi: Atma Ram, 1962. 203 p.

Rao, P. V. Geology and mineral resources of India. New Delhi, 1964. 44 p. (International Geological Congress. 22nd, India, 1964.)
The 22nd International Geological Congress has also

published 18 guidebooks on areas of geological interest in India.

Wadia, D. N. Geology of India. 3rd ed. London: Macmillan, 1957. 536 p.

IRAN

Rosen, Norman C. Bibliography of geology of Iran. Tehran: Geological Survey of Iran, 1969. 77 p. (Special Publication no. 2.)

* * *

Furon, Raymond. "Géologie du plateau iranien (Perse-Afghanistan-Beloutchistan)." In: Paris. Muséum national d'Histoire naturelle. Mémoires. v. 7, pt. 2, 1941. p. 177-411.

National Iranian Oil Company. Explanatory to the geologic map of Iran (1:2,500,000), with stratigraphical tables, index map and bibliography. 1959. 19 p.

IRAQ

Naqib, K. M. Al. Geology of the Arabian Peninsula--Southwestern Iraq. Washington: Government Printing Office, 1967. 54 p. (U.S. Geological Survey. Professional Paper no. 560-G.)

ISRAEL

Avnimelech, Moshe A. Bibliography of Levant geology, including Cyprus, Hatay, Israel, Jordania, Lebanon, Sinai and Syria. Jerusalem: Israel Program for Scientific Translations, 1965- .
V. 1, 1965. 192 p.; v. 2, 1969. 200 p.

Shaw, S. H. "The geology and mineral resources of Palestine." In: Imperial Institute (London). Bulletin. v. 46, 1948. p. 87-103.

JAPAN

Bibliography of Japanese palaeontology and related sciences, 1941-1950. Riuji Endo. Tokyo: Palaeontological Society of Japan, 1951. 71 p. (Special Paper no. 1.)

----------. 1951-1960. Fuyuji Takai. Tokyo: Palaeontological Society of Japan, 1962. 338 p. (Special Paper no. 9.)

Fujimoto, Haruyoshi. Bibliography of the geology of Japan, 1873-1955. Revised ed. Tokyo: Chijin Shokan, 1956. 711 p.

Isshiki, Naoki and others. Selected bibliography of Japanese volcanoes. Hisamoto-chō, Japan: Geological Survey of Japan, 1968. 78 p.

* * *

Association for Geological Collaboration of Japan. The geologic development of the Japanese Islands. Edited by Masao Minato and others. Tokyo: Tsukiji Shokan, 1965. 442 p.

"The earth sciences in Japan." In: Earth science reviews. v. 6, no. 3, 1970. Atlas section, p. A73-A156.

Japan. Geological Survey. Geology and mineral resources of Japan. 2nd ed. Hisamoto-chō, Japan, 1960. 304 p.

Takai, Fuyuji, ed. The geology of Japan. Berkeley: University of California Press, 1963. 279 p.

JORDAN

Avnimelech, Moshe A. Bibliography of Levant geology, including Cyprus, Hatay, Israel, Jordania, Lebanon, Sinai and Syria. Jerusalem: Israel Program for Scientific Translations, 1965- .
V. 1, 1965. 192 p.; v. 2, 1969. 200 p.

Bender, Friedrich. Geologie von Jordanien. Berlin: Borntraeger, 1968. 230 p. (Beiträge zur Regionalen Geologie der Erde, v. 7.)

---------- and others. Beiträge zur Geologie Jordaniens. Hannover: Bundesanstalt für Bodenforschung, 1969. 249 p. (Beihefte zum Geologischen Jahrbuch no. 81.)

Burdon, David J. Handbook of the geology of Jordan. Jordan: Government of the Hashemite Kingdom of Jordan, 1959. 82 p.

U. S. Geological Survey. Geology of the Arabian Peninsula. Part K: Jordan. Washington: Government Printing Office, in preparation. (Professional Paper no. 560-K.)

KOREA

Kobayashi, T. "Geologic structure of North Korea and south Manchuria." In: Ogura, Tsutomu, ed. Geology and mineral resources of the Far East, v. 2. Tokyo: University of Tokyo Press, 1969. p. 3-123.

KUWAIT

Milton, D. I. Geology of the Arabian Peninsula--Kuwait. Washington: Government Printing Office, 1967. 7 p. (U.S. Geological Survey. Professional Paper no. 560-F.)

LEBANON

Avnimelech, Moshe A. Bibliography of Levant geology, including Cyprus, Hatay, Israel, Jordania, Lebanon, Sinai and Syria. Jerusalem: Israel Program for Scientific Translations, 1965- .
V. 1, 1965. 192 p.; v. 2, 1969. 200 p.

Wolfart, Reinhard. Geologie von Syrien und dem Libanon. Berlin: Borntraeger, 1967. 326 p. (Beiträge zur Regionalen Geologie der Erde, v. 6.)

MALAYSIA

Gobbett, D. J. Bibliography and index of the geology of West Malaysia and Singapore. Kuala Lumpur: Geological Society of Malaysia, 1968. 152 p. (Bulletin no. 2.)

----------. Supplement. In: Geological Society of Malaysia. Bulletin. no. 3, 1970. p. 115-129.

Scrivenor, J. B. The geology of Malaya. London: Macmillan, 1931. 217 p.

OMAN

Morton, D. M. "The geology of Oman." In: World Petroleum Congress. 5th, New York, 1959. Proceedings: section 1, geology and geophysics. p. 277-294.

U.S. Geological Survey. Geology of the Arabian Peninsula: Part J: Oman, Trucial Coast and Dhufar. Washington: Government Printing Office, in preparation. (Professional Paper no. 560-J.)

PAKISTAN

Mehdiratta, R. C. Geology of India, Pakistan, and Burma. 2nd ed. Delhi: Atma Ram, 1962. 203 p.

PHILIPPINES

Daleon, Benjamin A. "Bibliography of Philippine palaeontology and stratigraphy, 1861-1957." In: Philippine geologist. v. 12, no. 1, 1957. p. 16-32.

Teves, Juan. Bibliography of Philippine geology, mining and mineral resources. Manila: Philippines Bureau of Mines, 1953. 155 p. (Bibliographical Series no. 1.)

* * *

Corby, Grant and others. Geology and oil possibilities of the Philippines. Manila: Department of Agriculture and Natural Resources, 1951. 363 p. (Technical Bulletin no. 21.)

Hashimoto, Wataru. "Paleontology of the Philippines." In: Kobayashi, T. and R. Toriyama, eds. Geology and palaeontology of southeast Asia. v. 6. Tokyo: University of Tokyo Press, 1969. p. 293-329.

PORTUGUESE TIMOR

Audley-Charles, M. G. The geology of Portuguese Timor. London: Geological Society of London, 1968. 76 p. (Memoir no. 4.)

SAUDI ARABIA

Powers, R. W. and others. Geology of the Arabian Peninsula--Sedimentary geology of Saudi Arabia. Washington: Government Printing Office, 1966. 147 p. (U.S. Geological Survey. Professional Paper no. 560-D.)

SYRIA

Avnimelech, Moshe A. Bibliography of Levant geology, including Cyprus, Hatay, Israel, Jordania, Lebanon, Sinai and Syria. Jerusalem: Israel Program for Scientific Translations, 1965- .
V. 1, 1965. 192 p.; v. 2, 1969. 200 p.

Wolfart, Reinhard. Geologie von Syrien und dem Libanon. Berlin: Borntraeger, 1967. 326 p. (Beiträge zur Regionalen Geologie der Erde, v. 6.)

TAIWAN

Yen, Tsang P. and others. Bibliography of the geology of Taiwan. Taipei: Taiwan Geological Survey, 1947. 58 p.

* * *

Bank of Taiwan. Economic Research Branch. Geology of Taiwan. Taipei, 1955. 164 p. (Taiwan Research Series no. 36.)

Juan, Vei Chow. Physiography and geology of Taiwan. Taipei: China Culture Publishing Foundation, 1954. 45 p.

THAILAND

Brown, Glen F. and others. Geologic reconnaissance of the mineral deposits of Thailand. Washington: Government Printing Office, 1951. 178 p. (U.S. Geological Survey. Bulletin no. 984.)

Kobayashi, T. "Geology of Thailand." and "Paleontology of Thailand." In: Kobayashi, T., ed. Geology and palaeontology of southeast Asia, v. 1. Tokyo: University of Tokyo Press, 1964. p. 1-30.

YEMEN

Geukens, F. Geology of the Arabian Peninsula--Yemen. Translated from the French by S. D. Bowers. Washington: Government Printing Office, 1966. 23 p. (U. S. Geological Survey. Professional Paper no. 560-B.)

AFRICA

"Géologie africaine." In: Chronique des mines et de la recherche minière. Paris, October, 1969- .
A new section devoted to "papers received in the office of the Association of African Geological Surveys relating to the geology of Africa, except those of purely local interest."

Haughton, S. H. Geological bibliography of Africa south of the Sahara. Bibliographie géologique de l'Afrique au Sud du Sahara. London: Commission for Technical Cooperation in Africa, 1957-1962.
Bibliography of the Karroo System. 1957; Bibliography of the Jurassic and Cretaceous systems of Africa south of the Sahara. 1959; Karroo System. Supplement no. 1. 1962.

Lexique stratigraphique international. Volume IV: Africa. Publie sur l'initative de la Commission de stratigraphie du Congress géologique international. Paris: Centre national de la recherche scientifique (CNRS), 1956- .
For a breakdown of the Lexique see entry in general section of this guide.

Rodier, J. Bibliography of African hydrology. Paris: UNESCO, 1963. 166 p. (Natural Resources Research, v. 2.)

* * *

De Kun, N. The mineral resources of Africa. New York: Elsevier, 1965. 750 p.
Includes geology of the African countries.

Furon, Raymond. Geology of Africa. Edinburgh: Oliver and Boyd, 1963. 377 p.
Translation of 2nd French edition, published in 1960. The 3rd French edition, Géologie de l'Afrique, published by Payot, Paris, in 1968. 374 p.

----------. Introduction a la géochronologie de l'Afrique.

Paris: Centre national de la recherche scientifique, 1969. 112 p. (Lexique Stratigraphique International, v. 4, no. 13.)

Haughton, S. H. The stratigraphic history of Africa south of the Sahara. Edinburgh: Oliver and Boyd, 1963. 365 p.

Krenkel, Erich. Geologie und Bodenschätze Afrikas. 2nd ed. Leipzig: Geest and Portig, 1957. 597 p.

UNESCO. Explanatory note, geological map of Africa (1/5,000,000). Paris, 1964. 39 p.
"African geological bibliography," p. 35-39.

----------. Explanatory note, international tectonic map of Africa, 1/5,000,000. Paris, 1969. 54 p.
"Bibliography," p. 51-54.

----------. A review of the natural resources of the African continent. Paris, 1963. 437 p. (Natural Resources Research, v. 1.)
Includes the geological sciences.

SPECIFIC REGIONS

ATLAS MOUNTAINS

Richter, Max von. "Das Atlas-System in Nordafrika." In: Deutschen Geologischen Gesellschaft. Zeitschrift. v. 120, 1970. p. 68-84.

FORMER BRITISH AFRICA

"Select bibliography on geology and the mineral industry." In: Overseas geology and mineral resources. London: Her Majesty's Stationery Office, 1950- .
Latest published is no. 37 (Overseas geology and mineral resources, v. 10, no. 1, 1966). Preface--"select bibliography on geology, mineral resources and the mineral industry relates principally to Commonwealth, Colonial and Dependent Territories, but also includes items from the world literature likely to be of interest to workers overseas."

* * *

Dixey, F. and E. S. Willbourne. "The geology of the British African colonies." In: International Geological Congress. 18th, London, 1948. Report. Pt. 14. London, 1951. p. 87-109.

FORMER FRENCH AFRICA

African Geological Surveys. Bibliographie géologique de l'Afrique centrale (Édition provisoire). Paris: Secretariat du Bureau d'études géologiques et minières coloniales; Liege, Secrétariat de la Société géologique de Belgique, Université de Liège, 1937. 298 p.

----------. Compléments a la bibliographie géologique de l'Afrique centrale publiée en 1937. Congo Belge, décade 1935-1944. Paris: Bureau d'études géologiques et minières coloniales, 1948. 37 p.

----------. Compléments a la bibliographie géologique de l'Afrique centrale publiée en 1937. Congo Belge, periode 1945-1949. Addenda et errata aux éditions de 1937 et 1948. Paris, 1951. 29 p.

Blondel, Fernand. Bibliographie géologique et minière de la France d'Outre-mer. Paris: Bureau d'etudes géologiques et minières coloniales, 1941-1952. 2 v.

--------- and G. Daumain. v. 1, 2nd ed. 1952. 568 p.

French Equatorial Africa. Direction des Mines et de la Géologie. Carte géologique de l'Afrique equitorial française au 1/2. 000. 000. Notice explicative. George Gérard. Paris: Imprimerie Typographique d'édition, 1958. 198 p.
Covers Chad, Central African Republic and Gabon. Bibliography, p. 173-197.

Marvier, L. Notice explicative de la carte géologique d'ensemble de l'Afrique occidentale française. Dakar: French West Africa. Direction des Mines, 1953. 99 p. (Bulletin no. 16.)
Geologic map published in 1952 at a scale of 1:6, 000, 000. Covers Mauritania, Senegal, Ivory Coast, Dahomey, Guinea, Mali, Upper Volta and Niger.

FORMER ITALIAN EAST AFRICA

Desio, Ardito. Bibliografia geologica italiana dell'Africa sino al 1948 incluso. Rome: Ufficio Studi Ministero dell'Africa Italiana, 1950. 83 p.
Covers Ethiopia and Italian Somaliland.

* * *

Dainelli, Giotto. Geologia dell'Africa orientale. Rome: Reale Accademia d'Italia, Centro Studi per l'Africa Orientale Italiana, 1943. 4 v.
Covers Ethiopia and Italian Somaliland.

Usoni, Luigi. Risorse minerarie dell'Africa Orientale: Eritrea, Ethiopia, Somalia. Rome: Jandi Sapi, 1952. 553 p.

SAHARA

Furon, Raymond. Le Sahara; géologie, ressources minérales, mise en valeur. Paris: Payot, 1957. 300 p.

SOUTHERN AFRICA

Haughton, S. H. Geological history of southern Africa. Johannesburg: Geological Society of South Africa, 1969. 535 p.

King, Lester C. South African scenery. 3rd ed. Edinburgh: Oliver and Boyd, 1963. 308 p.

Mountain, Edgar D. Geology of southern Africa. Cape Town: Books of Africa, 1968. 249 p.

Pelletier, R. A. Mineral resources of south-central Africa. London: Oxford, 1964. 277 p.
Includes geology of the countries from the Congo River south.

ALGERIA

"Bibliographie géologique annuelle de l'Algérie du Nord et du Sahara septentrional." In: Algeria. Service Géologique. Bulletin. 1953- .
Bibliography no. 11 for 1964 is in Bulletin no. n.s. 35, 1967. p. 127-130. Bibliography no. 12 for 1965-66 is

in Bulletin no. n. s. 39, 1969. p. 255- .

Merabet, Omar. Bibliographie de l'Algérie du Sud (Sahara) et des régions limitrophes. Algiers: Service Géologique de l'Algérie, 1968. 197 p. (Bulletin no. n. s. 37.)

* * *

International Geological Congress. 19th, Algiers, 1952. Monographies régionales. 1 Ser: Algérie. Algiers, 1952. 27 v.

ANGOLA

Mouta, Fernando. Esboço tectónico de Angola; notíca explicativa. Lisbon, 1956. 102 p. (Portugal. Junta de Investigações do Ultramar. Estudios de geologia. Anais. v. 10, no. 5.)

----------. Noticia explicativa do esboço geológico de Angola (1:2, 000, 000). Lisbon: Ministério do Ultramar, Junta de Investigações do Ultramar, 1954. 148 p.

BOTSWANA

Laughton, C. A. Annotated bibliography and index of the geology of Botswana to 1966. Gaberones: Botswana Geological Survey Department, n. d. 172 p.

* * *

Boocock, C. "Mineral resources of the Bechuanaland Protectorate." In: Overseas geology and mineral resources. v. 9, no. 4. London: Her Majesty's Stationery Office, 1965. p. 369-417.

McConnell, R. B. "Notes on the geology and geomorphology of the Bechuanaland Protectorate." In: International Geological Congress. 20th, Mexico, 1956. Asociación de Servicios Geológicos Africanos. Actas y Trabajos. Mexico, 1959. p. 175-186.

CAMEROUN

Gazel, J. and others. Carte géologique du Cameroun au 1/1. 000. 000; notice explicative. Paris, 1956. 62 p. (Cameroun. Direction des Mines et de la Géologie.

Bulletin no. 2.)

CONGO

Bibliographie géologique du Congo, du Rwanda et du Burundi. Tervuren, Belgium: Musée Royal de l'Afrique Centrale, 1952- . Biennial.
Formerly entitled Bibliographie géologique du Congo Belge et du Ruanda-Urundi. Covers from 1818.

Cahen, Lucien. Géologie du Congo Belge. Liege: H. Vaillant-Carmanne, 1954. 577 p.

Dadet, P. Notice explicative de la carte géologique de la République du Congo Brazzaville au 1/500, 000. Paris: Bureau de recherches géologiques et minières, 1969. 103 p. (Mémoires no. 70.)

ETHIOPIA

Abul-Haggag, Y. Physiography of northern Ethiopia. London: Althone Press, 1961. 153 p.

Mohr, Paul A. The geology of Ethiopia. Addis Ababa: University College of Addis Ababa Press, 1961. 268 p.

GABON

Hudeley, H. and Y. Belmonte. Carte géologique de la République gabonaise au 1/1, 000, 000, notice explicative. Paris: Bureau de recherches géologiques et minières, 1970. 187 p. (Mémoires no. 72.)

GHANA

James, W. T. A bibliography of Gold Coast geology, mining and archaeology to March 1937. Accra: Gold Coast Geological Survey, 1937. 55 p. (Bulletin no. 9.)

* * *

Bates, D. A. The geological evolution of the Gold Coast. London: Commission for Technical Cooperation in Africa, 1957. 13 p.

Junner, N. R. Geology of the Gold Coast and western Togoland. Gold Coast Geological Survey, 1940. 40 p. (Bulletin no. 11.)

GUINEA

Furon, Raymond. Géologie de la Guinée française. Paris, 1943. 46 p. (Bureau d'études géologiques et minières coloniales. Publication no. 19.)

KENYA

Dosaj, N. P. and J. Walsh. Bibliography of the geology of Kenya, 1859-1968. Nairobi: Geological Survey of Kenya, 1970. 65 p. (Bulletin no. 10.)

* * *

Cole, S. M. An outline of the geology of Kenya. Nairobi: Pittman, 1950. 58 p.

Pulfrey, William. The geology and mineral resources of Kenya. Nairobi: Geological Survey of Kenya, 1969. 34 p. (Bulletin no. 9.)

LIBERIA

Johnson, Donald H. and Richard W. White. Bibliography of the geology and mineral industry of Liberia and adjacent countries. U.S. Geological Survey Open File Report. 1968. (IR)-LI-32. 65 p.

----------. Supplement 1. 1970. (IR)-LI-44. 41 p.

LIBYA

Hill, R. W. A bibliography of Libya. Durham, England: University of Durham, 1959. 100 p. (Durham University. Durham Colleges. Department of Geography. Research Papers Series no. 1.)

* * *

Goudarzi, Gus H. Geology and mineral resources of Libya --a reconnaissance. Washington: Government Printing Office, 1970. (U.S. Geological Survey. Professional Paper no. 660.)

Hecht, Franz and others. "Zur geologie von Libyen." In: Geologische Rundschau. v. 53, no. 2, 1963. p. 413-469.

Klitzsch, E. "Outline of the geology of Libya." In: Petroleum Exploration Society of Libya. Geology and archaeology of Northern Cyrenaica, Libya. Tripoli, 1968. p. 71-78.
The Petroleum Exploration Society of Libya has published other guidebooks on the geology of Libya.

MALAGASY REPUBLIC

Besairie, Henri. Monographie géologique de Madagascar. Paris: Centre national de la recherche scientifique, 1960. 159 p. (Lexique Stratigraphic International. v. 4, fasc. 11, Supplement.)

Dixey, F. "Geology and geomorphology of Madagascar." In: Geological Society of London. Quarterly Journal. v. 116, 1960. p. 255-268.

MALAWI

Cooper, W. G. G. The geology and mineral resources of Nyasaland. Revised ed. Zomba: Nyasaland Geological Survey, 1957. 43 p. (Bulletin no. 6.)

MOROCCO

Morin, Philippe. Bibliographie analytique des sciences de la terre; Maroc et régions limitrophes (depuis le début des recherches géologiques à 1964). Morocco. Service géologique, 1965. 2 v. (Notes et mémoires no. 182.)
Continues "Bibliographie géologique du Maroc" published in the Notes et mémoires series since 1951 (covering 1949-).

* * *

International Geological Congress. 19th, Algiers, 1952. Monographies régionales. 3 Ser: Maroc. Casablanca, 1952. 6 v.
The sixth volume, Géologie du Maroc, pt. 1, by George Choubert and J. Marçais, has also been published as Morocco. Service géologique. Notes et Mémoires, no. 100.

Roch, Edouard. Histoire stratigraphique du Maroc. Morocco. Service géologique, 1950. 435 p. (Notes et Mémoires no. 80.)

MOZAMBIQUE

Dias, Manuel Bettencourt. Escôbo tectónico de Moçambique, 1:3.000.000; notícia explicativa. Lourenço Marques: Tip. Artes Gráficas, 1956. 34 p. (Provincia de Moçambique. Serviços de Indústria e Geologia. Série de geologia e minas--Memôrias e comunicações. Boletim no. 18.)

Freitas, Antonio J. de. A geologia e o desenvolvimento económico e social de Moçambique. Mozambique: Lourenço Marques, 1959. 396 p.
Geological bibliography, p. 253-293.

----------. Notícia explicativa do escôbo geológico de Moçambique, 1:2.000.000. Lourenço Marques: Tip. Artes Gráficas, 1957. 82 p. (Provincia de Moçambique. Servicos de Indústria e Geologia. Série de geologia e minas--Memôrias e comunicações. Boletim no. 23.)

NIGERIA

Reyment, R. A. Aspects of the geology of Nigeria; the stratigraphy of the Cretaceous and Cenozoic deposits. Ibadan: Ibadan University Press, 1965. 145 p.

Taham, C. M. "A review of Nigerian stratigraphy." In: Nigeria. Geological Survey. Annual report for 1943. p. 27-46.

PORTUGUESE GUINEA

Teixeira, João E. "Geologia da Guiné portuguesa." In: Curso de geologia do Ultramar. Lisbon: Junta de Investigações do Ultramar, 1968. v. 1, p. 53-103.

RHODESIA

Smith, Craig C. and H. E. Van der Heyde. "Rhodesian geology; a bibliography and brief index to 1968." In: National Museums of Rhodesia. Natural Sciences. Occasional paper. v. 4, no. 31B, 1971. p. 323-575.

Swift, W. H. An outline for the geology of Southern Rhodesia. Salisbury: Southern Rhodesia Geological Survey, 1961. 73 p. (Bulletin no. 50.)

SENEGAL

Senegal. Service des Mines et de la Géologie. Carte géologique du Sénégal, 1/500.000. Paris: Bureau des recherches géologiques et minières, 1962. 36 p.

SIERRA LEONE

Pollet, J. D. "The geology and mineral resources of Sierra Leone." In: Colonial geology and mineral resources. v. 2. London: Her Majesty's Stationery Office, 1951. p. 3-28.

SOMALIA (See also Former Italian East Africa)

Macfadyen, William A. The geology of British Somaliland. London: Published by the Government of the Somaliland Protectorate, 1933. 87 p.

Somaliland Oil Exploration Company, Ltd. A geological reconnaissance of the sedimentary deposits of the Protectorate of British Somaliland. London: Published by the Government of the Somaliland Protectorate, 1954. 41 p.

SOUTH AFRICA

South Africa. Geological Survey. Bibliography and subject index of South African geology. Pretoria, 1922- . Coverage from 1922-1936 was under the title Bibliography of South African geology by A. L. Hall and published as Geological Survey Memoirs 18, 22, 15, 27, 30 and 37. After a gap, 1936-1956 (for which an index is being prepared), coverage continued as Bibliography and subject index of South African geology 1957-1962 followed by annual supplements to date. The latest supplement is for the year 1966 published in 1969.

* * *

DuToit, Alex L. The geology of South Africa. 3rd ed. Edinburgh: Oliver and Boyd, 1954. 611 p.

SOUTH WEST AFRICA

Martin, Henno. A bibliography of geological and allied subjects, South West Africa. Cape Town: Precambrian Research Unit, Department of Geology, University of Cape Town, 1965. 87 p. (Bulletin no. 1.)
Covers through 1961.

South West Africa. Geological Survey. Geological literature on South West Africa, 1838-1961. Windhoek, n.d. 47 leaves. Mimeographed.

* * *

Frommurze, Herbert F. and T. W. Gevers. South West Africa. Pretoria, 1929. 46 p. (International Geological Congress. 15th, South Africa, 1929. Guide Book, Excursion no. C. 21.)

SPANISH GUINEA

Lizaur y Roldan, Juan de. "Geologîa y geografîa fîsica de la Guinea continental española." In: Sociedad Geográfico, Madrid. Boletin. v. 81, nos. 7-12, 1945. p. 503-561.

SPANISH SAHARA

Hernández-Pacheco, Eduardo. El sahara Español; estudio geológico, geográfico y botánico. Madrid: Consejo Superior de Investigaciones Cientificas, Instituto de Estudios Africanos, 1949. 808 p.

SUDAN

Andrew, Gerald. Sources of information on the geology of the Anglo-Egyptian Sudan. Khartoum: Sudan Geological Survey, 1945. 36 p. (Bulletin no. 3.)

* * *

Andrew, Gerald. "Geology of the Sudan." In: Agriculture of the Sudan. Neasden, England: Oxford University Press, 1948. p. 84-128.

Whiteman, A. J. The geology of the Sudan Republic. Oxford: Clarendon Press, 1971. 290 p.

SWAZILAND

Hunter, Donald R. The geology of Swaziland; an explanation of the 1:125, 000 geologic map. Mbabane: Swaziland Geological Survey and Mines Department, 1961. 104 p.

TANZANIA

Tanganyika. Geological Survey. Summary of the geology of Tanganyika. Dar Es Salaam, 1956- . (Memoir no. 1.)
Pt. 1. Introduction and stratigraphy. A. M. Quennell and others. 1956; pt. 2. Geological map. A. M. Quennell. 1960; pt. 4. Economic geology. J. F. Harris. 1961.

Tanzania. Bureau of Resource Assessment and Land Use Planning. Bibliography of the geology and mineral resources of Tanzania to December 1967. Dar Es Salaam, 1969. 249 p.

TOGO

Kachinsky, V. "Bibliographie géologique et minières du Togo." In: La chronique des mines. Année 2, no. 21. Paris: Bureau d'études géologiques et minières coloniales, 1933. p. 517-526.

TUNISIA

Bibliographie géologique de la Tunisie. Tunis: Service géologique, 1954- . Irregular.
Continues Blondel, F. and G. Daumain. Bibliographie géologique et minière de France d'outre-mer. Paris: B.R.G.M., 1952. Thus far 5 nos. have been published, the latest in 1968.

Castany, G. Carte géologique de la Tunisie au 1/500, 000; notice explicative. Tunis: Service des mines, de l'industrie et de l'energie, 1953. 143 p.

International Geological Congress. 19th, Algiers, 1952. Monographies régionales. 2 Ser: Tunisie. Tunis, 1952. 7 v.

Petroleum Exploration Society of Libya. Guidebook to the geology and history of Tunisia. Edited by Lewis Martin. Tripoli, 1967. 293 p. (9th Annual Field Conference.)

UGANDA

Barnes, J. W. Mineral resources of Uganda. Entebbe: Uganda Geological Survey, 1961. 89 p. (Bulletin no. 4.)

Davies, K. A. and Bisset, C. B. "Geology and mineral resources of Uganda." In: Imperial Institute (London). Bulletin. v. 45, no. 2, 1947. p. 161-180.

Swardt, A. M. J. de and A. F. Trendall. "The physiographic development of Uganda." In: Overseas geology and mineral resources. v. 10, no. 3. London: Her Majesty's Stationery Office, 1969. p. 241-288.

UNITED ARAB REPUBLIC

Egypt. Ministry of Finance. Department of Survey and Mines. A bibliography of geology and related sciences concerning Egypt up to the end of 1939. Elias H. Keldani. Cairo: Government Printer, 1941. 428 p.

Hume, William F. Geology of Egypt. Cairo: Egypt Geological Survey and Mineral Research Department, 1925-1937. 2 v. in 4.

Reilly, Frank A. Guidebook to the geology and archaeology of Egypt. Tripoli: Petroleum Exploration Society of Libya, 1964. (6th Annual Field Conference.)

Said, Rusdhi. The geology of Egypt. New York: Elsevier, 1962. 377 p.

---------- and E. M. el Shazly. "Review of Egyptian geology." In: Egyptian reviews of science. v. 1, 1957. p. 1-91.
Includes bibliography for Egyptian geology between 1939 and 1957.

ZAMBIA

Zambia. Geological Survey. Annotated bibliography and index of the geology of Zambia, 1931-1959. Compiled by G. J. Snowball. Lusaka, 1965. 121 p.

----------. 1960-1961. G. J. Snowball in the Records of the Geological Survey Dept. of Zambia, v. 9, 1963.

----------. 1962 to date. Published bi-ennially. Latest is for 1966-1967, published in 1969.

* * *

Mendelsohn, F., ed. The geology of the Northern Rhodesia copperbelt. London: Macdonald, 1961. 523 p.

Reeve, W. H. The geology and mineral resources of Northern Rhodesia. Lusaka: Northern Rhodesia Geological Survey, 1963. 2 v. (Bulletin no. 3.)

THE PACIFIC
(See Also Indonesian Archipelago and Hawaii)

Lexique stratigraphique international. Volume VI: Oceania. Publie sur l'initative de la Commission de stratigraphie du Congress géologique international. Paris: Centre national de la recherche scientifique (CNRS), 1956- . For a breakdown of the Lexique see entry in general section of this guide.

* * *

"The age and nature of the circum-Pacific orogenesis; papers presented at the 11th Pacific Science Congress, Tokyo, August, 1966." A special issue of Tectonophysics. v. 4, no. 4-6, 1967. p. 317-613.

Cullen, David J. "A tectonic analysis of the south-west Pacific." In: New Zealand journal of geology and geophysics. v. 13, no. 1, 1970. p. 7-20. (Tectonic map of s.w. Pacific in pocket.)

Menard, H. W. Marine geology of the Pacific. New York: McGraw-Hill, 1964. 271 p.

Pacific geology. Tokyo: Tsukiji Shokan Publishing Co., Ltd., 1968- . Irregular.
V. 1, 1968 devoted largely to papers presented in the Division of Geology, 11th Pacific Science Congress, Tokyo, 1966; v. 2, 1970.

Pacific Science Congress. 1st, 1920- .
Scientific meetings of the Pacific Science Association are held every 4 years and include, among other subjects, papers on the geology and geophysics of the Pacific Basin. Papers presented at symposiums appear in special issues of appropriate scientific journals or as monographs. For example, geological papers from the 11th Pacific Science Congress, Tokyo, 1966 have appeared in special issues of Tectonophysics, Tohoku University science reports, Osaka City University publications, and Pacific geology, a new journal. The American Geophysi-

cal Union has published a monograph entitled The crust and upper mantle of the Pacific area in 1968 as its Geophysical Monograph no. 12.

SPECIFIC ISLANDS

Baltzer, F. and others. "Geological and geophysical publications on New Caledonia." In: New Zealand journal of geology and geophysics. v. 10, no. 5, 1967. p. 1275-1279.

Coleman, P. J. "Geology of the Solomon and New Hebrides Islands, as part of the Melanesian re-entrant, southwest Pacific." In: Pacific science. v. 24, no. 3, 1970. p. 289-314.

Fiji. Geological Survey Department. Bibliography of the geology of Fiji, including published and unpublished references to November 1968. Suva: Ministry of Natural Resources, 1968. 81 p.

Karig, Daniel E. "Structural history of the Mariana Island Arc System." In: Geological Society of America. Bulletin. v. 82, no. 2, 1971. p. 323-344.

Lillie, A. R. and R. N. Brothers. "The geology of New Caledonia." In: New Zealand journal of geology and geophysics. v. 13, no. 1, 1970. p. 145-183.

McBirney, A. R. and Howel Williams. Geology and petrology of the Galápagos Islands. Boulder, Colorado: Geological Society of America, 1969. 197 p. (Memoir no. 118.)

Mitchell, A. H. G. "Geological and geophysical publications on the New Hebrides." In: New Zealand journal of geology and geophysics. v. 10, no. 5, 1967. p. 1274.

U.S. Geological Survey. Bikini and nearby atolls, Marshall Islands. Washington: Government Printing Office, 1954-1964. (Professional Paper no. 260.)

----------. Geology of Saipan, Mariana Islands. 1956-1959. (Professional Paper no. 280.)

----------. Geology and hydrology of Guam, Mariana Islands. 1962-1966. (Professional Paper no. 403.)

Wood, Bryce L. and R. F. Hay. Geology of the Cook Islands. Lower Hutt: New Zealand Geological Survey, 1970. 103 p. (Bulletin no. n.s. 82.)

AUSTRALIA

Publications of the Bureau of Mineral Resources, Geology and Geophysics. Canberra: Government Printing Office, 1969. 107 p.

In addition to the work of the Australian Bureau of Mineral Resources, Geology and Geophysics, the 6 Australian state geological surveys--Queensland, New South Wales, Victoria, Tasmania, South Australia, Western Australia --carry out geological investigations and mapping in their own areas. Results are published through these state surveys, who furnish lists of reports and publications available. Headquarters of the surveys are located in the capital city of each state.

* * *

Brown, D. A., K. S. W. Campbell and K. A. W. Crook. The geological evolution of Australia and New Zealand. Oxford: Pergamon, 1968. 409 p. (Commonwealth and International Library.)

Commonwealth Mining and Metallurgical Congress. 8th, Australia and New Zealand, 1965. Publications. Melbourne: Australasian Institute of Mining and Metallurgy, 1965. 6 v.

Includes, v. 1: Geology of Australian ore deposits and v. 4: Economic geology of New Zealand.

David, T. W. E. The geology of the Commonwealth of Australia. London: Arnold, 1950. 3 v.

Geological Society of Australia. Tectonic Map Committee. Geological notes in explanation of the tectonic map of Australia. Canberra: Bureau of Mineral Resources, Geology and Geophysics, 1962. 72 p.

Accompanies the tectonic map published in 1960 at a scale, 1:2,534,400. Includes a bibliography of the Australian states. A revised map is in preparation.

Jennings, J. N. and J. A. Mabbutt, eds. Landform studies from Australia and New Guinea. Canberra: Australian National University Press, 1967. 434 p.

Australian geomorphology also predominates in the following textbooks: Twidale, C. K. Geomorphology, with special reference to Australia. London: Nelson, 1968. 406 p., and Introduction to systematic geomorphology, a series of textbooks published by the Australian National University Press, beginning in 1968 with Coasts, by E. C. F. Bird.

Joplin, Germaine A. A petrography of Australian igneous rocks. 2nd ed. New York: Elsevier, 1968. 210 p.

----------. A petrography of Australian metamorphic rocks. New York: Elsevier, 1968. 262 p.

Raggatt, H. G. Mountains of ore. Melbourne: Lansdowne Press, 1968. 406 p.
Authoritative descriptions of important mineral deposits in Australia.

THE STATES

A series of geological monographs on the Australian states were published by the Geological Society of Australia as part of its Journal. Published thus far: The stratigraphy of Western Australia. J. R. H. Whae and others. 1958. 161 p. (Geological Society of Australia. Journal. v. 4, pt. 2); The geology of South Australia. M. F. Glaessner and L. W. Parkin, eds. 1958. 163 p. (Journal. v. 5, pt. 2); The geology of Queensland. D. Hill and A. K. Denmead, eds. 1960. 474 p. (Journal. v. 7); The geology of Tasmania. A. Spry and M. R. Banks, eds. 1962. (Journal. v. 9, pt. 2, p. 107-362); The geology of New South Wales. G. H. Packham, ed. 1969. 654 p. (Journal. v. 16, pt. 1.)

SOUTH AUSTRALIA

Teesdale-Smith, E. N. Bibliography of South Australian geology. Adelaide: Geological Survey of South Australia, 1959. 235 p.

* * *

South Australia. Geological Survey. Handbook of South Australian geology. Edited by L. W. Parkin. Adelaide, 1969. 268 p.

TASMANIA

Jennings, I. B. and E. Williams. Geology and mineral resources of Tasmania. Hobart: Tasmania Department of Mines, 1967. 90 p. (Tasmania Geological Survey. Bulletin no. 50.)

VICTORIA

McAndrew, John and Marcus A. H. Marsden, eds. A regional guide to Victorian geology. Parkville: University of Melbourne, Geology Department, 1968. 219 p.

GREAT BARRIER REEF

Maxwell, W. G. H. Atlas of the Great Barrier Reef. Amsterdam: Elsevier, 1968. 258 p.

NEW ZEALAND

Adkin, G. L. and B. W. Collins. A bibliography of New Zealand geology to 1950. Lower Hutt: New Zealand Geological Survey, 1967. 244 p. (Bulletin no. n.s. 65.) Subject index to this bibliography is being prepared by D. L. Jenkins. For literature subsequent to 1950, refer to series "Abstracts of papers on New Zealand geology" in New Zealand journal of science and technology, Section B for the years 1950-1955, and to "New Zealand geological abstracts" in the New Zealand journal of geology and geophysics for the years 1956- .

Grindley, G. W., H. J. Harrington and B. L. Wood. The geological map of New Zealand, 1:2,000,000. Lower Hutt: New Zealand Geological Survey, 1961. 111 p. (Bulletin no. n.s. 66.)
Descriptive material on the geology of New Zealand accompanying the map.

* * *

Brown, D. A., K. S. W. Campbell and K. A. W. Crook. The geological evolution of Australia and New Zealand. Oxford: Pergamon, 1968. 409 p. (Commonwealth and International Library.)

Suggate, R. P., ed. The geology of New Zealand. Wellington: Department of Scientific and Industrial Research, in preparation. 2 v.

Williams, G. J. Economic geology of New Zealand. Melbourne: Australasian Institute of Mining and Metallurgy, 1965. 384 p. (Commonwealth Mining and Metallurgical Congress. 8th, Australia and New Zealand, 1965. Publications, v. 4.)

ANTARCTICA

Lexique stratigraphique international. Volume 6, Fasc. 6: Antarctica. Publie sur l'initative de la Commission de stratigraphie du Congress géologique international. Paris: Centre national de la recherche scientifique (CNRS), 1956- .

"Recent polar literature." In: Polar record, publication of the Scott Polar Research Institute. 3/yr.
A list of recent accessions to the Scott Polar Research Institute library. In 1969, the Polar record listed over 1800 items, including those in many branches of geology. Coverage includes both arctic and antarctic areas.

Splettstoesser, John R. "Antarctic geological literature." In: Geotimes. v. 11, no. 4, 1966. p. 33-35.

Stefansson Collection. Dictionary catalog of the Stefansson collection on the polar regions in the Dartmouth College Library. Boston: G. K. Hall, 1967. 8 v.

U.S. Library of Congress. Antarctic bibliography. Edited by Geza T. Thuronyi. Sponsored by the Office of Antarctic Programs, National Science Foundation. Washington: Government Printing Office, 1965- . Irregular. Coverage of biological, geoscience, and geographical subjectas as well as others. V. 1, 1965; v. 2, 1966; v. 3, 1968; v. 4, 1971. A retrospective complement to the current volumes is Antarctic bibliography, 1951-1961, published by the Library of Congress in 1970. 349 p.

* * *

Adie, Raymond J. "The geology of Antarctica." In: Pacific Science Congress. 10th, Honolulu, 1961. Antarctic research. Washington: American Geophysical Union, 1962. (Geophysical Monograph no. 7.) (NAS-NRC Publication no. 1036.) p. 26-39.

American Geographical Society of New York. Antarctic map folio series. Edited by Vivian C. Bushnell. New York, 1964- .

American Geophysical Union. Antarctic research series. Washington, 1964- .
Selected titles in the series: Antarctic snow and ice studies, no. 2; Geomagnetism and aeronomy, no. 4; Geology and paleontology of the Antarctic, no. 6; Antarctic soils and soil forming processes, no. 8.

Cailleux, André. Géologie de l'Antarctique. Paris: Centre de documentation universitaire et Société d'édition d'enseignement supérieur, 1963. 201 p.

Cordini, I. Rafael. El conocimiento geológico de la Antártida. Buenos Aires: Instituto Antártico Argentino, 1959. 159 p. (Publicación no. 6.)

Ford, Arthur B. "Review of Antarctic geology." In: American Geophysical Union. Transactions. v. 45, no. 2. Washington, 1964. p. 363-381.

International Symposium on Antarctic Geology. 1st, Capetown, 1963. Antarctic geology; proceedings. Edited by Raymond J. Adie. New York: Interscience, 1964. 758 p.

National Research Council. Committee on Polar Research. Polar research; a survey. Washington: National Academy of Sciences, 1970. 204 p. (NAS-NRC Publication no. 1753.)

Warren, Guyon. "Geology of Antarctica." In: Hatherton, Trevor, ed. Antarctica. New York: Praeger, 1965. p. 279-320.

Woollard, G. P. The geology of Antarctica. Honolulu: Hawaii Institute of Geophysics, University of Hawaii, 1970. 99 p. (HIG no. 70-8.)

GEOLOGIC MAPS

A Contribution to Geologic Reference Sources

by
Mark W. Pangborn, Jr.
U.S. Geological Survey Library

1971

MAP AND CARTOGRAPHIC LITERATURE

This is a selected bibliography of modern books and articles on maps and cartography. The geosciences are emphasized; bibliographic works on topographic mapping are largely omitted.

MAP LIBRARIES

Hill, J. Douglas. "Map and atlas cases." In: Library trends. v. 13, no. 4, 1965. p. 481-487.

Le Gear, Clara E. Maps; their care, repair, and preservation in libraries. Revised ed. Washington: Library of Congress, Map Division, 1956. 75 p.

Ristow, Walter W. "The emergence of maps in libraries." In: Special libraries. v. 58, no. 6, 1967. p. 400-419.
Includes a bibliography of 118 items which cites nearly all of the important literature on the subject of maps in libraries.

Special Libraries Association. Geography and Map Division. Directory Revision Committee. Map collections in the United States and Canada; a directory. 2nd ed. David K. Carrington, Chm. New York: Special Libraries Association, 1970. 159 p.

Woods, Bill M. Map librarianship; a selected bibliography. Revised ed. N.p.: New Jersey Library Association, 1970. 20 leaves.

BIBLIOGRAPHIES AND INDEXES, JOURNALS

American Geographical Society. Current geographical publications. New York, v. 1, 1938- .
Lists new books, articles and maps.

----------. Map Department. Index to maps in books and periodicals. Boston: G. K. Hall, 1968. 10 v.

Association of American Geographers. High School Geography Project. Sources of information and materials: maps and aerial photographs. Boulder, Colo., 1970. 159 p.
Lists sources of cartographic information, ideas and materials.

Bibliography and index of engineering geological mapping. Prague: GEOFOND, no. 1, 1970- .

Bibliotheca Cartographica. Herausgegeben vom Institut für Landeskunde. Bad Godesberg: Bundesforschungsanstalt für Landeskunde und Raumordnung, 1957- . Annual.
Classed bibliography of literature, not maps.

Commission for the Geological Map of the World. Bulletin. Paris, no. 1, 1963- .
Contains information on new geologic maps.

----------. Selected geologic bibliography; regional descriptions and maps. Paris, 1964- .
1. Africa; 3. South America; 4. Middle East.

Drazniowsky, Roman. "Bibliographies as tools for map acquisition and map compilation." In: Cartographer. v. 3, no. 2, 1966. p. 138-144.
Lists 52 sources.

International yearbook of cartography. London: Philip, v. 1, 1961- .
Annual volumes include summary articles.

Lock, Clara Beatrice Muriel. Modern maps and atlases; an outline guide to twentieth century production. London: Bingley, 1969. 619 p.
Encyclopedia of information on maps and atlases, arranged geographically.

Royal Geographical Society. New geographical literature and maps. London, v. 1, 1951- .

Special Libraries Association. Geography and Map Division. Bulletin. New York, no. 1, 1950- .
Lists new books and maps; there are occasional articles and bibliographies dealing with maps.

Stephenson, Richard W. "Published sources of information about maps and atlases." In: Special libraries. v. 61, no. 2, 1970. p. 87-98, 110-112.
Useful article on how to go about building a map collection. Lists review journals, out-of-print dealers, publishers and sellers.

Surveying and mapping. Washington: American Congress of Surveying and Mapping, 1941- . Quarterly.
Lists new books and maps.

World cartography. New York: United Nations, v. 1, 1951- . Annual.
Contains useful summary articles on cartography, e.g., "The status of world topographic mapping," in v. 10.

CARTOGRAPHY

Arnberger, Erik. Handbuch der thematischen Kartographie. Wien: Franz Deuticke, 1966. 554 p.

Baldock, E. D. Manual of map reproduction techniques. Ottawa: Canada Dept. of Mines and Technical Surveys, Surveys and Mapping Branch, n.d. 32 p.

Birch, Thomas William. Maps: topographical and statistical. 2nd ed. Fair Lawn, N. J.: Oxford University Press, 1964. 240 p.

Blackadar, Robert Gordon. Guide for the preparation of geological maps and reports. Revised ed. Ottawa: Geological Survey of Canada, 1968. 147 p. (Miscellaneous Report no. 16.)

Cramer, Robert E. Workbook in essentials of cartography and mapping. Dubuque, Iowa: W. C. Brown, 1963. 215 p.

Dickinson, Gordon C. Maps and air photographs. London: Edward Arnold, 1969. 286 p.

Greenwood, David. Mapping. Chicago: University of Chicago Press, 1963. 289 p.
Revision of his Down to earth; mapping for everybody, 1951.

Hodgkiss, A. G. Maps for books and theses. New York: Pica Press, 1970. 267 p.

Imhof, Eduard. Gelände und Karte. 3rd ed. Stuttgart: Rentsch, 1968. 259 p.

----------. Kartographische Geländedarstellung. Berlin: de Gruyter, 1965. 425 p.

Lobeck, Armin K. Block diagrams and other graphic methods used in geology and geography. 2nd ed. Amherst, Mass.: Emerson-Trussell, 1958. 212 p.

Low, Julian W. Plane table mapping. New York: Harper, 1952. 365 p.

Monkhouse, Francis J. and H. R. Wilkinson. Maps and diagrams; their compilation and construction. 3rd ed. London: Methuen, 1971. 522 p.

Raisz, Erwin J. General cartography. 2nd ed. New York: McGraw-Hill, 1948. 354 p.
College text.

----------. Principles of cartography. New York: McGraw-Hill, 1962. 315 p.
High school text.

Robinson, Arthur Howard. The look of maps; an examination of cartographic design. Madison: University of Wisconsin Press, 1952. 105 p.

---------- and Randall D. Sale. Elements of cartography. 3rd ed. New York: Wiley, 1969. 415 p.
College text.

Schuster, Matthäus. Das geographische und geologische Blockbild; eine Einführung in dessen Erzeichnung. Berlin: Akademie-Verlag, 1954. 222 p.

Steward, Harry. Education and training in the mapping sciences; a working bibliography. New York: American Geographical Society, 1969. 60 p.

Tennessee Valley Authority. Maps and Surveys Branch. Division of Water Control Planning. Maps and surveys. Chattanooga, Tenn.: TVA Map and Records Unit, 1966. 56 p.

U.S. Army Topographic Command. Glossary of mapping, charting, and geodetic terms. 2nd ed. Prepared for the Dept. of Defense. Washington: Government Printing Office, 1969. 281 p.

U.S. Department of the Army. Map compilation, color separation and revision. Washington, 1962. 152 p. (Technical Manual TM5-240.)

Witt, Werner. Thematische Kartographie. Hannover: Gebrüder Jänecke, 1967. 762 col. (Veröffentlichungen der Akademie für Raumforschung und Landsplanung. Abhandlungen, v. 49.)

SURVEYING AND CHARTING

Bouchard, Harry. Surveying. 5th ed. Revised by Francis H. Moffitt. Scranton, Pa.: International Textbook Co., 1965. 754 p.

Cohen, Philip M. Bathymetric navigation and charting. Annapolis, Md.: United States Naval Institute, 1970. 138 p.

Davis, Raymond E. and Joe W. Kelly. Elementary plane surveying. 4th ed. New York: McGraw-Hill, 1967. 574 p.

Kissam, Philip. Surveying practice; the fundamentals of surveying. New York: McGraw-Hill, 1966. 460 p. Elementary text.

Surveying and mapping. Washington: American Congress of Surveying and Mapping, v. 1, 1941- . Quarterly.

MAP READING AND INTERPRETATION

DeBruin, Richard. 100 topographic maps illustrating physiographic features. Northbrook, Ill.: Hubbard Press, 1970. 128 p.

Dury, George H. Map interpretation. 3rd ed. London: Pitman, 1967. 215 p.

Lobeck, Armin K. Things maps don't tell us--an adventure into map interpretation. New York: Macmillan, 1957. 159 p.

Elementary text.

U.S. Department of the Army. Map reading. Washington: Government Printing Office, 1969. Various paging. (Field Manual FM21-26.)
Military slant.

U.S. Military Academy. Department of Earth, Space, and Graphic Sciences. Atlas of landforms. James L. Scovel and others. New York: Wiley, 1966. 164 p.

Upton, William Bayly, Jr. Landforms and topographic maps, illustrating landforms of the continental United States. New York: Wiley, 1970. 134 p.

GEOLOGIC MAPS

Aubouin, Jean and others. Manuel de travaux pratiques de cartographie; 1er cycle et maitrise. Paris: Dunod, 1970. 332 p.

Bennison, George M. An introduction to geological structures and maps. London: Arnold, 1965. 64 p.

Bishop, Margaret S. Subsurface mapping. New York: Wiley, 1960. 198 p.

Blyth, Francis F. H. Geological maps and their interpretation. London: Arnold, 1965. 48 p.

Bonté, Antoine. Introduction à la lecture des cartes géologiques. 4th ed. Paris: Masson, 1969. 277 p.

Bradshaw, Michael J. and E. A. Jarman. Reading geological maps. London: English Universities Press, 1969. 48 p.

Currier, Louis W. Geologic map portfolio; a laboratory study of geologic maps and sections. Washington: Williams and Heintz Map Corp., 1965. Various paging.
With text.

Falke, Horst. Die Geologische Karte, ihre Anfertigung und Ausdeutung. Berlin: de Gruyter, 1969. 180 p.

Foucault, Alain and J. F. Raoult. Coupes et cartes géologiques. Paris: Société d'Edition d'Enseignement Supérieur, 1966. 147 p.

Ireland, H. Andrew. "History of the development of geologic maps." In: Geological Society of America. Bulletin. v. 54, no. 9, 1943. p. 1227-1280.

Levorsen, Arville I. Paleogeologic maps. San Francisco: Freeman, 1960. 174 p.

Platt, John Isaac. Selected exercises upon geological maps. London: Murby, 1951. Various paging.

----------. A series of elementary exercises upon geological maps. 3rd ed. London: Murby, 1952. 32 p.

---------- and John Challinor. Simple geological structures; a series of notes and map exercises. 4th ed. London: Murby, 1968. 56 p.

Simpson, Brian. Geological maps. New York: Pergamon, 1968. 98 p.

----------. Geological map exercises. 2nd ed. London: Philip, 1968. 68 p.

Spiridonow, A. I. Geomorphologische Kartographie. Berlin: VEB Deutscher Verlag der Wissenschaften, 1956. 160 p.

"Stratigraphic maps." In: Krumbein, William C. and L. L. Sloss. Stratigraphy and sedimentation. 2nd ed. San Francisco: Freeman, 1963. p. 432-500.
Among other topics, discusses structural contour, isopach, paleogeologic, facies, and biofacies maps.

Thomas, John A. G. An introduction to geological maps. London: Murby, 1966. 64 p.

Vacquier, Victor. Interpretation of aeromagnetic maps. New York: Geological Society of America, 1951. 151 p. (Memoir no. 47.)

MAPS AND ATLASES

This is a selected bibliography of modern maps and atlases of interest to the geoscientist. It includes small- and medium-scale maps of the world, the continents and ocean basins, countries, provinces, and states. Emphasis is on areal geology, tectonics, and geomorphology, but some interesting physiographic diagrams, soil, ground water, and mineral maps have been included. Gravity and other geophysical maps, as well as topographic maps, have generally been omitted.

Bibliographic information on geoscience maps has been limited to publications that deal with maps only. Bibliographies and publication lists which describe geological publications in general (including maps) are found in the regional section.

WORLD

Akademiia Nauk. Geologicheskii Institut. Tektonicheskaia karta Evrazii; Tectonic map of Eurasia, 1:5,000,000. A. L. IAnshin, editor. Moskva, Glavnoe Upravlenie Geodezii i Kartografii Ministerstva Geologii SSSR, 1966. 12 sheets. Structure shown in some detail.

Akademiia Nauk SSR. Institut Geologicheskikh Nauk. Skhema struktury zemnoi kory, 1:27,000,000. N.p., 1962. 2 sheets. Generalized structure.

Bartholomew, John, ed. Times atlas of the world. Boston, Houghton Mifflin, 1955-58. 5 vols.

----------. Times atlas of the world. Comprehensive edition. Boston, Houghton Mifflin, 1967. 272 p.

Bederke, Erich, and Wunderlich, Hans-Georg. Atlas zur Geologie. Mannheim, Bibliographisches Institut, Kartographisches Institut Meyer, 1968. 75 p. (Meyers Grosser physischer Weltatlas, Bd. 2.) Contains rather generalized, small scale geology, tectonic, drift, mineral, and other types of maps, covering the continents and the world as a whole.

Drecka, Jolanta, and Tuszynska-Rekawek, Halina. National and regional atlases; sources, bibliography, articles.

Warsaw, 1964. 155 p. (Polish Academy of Sciences. Institute of Geography. Dokumentacja geograficzna nr. 1.)

Fahlbusch, Klaus. Geologie der Erde; Geology of the earth; Geologí a del mundo, 1:16,000,000. Darmstadt, Justus Perthes, 1969? 2 sheets. Generalized geology.

Food and Agricultural Organization of the United Nations. World Soil Resources Office. Catalogue of maps. 3rd ed. Rome, FAO, 1965. 165 p.

Ganssen, Robert, and Hädrich, Friedhelm. Atlas zur Bodenkunde. Mannheim, Bibliographisches Institut, Kartographisches Institut Meyer, 1965. 85 p. (Meyers Grosser Physischer Weltatlas, Bd. 1.) Includes world soil map, 1:80,000,000, and regional maps at various scales.

Laffitte, P., and Rouveyrol, P. Carte minière du globe sur fond tectonique, 1:20,000,000. Edition provisoire. 1 ère partie. Paris, Bureau de Recherches Géologiques et Minières, 1963. 16 sheets.

----------. Carte minière du globe sur fond tectonique, 1:20,000,000. Édition provisoire. 2 ème partie. Paris, Bureau de Recherches Géologiques et Minières, 1965. 2 sheets.

McGill, John T. Coastal landforms of the world, 1:25,000,000. N.p., American Geographical Society, 1958. (Plate in Geographical review, vol. 48, 1958.)

Murphy, Richard E. Landforms of the world, 1:50,000,000. Washington, Association of American Geographers, 1968. (Map supplement no. 9, Annals, vol. 58, no. 1, March 1968.)

National Geographic Society. National Geographic atlas of the world. 3rd ed. Washington, 1970. 331 p.

Orvedal, A. C. Great soil groups, 1:22,800,000. Chicago, Rand McNally, 1951?

Petzold, G., and Pardeyke, K. World geology, 1:15,000,000. Edited by R. Jubelt and G. Olszak. Gotha/Leipzig, VEB Hermann Haack, Geographisch-Kartographische Anstalt, 1969? 4 sheets. Generalized geology.

Russia. Glavnoe Upravlenie Geodezii i Kartografii. Fiziko-geograficheskii atlas mira. I. P. Gerasimov, ed. Moskva, Akademiia Nauk SSSR, 1964. 298 p. Includes world maps, 1:60, 000, 000, showing geology, structure, earthquakes and volcanoes, soils, vegetation, geomorphology, drift, sea-bottom sediments; similar maps of continents at various scales.

Russia. Ministerstvo Geologii SSSR. Geologicheskaia karta mira, 1:15, 000, 000. Moskva, Glavnoe Upravlenie Geodezii i Kartografii, 1966. 6 sheets. Geology in some detail.

Schubnel, H. J. World map of gemstone deposits, 1:40, 000, 000. Orleans, Bureau de Recherches Géologiques et Minières, 1970? With text.

Shatskii, N. S. Geologicheskaia karta Evrazii, 1:6, 000, 000. Moskva, Glavnoe Upravlenie Geodezii i Kartografii, MVD, SSSR, 1956. 6 sheets.

Termier, Henri. Atlas de paleogeographie. Paris, Masson, 1960. 100 p.

U. S. Coast and Geodetic Survey. National Earthquake Information Center. World seismicity, 1961-1969, 1:35, 000, 000. Washington, 1970. (Chart NEIC 3005.) Sheets 3008 through 3019 of this series cover parts of the world in detail; 3012 covers the United States.

Van Royen, William, and Bowles, Oliver. The mineral resources of the world. New York, Prentice-Hall, 1952. 181 p. (Maryland. University. Dept. of Geography. Atlas of the world's resources, vol. 2.)

Wolfanger, Louis A. World soils, 1:28, 828, 800, 455 miles to the inch. Chicago, Denoyer-Geppert, 1960. (Map P79s.)

OCEANS

Chase, T. E. Bathymetry of the North Pacific, 1:6, 000, 000. La Jolla, Calif., Scripps Institution of Oceanography, and Institute of Marine Resources, 1970. (IMR Technical report series, TR-3, TR-6, TR-7, TR-9, TR-10, TR-11, TR-12, TR-13, TR-14, TR-15.) 10 sheets.

Dietrich, Günter, and Ulrich, Johannes. Atlas zur Ozeanographie. Mannheim, Bibliographisches Institut, 1968. 76 p. (Meyers Grosser Physicher Weltatlas, Bd. 7.)

Heezen, Bruce C., and Tharp, Marie. Physiographic diagram, Atlantic Ocean (sheet 1)...South Atlantic Ocean, the Caribbean Sea, the Scotia Sea, and the eastern margin of the South Pacific Ocean, ca. 1:10,000,000. N.Y., Geological Society of America, 1957-61. (Sheet 1 accompanied by Geological Society of America. Special paper 65, The floors of the oceans, by Bruce C. Heezen and others. N.Y., 1959. 122 p.)

----------. Physiographic diagram of the Indian Ocean, the Red Sea, the South China Sea, the Sulu Sea and the Celebes Sea, ca. 1:10,000,000. N.Y., Geological Society of America, 1964. 2 sheets.

----------. Physiographic diagram of the western Pacific Ocean, 1:10,000,000. Boulder, Colo., Geological Society of America, 1971.

Kropotkin, P. N. Tektonicheskaia karta Tikhogo Okeana i Tikhookeanskogo podvizhnogo poiasa, 1:15,000,000. N.p., 1962. 4 sheets.

McKelvey, Vincent E., and Wang, F. F. H. Preliminary map...World subsea mineral resources, 1:39,293,200. Washington, 1969. (U.S. Geological Survey. Miscellaneous geologic investigations map I-632.) 4 sheets.

Menard, H. W., and others. Physiography of the northeastern Pacific, ca. 1:6,000,000. N.Y., McGraw-Hill, 1964. (Plate in Menard, H. W. Marine geology of the Pacific, N.Y., McGraw-Hill, 1964. Separate map also available from C. S. Hammond and Co., Maplewood, N.J.)

Pepper, James F. The Indian Ocean: the geology of its bordering lands and the configuration of its floor, 1:13,655,000. Washington, 1963. (U.S. Geological Survey. Miscellaneous geological investigations map, I-380.)

Russia. Akademiia Nauk. Mezhduvedomstvennyi Geofizicheskii Komitet. Atlanticheskii Okean: Geomorfologiia, 1:20,000,000. Moskva, Glavnoe Upravlenie Geodezii i

Kartografii pri Sovete Ministrov SSSR, 1969.

----------. Atlanticheskii Okean: Tipy donnykh osadkov, veshchestvenno-geneticheskii sostav... 1:20, 000, 000. Moskva, Glavnoe Upravlenie Geodezii i Kartografii pri Sovete Ministrov SSSR, 1969. 1 sheet, shows bottom deposits.

THE ARCTIC

Atlasov, I. P. Tektonicheskaia karta Arktiki i Subarktiki, 1:5, 000, 000. N.p., Ministerstvo Geologii SSSR, Nauchno-Issledovatel'skii Institut Geologii Arktiki, 1964. 6 sheets, with text.

International Symposium on Arctic Geology, 1st, Calgary, 1960. Geological map of the Arctic, 1:7, 500, 000. Calgary, Alberta Society of Petroleum Geologists, 1960.

Markov, F. G. Geologicheskaia karta Arktiki i Subarktiki, 1:5, 000, 000. Moskva, Vsesoiuznyi Aerogeologicheskii Trest Ministerstva Geologii SSSR, 1969.

Pushcharovskii, IU. M. Tektonicheskaia karta Arktiki, 1:10, 000, 000. Moskva, Akademiia Nauk SSSR, Geologicheskii Institut, 1963.

Russia. Ministerstvo Geologii SSSR. Nauchno-Issledovatel'skii Institut Geologii Arktiki. Tektonicheskaia karta poliarnykh oblastei Zemli; Tectonic map of polar regions of the earth, 1:10, 000, 000. Glavnyi redaktor: B. Kh. Egiazarov. Moskva, Vsesoiuznyi Aerogeologicheskii Trest Ministerstva Geologii SSSR, 1971. 4 sheets.

NORTH AMERICA

American Association of Petroleum Geologists, and U. S. Geological Survey. Basement map of North America, between latitudes 24° and 60° N, 1:5, 000, 000. Prepared by the Basement Rock Project Committee, Peter T. Flawn, Chairman. Douglas M. Kinney, Editor. Washington, U. S. Geological Survey, 1967.

Flint, Richard Foster, and others. Glacial map of North America, 1:4, 555, 000. New York, Geological Society of America, 1945. (Part 1 of its Special papers, no. 60.) 2 sheets.

King, Philip B. Tectonic map of North America, 1:5, 000, 000. Washington, U.S. Geological Survey, 1969. 2 sheets.

North American Geologic Map Committee. Geologic map of North America, 1:5, 000, 000. Washington, U.S. Geological Survey, 1965. 2 sheets.

Oxford University Press. Oxford regional economic atlas, United States and Canada. Editors: John D. Chapman and John C. Sherman. Oxford, Clarendon Press, 1967. 128 p. Includes permafrost, solid geology, drift, soil, and mineral maps, generally at scale 1:15, 840, 000.

Prest, V. K. Retreat of Wisconsin and recent ice in North America; speculative ice-marginal positions during recessions of the last ice-sheet complex, 1:5, 000, 000. Ottawa, 1969. (Canada. Geological Survey. Map 1257A.)

GREENLAND

Escher, A., and others. Tectonic/geological map of Greenland, 1:2, 500, 000. Copenhagen, Geological Survey of Greenland, 1970.

CANADA

Brown, Roger James Evan. Permafrost in Canada, 1:7, 603, 200. Ottawa, Geological Survey of Canada, 1967. (Map 1246A.)

Canada. Dept. of Mines and Technical Surveys. Geographical Branch. Atlas of Canada. Ottawa, 1957- . Loose-leaf; includes glacial geology, bedrock geology, iron and steel, and mineral fuel maps, 1:10, 000, 000; metal and industrial mineral maps, 1:5, 000, 000.

Canada. Geological Survey. Geological map of Canada, 1:5, 000, 000. Ottawa, 1969. (Map 1250 A.)

----------. Geological map of Canada. 1:1, 000, 000. Ottawa, 1965- . 7 of about 70 sheets published.

----------. Geological maps published by the Geological Survey of Canada; index sheet 1-- . Ottawa, 1960- . Set of about 70 index diagrams, covering Canada, nearly all published.

----------. Physiographic regions of Canada, 1:5, 000, 000. Ottawa, 1970. (Map 1254 A.)

----------. Principal mineral areas of Canada, 1:7, 603, 200. 20th ed. Ottawa, 1970. (Map 900 A.)

----------. Mineral deposits of Canada, 1:5, 000, 000. Ottawa, 1969. (Map 1252 A.)

----------. Tectonic map of Canada, 1:5, 000, 000. Ottawa, 1969. (Map 1251 A.)

----------. Tectonic Map of Canada Committee. Tectonic map of the Canadian Shield, 1:5, 000, 000. Ottawa, 1965. (Map 4-1965.)

Prest, V. K., and others. Glacial map of Canada, 1:5, 000, 000. Ottawa, Geological Survey of Canada, 1968. (Map 1253 A.)

Raisz, Erwin. Landform map of Canada, prepared for the Environmental Protection Section of the Office of the U. S. Quartermaster General, 1:6, 336, 000. Cambridge, Mass., Institute of Geographical Exploration, Harvard University, 1950.

Wanless, R. K. Isotopic age map of Canada, 1:5, 000, 000. Ottawa, Geological Survey of Canada, 1970. (Map 1256 A.)

Winearls, Joan, and Tessier, Yves. Directory of Canadian map collections. Ottawa, Association of Canadian Map Libraries, 1969. 72 p.

ALBERTA

Alberta. University. Department of Geography. Atlas of Alberta. Edmonton, University of Alberta Press, 1969. 158 p. Includes bedrock geology, glacial, groundwater, and mineral resource maps, scales not given.

Canada. Geological Survey. Geological map of Alberta, 1:1, 267, 200. Ottawa, 1951. (Map 1002 A.)

----------. Oil and gas pools of western Canada: northern Alberta and northeastern British Columbia, 1:760, 320. Ottawa, 1970. (Map 1317 A.)

----------. Oil and gas pools of western Canada: southern Alberta, 1:760, 320. Ottawa, 1970. (Map 1316 A.)

BRITISH COLUMBIA

Canada. Geological Survey. Geological map of British Columbia, 1:1, 267, 200. 2nd ed. Ottawa, 1962. (Map 932 A.)

----------. Mineral map of British Columbia, 1:1, 267, 200. N.p., 1951. (Map 1008 A.) 2 sheets.

Chapman, J. D., and others. British Columbia atlas of resources. N.p., British Columbia Natural Resources Conference, 1956. 48 p. Includes bedrock geology, glacial geology, and mineral resources maps, 1:3, 500, 000.

MANITOBA

Davies, J. F. Metallogenic map of Manitoba, 1:1, 267, 200. N.p., Manitoba Department of Mines and Natural Resources, 1966. (Map 66-2.)

Manitoba. Department of Mines and Natural Resources. Mines Branch. Geological index map of Manitoba, 1:1, 267, 200. Winnipeg, 1960. (Map 59-5.)

----------. Geological Division. Geological map of Manitoba, 1:1, 267, 200. Winnipeg, 1964. (Map 65-1.)

Weir, Thomas Robert. Economic atlas of Manitoba. N.p., Department of Industry and Commerce, Province of Manitoba, 1960. 81 p. Includes bedrock geology, glacial geology, and mineral resources maps, 1:2, 500, 000.

NEW BRUNSWICK

Potter, R. R., and others. Geological map, New Brunswick... 1:500, 000. N.p., New Brunswick Department of Natural Resources, 1968. (Map no. NR-1.)

NEWFOUNDLAND

Canada. Geological Survey. Island of Newfoundland, 1:1, 000, 000. Ottawa, 1967. (Map 1231 A; Geological map of Canada, 1:1, 000, 000, sheet 1/2.)

Mullins, W. J. Map index, geological, geophysical and related maps, to December 31, 1969. N.p., 1969. 59 p. (Newfoundland. Dept. of Mines, Agriculture and Resources. Mineral Resources Division. Information circular no. 13.)

NOVA SCOTIA

Buehren, Ernest A., comp. Industrial mineral map of the Province of Nova Scotia, 1:506, 880. Halifax, Department of Mines, Nova Scotia, 1966. With text.

----------. Metallic mineral map of the Province of Nova Scotia, 1:506, 880. Halifax, Department of Mines, Nova Scotia, 1966. With text.

Nova Scotia. Department of Mines. Geological map of the Province of Nova Scotia, 1:506,880. Halifax, 1965.

NORTHWEST TERRITORIES

Canada. Geological Survey. Geological map of the District of Mackenzie, Northwest Territories, 1:1,267,200. N.p., 1958. (Map 1055 A.) 2 sheets.

ONTARIO

Ayers, L. D., and others. Ontario geological map, 1:1,013,760. Toronto, Ontario Department of Mines and Northern Affairs, 1971. (Sheets 2197, 2198, 2199, 2200, 2201.) 5 sheets.

Dean, W. G., ed. Economic atlas of Ontario; Atlas économique de l'Ontario. N.p., University of Toronto Press, 1969. 113 pl. Includes metallic mineral production maps, 1:4,400,000.

Ontario. Department of Mines. Index of preliminary maps ...Maps in print...Out of print and miscellaneous maps, 1:1,013,760. Toronto, 1965. (Maps 2079 through 2093.)

----------. Ontario mineral map, 1:1,584,000. N.p., 1968. (Map 2148.)

Ontario. Department of Mines and Northern Affairs. Ontario, geology and principal mineral collecting localities, ca. 1:4,200,000. N.p., 1971. (Map 2211.)

QUEBEC

Bergeron, R. Geology north of 50th parallel in the Province of Quebec, 1:2,027,520. N.p., Province of Quebec, Department of Mines, 1956. (Map no. 1060; accompanied by its Geological report 75.)

Quebec. Department of Mines. Geological map, Province of Quebec, 1:3,168,000. N.p., 1957. (Map no. 1129.)

----------. Mineral map, Province of Quebec, 1:3,168,000. N.p., 1957. (Map no. 1130.)

Quebec. Direction General des Mines. Carte géologique du Quebec; Geological map of Quebec, 1:1,013,760...

Coordinator: André F. Laurin. N.p., 1969. 2 sheets.

Quebec. Ministère des Richesses Naturelles. Index des cartes géologiques publiees, 1:2, 000, 000. Quebec, 1966. (Map 1630; accompanied by: Liste des cartes publiees par les Services Géologiques. Quebec, 1966. 15 p.)

Sabourin, R. J. E. Carte glaciaire de Québec, 1:2, 027, 520. Quebec, 1957. (Université de Laval, Faculté des Sciences. Géologie et mineralogie, Contributions no. 128.)

SASKATCHEWAN

Canada. Geological Survey. Geological map of Saskatchewan, 1:1, 267, 200. Rev. N.p., 1954. (Map 895 A.)

----------. Mineral map of Saskatchewan, 1:1, 267, 200. N.p., 1947. (Map 896 A.)

----------. Oil and gas pools of western Canada; southern Saskatchewan and southwestern Manitoba, with Yukon Territory and part of Northwest Territories, 1:760, 320. Ottawa, 1970. (Map 1318 A.)

Richards, J. Howard, and Fung, K. I. Atlas of Saskatchewan. Saskatoon, University of Saskatchewan, 1969. 236 p. Includes bedrock geology, glacial geology, groundwater, and mineral resource maps.

YUKON

Canada. Geological Survey. Geological map of Yukon Territory, 1:1, 267, 200. N.p., 1957. (Map 1048 A.)

UNITED STATES

American Association of Petroleum Geologists. Geological highway Map Committee. Geological highway map, 1:1, 875, 000, no. 1-- . Tulsa, Okalhoma, 1966- . Eleven sheets will cover the U. S., of which four have been issued: 1, Mid-continent region; 2, Southern Rocky Mountain region; 3, Pacific southwest region; and 4, Mid-Atlantic region.

Bayley, Richard W., and Muehlberger, William R. Basement rock map of the United States, exclusive of Alaska and Hawaii, 1:2, 500, 000. Washington, U. S. Geological Survey, 1968. 2 sheets.

Belcher, Donald J., and others. Origin and distribution of United States soils, 1:4, 000, 000. Prep. cooperatively by the Technical Development Service, Civil Aeronautics Administration, and the Engineering Experiment Station, Purdue University. Ithaca, N. Y., Donald J. Belcher and Associates, Inc., 1946. (Similar map in: U. S. Civil Aeronautics Administration. Technical Development report no. 52, The origin, distribution, and airphoto identification of United States soils, by D. S. Jenkins and others. Washington, 1946. 2 v.)

Cox, Dennis P. Geologic map of Appalachian region, 1:2, 500, 000. Washington, 1967. (Plate 2 of U. S. Geological Survey. Professional paper 580.)

Fenneman, Nevin M., and Johnson, Douglas W. Physical divisions of the United States, 1:7, 000, 000. 1946. Washington, U. S. Geological Survey, 1957.

Feth, J. H., and others. Preliminary map of the conterminous United States showing depth to and quality of shallowest ground water... 1:3, 168, 000. Washington, 1966. (U. S. Geological Survey. Hydrogeologic investigations atlas, HA-199.) 2 sheets.

Flint, Richard Foster, and others. Glacial map of the

United States east of the Rocky Mountains, 1:1, 750, 000. N. Y., Geological Society of America, 1959. 2 sheets.

Hammond, Edwin H. Classes of land-surface form in the forty eight states, U. S. A., 1:5, 000, 000. Washington, 1964. (Association of American Geographers, Annals, vol. 54, no. 1, March 1964, Map supplement no. 4.)

Industrial Atlas Corporation. United States mines; properties in operation, 1:7, 500, 000. N. Y., 1967.

Lobeck, Armin Kohl. Geologic map of the United States, 1:5, 000, 000. N. Y., Geographical Press, Columbia University, 1941. Now distributed by C. S. Hammond & Co.

----------. Physiographic diagram of the United States, 1:9, 000, 000. Rev. Maplewood, N. J., Geographical Press, C. S. Hammond & Co., 1957. 2 sheets, text on verso.

Marbut, Curtis F. Soil map of the United States, 1:2, 500, 000. N. p., 1931. (Also issued as plate 5 of U. S. Department of Agriculture. Atlas of American agriculture, Pt. 3, Soils of the United States. Washington, G. P. O., 1935. 98 p.) 12 sheets.

McGuinness, C. L. Generalized map showing annual runoff and productive aquifers in the coterminous United States, 1:5, 000, 000. Washington, 1964. (U. S. Geological Survey. Hydrologic investigations atlas, HA-194; similar map is Plate 1 in Water supply paper 1800.)

Miller, David William and others. Water atlas of the United States; basic facts about the Nation's water resources. 2nd ed. Port Washington, N. Y., Water Information Center, 1963. 8 p., 40 pl.

National Research Council. Division of Geology and Geography. Committee for the Study of Eolian Deposits. Pleistocene eolian deposits of the United States, Alaska, and parts of Canada, 1:2, 500, 000. N. p., Geological Society of America, 1952. 2 sheets.

Raisz, Erwin. Landforms of the United States, 1:4, 500, 000. 6th ed. Cambridge, Mass., The Author, 1957.

Riggs, E. A. Major basins and structural features of United States, 1:5,000,000. Maplewood, N.J., Geographical Press, C. S. Hammond & Co., 1960.

Stephenson, Richard W., and Galneder, Mary. Anglo-American state and provincial thematic atlases: a survey and bibliography. (Canadian cartographer, vol. 6, no. 1, June 1969, p. 15-45.)

Stose, George Willis, and Ljungstedt, O. A. Geologic map of the United States, 1:2,500,000. 1932. Washington, U.S. Geological Survey, 1960. 4 sheets.

Trumbull, James Van Alen. Coal fields of the United States... 1:5,000,000. 1960. Washington, U.S. Geological Survey, 1967. 2 sheets.

U.S. Bureau of the Census. Geography Division. Mineral industries in the United States: 1963, 1:5,000,000. Washington, G.P.O., 1968. (United States maps, GE-50, no. 23.)

U.S. Geological Survey. Geologic map index of...(state). Washington, 194 - . A series (1 for each state) of "geologic map indexes outlining in color the areas in (each) state for which geologic maps have been published..."

----------. The national atlas of the United States of America, Arch C. Gerlach, editor. Washington, 1970. 417 p. Includes landform, earthquake, tectonic, geologic, soils, water, and mineral maps, 1:7,500,000; gazetteer. Includes Alaska.

----------. Paleotectonic maps...system, edited by E. D. McKee. Washington, 1956- . Three folios have so far appeared: Miscellaneous geologic investigations map, I-175, Jurassic system; I-300, Triassic system; I-450, Permian system.

----------. Water resources of the Appalachian region, Pennsylvania to Alabama, 1:2,500,000, by William J. Schneider and others. Washington, 1965. 11 p. (Hydrologic investigations atlas, HA-198.)

----------. Geologic Inquiries Group. (List of) State geologic maps. Washington, 1969. 20 p. Revised occasionally.

----------, and American Association of Petroleum Geologists. *Tectonic map of the United States*, prep. by a Committee composed of George V. Cohee and others, 1:2,500,000. 1962. Washington, 1968. 2 sheets.

Vlissides, Sophie D., and Quirin, Barbara A. *Oil and gas fields of the United States, exclusive of Alaska and Hawaii*, 1:2,500,000. 1964. Washington, U.S. Geological Survey, 1967. 2 sheets.

ALABAMA

MacNeil, F. Stearns. *Geologic map of the Tertiary formations of Alabama*, 1:500,000. 1946, rev. 1955. Washington, 1955. (U.S. Geological Survey. Oil and gas investigations, Preliminary map 45.)

Powell, W. J., and Duncan, A. C. *Water-level fluctuations in observation wells and radiochemical and chemical quality of ground water in Alabama*, 1:500,000. N.p., 1964. (Alabama. Geological Survey. Map 29.)

ALASKA

Cobb, Edward H., and Kennedy, Della L. *Geologic map index of Alaska*, 1:2,500,000. 5th ed. Washington, U.S. Geological Survey, 1967.

Dutro, J. Thomas, and Payne, T. G. *Geologic map of Alaska*, 1:2,500,000. Washington, U.S. Geological Survey, 1957.

Ferrians, Oscar J., Jr. *Permafrost map of Alaska*, 1:2,500,000. Washington, 1970. (U.S. Geological Survey. Miscellaneous geological investigations map, I-445.)

Johnson, Philip R. *Environmental atlas of Alaska*. College, Alaska, Institute of Arctic Environmental Engineering, Institute of Water Resources, University of Alaska, 1969. 111 p. Include general maps covering physiography, geology, tectonics, earthquakes, permafrost, glaciation; emphasis on light, water, climate.

Karlstrom, Thor N. V., and others. Surficial geology of Alaska, 1:1,584,000. Washington, 1964. (U.S. Geological Survey. Miscellaneous geologic investigations map, I-357.)

Payne, Thomas Gibson. Mesozoic and Cenozoic tectonic elements of Alaska, 1:5,000,000. Washington, 1955. (U.S. Geological Survey. Miscellaneous geologic investigations, Map I-84.)

----------, and Dutro, J. Thomas, Jr. Mesozoic and Cenozoic tectonic elements of Alaska, 1:2,500,000. Washington, 1958. (Plate 2 of U.S. Geological Survey. Bull. 1094.)

Raisz, Erwin. Landform map of Alaska, 1:2,500,000, for Office of the Quartermaster General. N.p., 1948. (A.M.S. 6-48.)

----------. Landform map of Alaska, 1:3,100,000, originally prep. for the Environmental Protection Section, Office of the Quartermaster General. N.p., Institute of Geographical Exploration, Harvard University, 1948.

U.S. Geological Survey. Alaska Glacial Map Committee. Map showing extent of glaciations in Alaska, 1:2,500,000, by H. W. Coulter and others. Washington, 1965. (Miscellaneous geologic investigations map, I-415.)

----------. Military Geology Branch. Terrain study of Alaska, Part 1- Washington, 1961- . (U.S. Army. Corps of Engineers. Engineer intelligence study, EIS 301, Part 1- . Several sheets have been issued, including 2, Physiographic regions; 3, Soils; 7, Soil moisture; 8, Permafrost, all 1:2,500,000.)

Wahrhaftig, Clyde. Maps of physiographic divisions of Alaska, 1:2,500,000. Washington, 1965. (Plate 1 of: U.S. Geological Survey. Professional paper 482.)

ARIZONA

Cooley, M. E. Arizona highway geologic map, 1:1,000,000. N.p., Arizona Geological Society, 1967.

Forrester, J. D., and Moore, R. T. Maps, with special reference to the geologic and mineral maps of Arizona.

Tucson, 1962. 11 p. (Arizona. Bureau of Mines. Bull. 171.)

Wilson, Eldred D., and others. Geologic map of Arizona, 1:500,000. Washington, U.S. Geological Survey, 1969.

ARKANSAS

Arkansas. Resources and Development Commission, Division of Geology. Arkansas mineral resources, ca. 1:1,000,000. Little Rock, 1952.

Hess, Albert. Geology of Arkansas, surface and below surface, 1:1,500,000. Rev. N.p., Arkansas Geological and Conservation Commission, 1959. Geological block diagram, with text.

Miser, Hugh D., and Stose, George W. Geologic map of Arkansas, 1:500,000. 1929. N.p., Arkansas Geological Survey, 1949.

CALIFORNIA

California. Division of Mines and Geology. Geologic atlas of California, 1:250,000. N.p., 1958-71. 28 sheets, each with Explanatory data sheet.

Proctor, R. J. Index map of selected major geologic and engineering features in southern California, n.p., Association of Engineering Geologists, 1966. (Plate in: Engineering geology in southern California, ed. by Richard Lung and Richard Proctor. Glendale, Calif., Association of Engineering Geologists, Los Angeles Section, 1966. 389 p.) Includes faults.

Smith, Merritt B. Map showing distribution and configuration of basement rocks in California... Washington, 1964. (U.S. Geological Survey. Oil and gas investigations map, OM-215.)

----------, and others. Reported occurrences of selected minerals in...California, 1:500,000. Washington, 1967- . (U.S. Geological Survey. Mineral investigations resource maps, MR-47, Northern third, 1967; MR-48, Central third, 1971.) With text. Sheet MR-49, Southern third, not yet issued.

Strand, Rudolph G., and others. Index to geologic maps of California, to December 31, 1956. San Francisco, 1958. 128 p. (California. Division of Mines. Special report 52.) Supplemented by Special report 52A, 1957-1960, and 52B, 1961-1964.

U.S. Geological Survey. Geologic map of California, 1:2,500,000. 1966. Washington, 1970. (Miscellaneous geologic investigations map, I-512.)

COLORADO

Burbank, W. S. Geologic map of Colorado, 1:500,000. 1935. U.S. Geological Survey, in cooperation with Colorado State Geological Survey Board and Colorado Metal Mining Fund, 1967.

Finley, E. A., and others. Preliminary structure contour map of the Colorado plains, 1:500,000. Washington, 1955. (U.S. Geological Survey. Oil and gas investigations map, OM 176.)

Oriel, Steven S. Tectonic map of Colorado, 1:1,000,000. N.p., 1954. (Plate 1 of: The oil and gas fields of Colorado, a symposium edited by Fred S. Jensen and others. Denver, Rocky Mountain Association of Geologists, 1954. 302 p.)

U.S. Geological Survey. Conservation Division. Branch of Mineral Classification. Reported occurrences of selected minerals in Colorado; includes most metals and nonmetals; does not include mineral fuels, sodium and potassium salts, or phosphate, 1:500,000. Washington, 1971. (Mineral investigation resource map, MR-57.)

CONNECTICUT

Rodgers, John, and others. Preliminary geological map of Connecticut, 1:253,440. N.p., 1956. (Accompanied by: Connecticut. Geological and Natural History Survey. Bull. 84, Explanatory text for preliminary geological map of Connecticut, by John Rodgers and others. Storrs, 1959. 64 p.)

DELAWARE

Spoljaric, Nenad, and Jordan, Robert R. Generalized geologic map of Delaware, 1:300, 000. N.p., Delaware Geological Survey, 1966.

FLORIDA

Cooke, Charles Wythe. Geologic map of Florida, 1:1, 000, 000. N.p., 1945. (Plate 1 of: Florida. Geological Survey. Bull. no. 29.)

Raisz, Erwin, and Dunkle, John R. Atlas of Florida. Gainesville, University of Florida Press, 1964. 52 p.

Vernon, R. O., and Puri, H. S. Geologic map of Florida, 1:2, 000, 000. N.p., 1965. (Florida. Division of Geology. Map series, no. 18.)

GEORGIA

Furcron, A. S., and others. Mineral resource map, 1:500, 000. N.p., Georgia Department of Mines, Mining, and Geology, 1969.

----------. Mineral resource map of Georgia, ca. 1:2, 200, 000. N.p., Georgia Department of Mines, Mining, and Geology, 1969.

Georgia. Division of Mines, Mining, and Geology. Geologic map of Georgia, 1:500, 000. N.p., 1939.

MacNeil, Francis Stearns. Geologic map of the Tertiary and Quaternary formations of Georgia, 1:500, 000. Washington, 1947. (U.S. Geological Survey. Oil and gas investigations, Preliminary map no. 72.)

Smith, James W., and Green, Martha A. Geologic map of Georgia, ca. 1:2, 200, 000. N.p., Georgia Department of Mines, Mining, and Geology, 1968.

IDAHO

Hubbard, Charles R. Mineral resources of the State of Idaho, 1:1, 000, 000. N.p., Idaho Bureau of Mines and Geology, 1955. (Accompanied by its Bull. no. 15 and Pamphlet no. 105.)

Ross, Clyde P., and Forrester, J. D. Geologic map of the State of Idaho, 1:500, 000. 1947. Washington, U.S.

Geological Survey, 1959.

ILLINOIS

Bier, James A. Landforms of Illinois, 1:1,000,000. Urbana, Illinois State Geological Survey, 1956.

Illinois. Geological Survey. Geologic map of Illinois, showing bedrock below the glacial drift, 1:1,000,000. N.p., 1961. (Plate 1 of its Educational series, 7.)

----------. Mineral industries of Illinois, 1:500,000. Urbana, 1961.

Illinois. University. Dept. of Geography. Atlas of Illinois resources. N.p., 1958-1962. 5 vols. 1, Water resources and climate; 2, Mineral resources.

Meents, Wayne F. Oil and gas industry in Illinois, 1:500,000. Urbana, Illinois State Geological Survey, 1968.

Thornburn, Thomas H. Map of surface deposits of Illinois, 1:1,000,000. N.p., University of Illinois Engineering Experiment Station, 1960. (Plate in its Circular no. 80.)

Willman, H. B., and Frye, John C. Glacial map of Illinois, 1:2,500,000. N.p., 1970. (Plate 2 of Illinois. Geological Survey. Bull. 94.)

----------, and others. Geologic map of Illinois, 1:500,000. Urbana, Illinois State Geological Survey, 1967.

INDIANA

Indiana. Geological Survey. Geological map of Indiana, 1:253,440. N.p., 1932.

----------. Map of Indiana showing bedrock geology, 1:1,800,000. N.p., 1970. (Its Miscellaneous map no. 16.)

----------. Regional geologic map, 1:253,440. N.p., 1961- . Two of 8 sheets have been issued: each sheet has text.

Patton, John B. Geologic map of Indiana, 1:1, 000, 000. N. p., 1956. (Indiana. Geological Survey. Atlas of mineral resources of Indiana, Map no. 9.)

Wayne, William J. Glacial geology of Indiana, 1:1, 000, 000. N.p., 1958. (Indiana. Geological Survey. Atlas of mineral resources of Indiana, Map no. 10; 1960 revision appears as Plate 1 of its Report of progress, no. 17.)

IOWA

Dorheim, Fred H. Mineral resources of Iowa, 1:500, 000. Iowa City, Iowa Geological Survey, 1970.

Iowa. Geological Survey. Geologic map of Iowa, 1:500, 000. N.p., 1969.

KANSAS

Bayne, Charles K., and Ward, John R. General availability of ground water in Kansas, 1:500, 000. N.p., State Geological Survey of Kansas, 1967. (Map M-4.)

----------. Saturated thickness and specific yield of Cenozoic deposits in Kansas, 1:500, 000. N.p., 1969. (Kansas. Geological Survey. Map M-5.)

Cole, Virgil B. Configuration on top of Precambrian basement rocks in Kansas, 1:625, 000. N.p., 1962. (Kansas. Geological Survey. Oil and gas investigations, no. 26.) Nos. 22, 23, and 25 show contours on tops of other formations.

Kansas. Geological Survey. Geologic map of Kansas, 1:500, 000. N.p., 1964. (Map M-1.)

----------. Kansas mineral resources, 1:600, 000. N.p., 1951.

Oros, Margaret O. Map of oil and gas pipelines and industries in Kansas, 1:500, 000. Lawrence, State Geological Survey of Kansas, 1963. (Map M-2.)

KENTUCKY

Hopkins, H. T. Fresh-saline water interface map of Kentucky, 1:500, 000. N.p., Kentucky Geological Survey, 1966.

Jillson, Willard Rouse. Geologic map of Kentucky, 1:1, 000, 000. Lexington, Kentucky Geological Survey, 1954.

Johnson, J. H., and others. Mineral resources and mineral industries of Kentucky, 1:500, 000. Frankfort, Kentucky Dept. of Commerce, Maps and Minerals Division, 1962.

LOUISIANA

Coignet, C. O. Oil and gas map of Louisiana, 1:500, 000. Baton Rouge, Louisiana Geological Survey, 1964.

Louisiana. Geological Survey. General availability of ground water in Louisiana, ca. 1:1, 275, 000. N.p., 1960? (Plate 4 in its Water resources bull. no. 4.)

----------. Generalized geological map of Lousiana, ca. 1:1, 250, 000. N.p., 1959.

Winslow, A. G., and others. Saline ground water in Louisiana, 1:750, 000. Washington, 1968. (U.S. Geological Survey. Hydrologic investigations atlas, HA-310.) 4 sheets.

MAINE

Hussey, Arthur M., II, and others. Preliminary geologic map of Maine, 1:500, 000. N.p., Maine Geological Survey, 1967.

Perkins, Edward Henry. Glacial deposits, State of Maine, ca. 1:450, 000. Chester, Vt., National Survey Co., 1934. (Supplement to Maine. Technology Experiment Station. Bulletin 30, vol. 2.)

MARYLAND

Cleaves, Emery T., and others. Geologic map of Maryland, 1:250, 000. N.p., Maryland Geological Survey, 1968.

Pearre, Nancy G. Mineral deposits of Maryland, excluding fuels, sand and gravel. Washington, 1961. (U.S. Geological Survey. Mineral investigations resource map, MR 12.)

MASSACHUSETTS

Pearre, Nancy C. Mineral deposits and occurrences in Massachusetts and Rhode Island, 1:500,000. Washington, 1956. (U.S. Geological Survey. Mineral investigations resource map, MR-4.)

MICHIGAN

Kirkby, Edward A. Geologic map index of Michigan, 1843-1962. Lansing, 1970. 56 p. (Michigan. Geological Survey. Circular 8.)

Martin, Helen M. The Centennial geological map of the Northern Peninsula of Michigan, 1:500,000. N.p., 1936. (Michigan. Geological Survey. Publ. 39, Geological series 33.)

----------. The Centennial geological map of the Southern Peninsula of Michigan, 1:500,000. N.p., 1936. (Michigan. Geological Survey. Publ. 39, Geological series 33.)

----------. Map of the surface formations of the Northern Peninsula of Michigan, 1:500,000. N.p., 1957. (Michigan. Geological Survey. Publ. 49.)

----------. Map of the surface formations of the Southern Peninsula of Michigan, 1:500,000. N.p., 1957. (Michigan. Geological Survey. Publ. 49.)

Twenter, Floyd R. General availability and quality of groundwater in the bedrock deposits in Michigan, 1:1,100,000. N.p., State Resources Planning Division, Michigan Dept. of Commerce, and Water Resources Commission, Michigan Dept. of Conservation, n.d.

MINNESOTA

Minnesota. Geological Survey. Geologic map of Minnesota, 1:250,000. N.p., 1966- . Three of eleven sheets have been published thus far.

Schwartz, George Melvin, and Prokopovich, N. Map of mineral resources of Minnesota, 1:1,000,000. Rev. ed. N.p., Minnesota Geological Survey, 1966.

Sims, P. K. Geologic map of Minnesota, bedrock geology, 1:1,000,000. N.p., Minnesota Geological Survey, 1970. (Miscellaneous map series, Map M-14.)

MISSISSIPPI

Bicker, Alvin R., Jr. Geologic map of Mississippi, 1:500,000. N.p., Mississippi Geological Survey, 1969.

Mississippi. State Oil and Gas Board. Oil and gas map of Mississippi, ca. 1:950,000. Rev. by R. S. Myers, Jr. N.p., 1968. (Plate 2 of Mississippi. Geological, Economic and Topographical Survey. Bull. 112.)

Vestal, Franklin Earl. Mississippi...mineral resources, 1:1,000,000. N.p., Mississippi State Geological Survey, 1948.

MISSOURI

Fuller, Dale L. Groundwater quality map of deep aquifers in Missouri, 1:750,000. N.p., Division of Geological Survey and Water Resources, 1962.

Kisvarsanyi, Eva B., and Searight, Walter V. Mineral resources and industry map of Missouri, 1:500,000. N.p., Division of Geological Survey and Water Resources, 1965.

Knight, Robert D. Groundwater areas of Missouri, 1:750,000. N.p., Division of Geological Survey and Water Resources, 1962.

McCracken, Earl, and McCracken, Mary. Subsurface maps of the Lower Ordovician (Canadian series) of Missouri, 1:1,000,000. Rolla, Division of Geological Survey and Water Resources, 1965. 6 sheets.

McCracken, Mary H. Geologic map of Missouri, 1:500,000. N.p., Division of Geological Survey and Water Resources. 1961.

Robertson, Charles E. Water well yield map of Missouri, 1:750,000. N.p., Division of Geological Survey and Water Resources, 1962.

Stout, Larry N. Index to Missouri areal geologic maps, 1890-1969. Rolla, 1969. 67 p. (Missouri. Geological Survey and Water Resources. Information Circular 22.)

MONTANA

Bentley, Craig B., and Mowat, George D. Reported occurrences of selected minerals in Montana. 1:500,000. Washington, 1967. (U.S. Geological Survey. Mineral investigations resource map, MR-50.)

Colton, Roger B., and others. Glacial map of Montana east of the Rocky Mountains, 1:500,000. Washington, 1961. (U.S. Geological Survey. Miscellaneous geologic investigations map, I-327.)

Dobbin, Carroll Edward, and Erdmann, C. E. Structure contour map of the Montana plains, 1:500,000. Rev. ed. Washington, 1955. (U.S. Geological Survey. Oil and gas investigations, Map OM 178 A.)

Ross, Clyde P., and others. Geologic map of Montana, 1:500,000. 1955. Washington, U.S. Geological Survey, 1958.

Vine, J. D., and others. Map of Montana showing oil and gas fields and test wells for oil and gas, 1:500,000. Washington, 1956. (U.S. Geological Survey. Oil and gas investigations, Map OM 170.)

NEBRASKA

Burchett, Raymond R. Geologic bedrock map of Nebraska, 1:1,000,000. Lincoln, Nebraska Geological Survey, 1969.

Carlson, Marvin P. Contour map of Precambrian surface in Nebraska, 1:1,000,000. N.p., Nebraska Geological Survey, 1961.

Nebraska. University. Conservation and Survey Division. Ground water atlas of Nebraska. N.p., 1969. 7 pl.

Reed, Eugene Clifton, and others. Map of Nebraska showing areal distribution of pre-Pennsylvanian rocks, anticlines and basins, oil and gas fields, pipelines, and unsuccessful test wells, 1:500,000. Washington, 1958. (U.S. Geological Survey. Oil and gas investigations map OM 198.)

Schreurs, Raymond L. Configuration of the water table in Nebraska, 1:1, 250, 000. Washington, 1954. (U.S. Geological Survey. Hydrologic investigations atlas, HA 4.)

NEVADA

Horton, Robert C. Industrial minerals of Nevada, 1:1, 000, 000. Reno, 1964. (Nevada. Bureau of Mines. Map 27.)

Nevada. Bureau of Mines. Geologic source index map, showing sources used in compiling Nevada Bureau of Mines Map 16, Progress geologic map of Nevada, 1:500, 000. N.p., 1962? (Map 16A.)

Schilling, John H. Metal mining districts of Nevada, 1:1, 000, 000. 2nd ed. Reno, 1969. (Nevada. Bureau of Mines. Map 37.)

Webb, Barbara, and Wilson, Roland V. Progress geologic map of Nevada, 1:500, 000. Reno, 1962. (Nevada. Bureau of Mines. Map 16.)

NEW HAMPSHIRE

Billings, Marland Pratt. Geologic map of New Hampshire, 1:250, 000. Washington, U.S. Geological Survey, 1955. (Accompanied by: The geology of New Hampshire, Part 2, Bedrock geology, by Marland P. Billings, Concord, New Hampshire State Planning and Development Commission, 1956. 203 p.)

Goldthwait, James Walter, and others. Surficial geology of New Hampshire, 1:250, 000. N.p., New Hampshire State Planning and Development Commission, 1950. (Accompanied by: The geology of New Hampshire, Part 1, Surficial geology, by James Walter Goldthwait and others, Concord, New Hampshire State Planning and Development Commission, 1963. 84 p.)

Pearre, Nancy C., and Calkins, James A. Mineral deposits and occurrences in New Hampshire, exclusive of clay, sand and gravel, and peat. Washington, 1957. (U.S. Geological Survey. Mineral investigations resource map, MR 6.)

NEW JERSEY

Lewis, Joseph Volney, and others. Geologic map of New Jersey, 1:250,000. Rev. N.p., 1950. (New Jersey. Dept. of Conservation and Economic Development. Atlas sheet no. 40.)

New Jersey. Bureau of Geology and Topography. Geologic map of New Jersey, 1:1,000,000. N.p., 1959. (Also appears in: Geological Society of New Jersey. Reports no. 1 and 2.)

NEW MEXICO

Dane, Carle H., and Bachman, George O. Geologic map of New Mexico, 1:500,000. Washington, U.S. Geological Survey, 1971. 2 sheets.

Foster, R. W., and Stipp, T. F. Preliminary geologic and relief map of the Precambrian of New Mexico, 1:600,000. N.p., 1961. (Plate in: New Mexico. Bureau of Mines and Mineral Resources. Circular 57.)

Haigler, Leon B., and Sutherland, Helen L. Reported occurrences of selected minerals in New Mexico, 1:500,000. Washington, 1965. (U.S. Geological Survey. Mineral investigations resource map, MR-45.)

New Mexico Geological Society. Geologic highway map of New Mexico, ca. 1:1,150,000. N.p., 1961.

Vlissides, Sophie D., and Bieberman, Robert A. Map of New Mexico, showing oil and gas fields, unsuccessful test wells, Precambrian rocks, and pipelines. 1:500,000. 1961. Washington, 1969. (U.S. Geological Survey. Oil and gas investigations map, OM-207.)

NEW YORK

Bier, James A. Landforms and bedrock geology of New York State, 1:1,000,000. N.p., New York State Geological Survey, 1966. (Accompanied by: New York State Museum and Science Service. Educational leaflet 20.)

Heath, Ralph C. Ground water in New York, 1:1,000,000. Albany, 1964. (New York (State). Water Resources

Commission. Bull. GW-51.)

Luedke, Elaine M., and others. Map of New York State showing mineral occurrences, 1:500,000. Washington, 1959? (Plate 9 of: U.S. Geological Survey. Bull. 1072-F, Mineral occurrences of New York State, by Elaine M. Luedke and others. Washington, G.P.O., 1959.)

New York (State). Geological Survey. Geologic map of New York, 1:250,000. Albany, 1962. 10 sheets. (Map and chart series, no. 5.)

NORTH CAROLINA

North Carolina. Dept. of Conservation and Development. Division of Mineral Resources. Geologic map of North Carolina, 1:500,000. N.p., 1958. (Accompanies its Bull. no. 71.)

NORTH DAKOTA

Anderson, S. B., and Mendoza, H. A. Contour map of the Pre-Mesozoic surface in North Dakota, ca. 1:950,000. N.p., 1960. (North Dakota. Geological Survey. Miscellaneous map no. 6.)

Carlson, Clarence G. Bedrock geologic map of North Dakota, 1:1,000,000. N.p., 1969. (North Dakota. Geological Survey. Miscellaneous map no. 10.)

Colton, Roger B., and others. Preliminary glacial map of North Dakota, 1:500,000. Washington, 1963. (U.S. Geological Survey. Miscellaneous geologic investigations map, I-331.)

OHIO

Bier, James A. Ohio landform map, 1:1,250,000. N.p., Ohio Division of Geological Survey, 1956.

Bownocker, J. A. Geologic map of Ohio, 1:500,000. N.p., Ohio Division of Geological Survey, 1965.

Calvert, Warren L. Oil and gas fields of Ohio, 1:380,160. Rev. ed. N.p., Ohio Division of Geological Survey, 1964.

Goldthwait, Richard P., and others. Glacial map of Ohio, 1:500,000. Rev. ed. Washington, 1967. (U.S. Geological Survey. Miscellaneous geologic investigations map I-316.)

Sedam, A. C., and Stein, R. B. Saline ground-water resources of Ohio, 1:500,000. Washington, 1970. (U.S. Geological Survey. Hydrologic investigations atlas, HA-366.) 2 sheets.

Smith, Robert W. Mineral industry map of Ohio, 1:500,000. N.p., Ohio Division of Geological Survey, 1956.

OKLAHOMA

Arbenz, J. Kaspar, Tectonic map of Oklahoma, showing surface structural features, 1:750,000. N.p., 1956. (Oklahoma. Geological Survey. Map GM-3.) Maps GM-5, GM-9, and GM-14 also indicate structure.

Branson, Carl C., and Jordan, Louise. Index to geologic mapping in Oklahoma, 1:1,000,000. N.p., Oklahoma Geological Survey, 1961. 5 sheets. Supplement 1, 1964; Supplement 2, 1967.

Johnson, Kenneth S. Mineral map of Oklahoma (exclusive of oil and gas fields). N.p., 1969. (Oklahoma. Geological Survey. Map GM-15.)

Jordan, Louise, and others. Oil and gas fields of Oklahoma. 1:750,000. N.p., 1966. (Oklahoma. Geological Survey. Map GM-10.)

Miser, Hugh D. Geologic map of Oklahoma, 1:500,000. 1954. Washington, U.S. Geological Survey, 1967.

Schoff, Stuart L. Map showing ground-water reservoirs in Oklahoma, 1:720,000. N.p., 1965. (Oklahoma. Geological Survey. Map 72-2.)

OREGON

Corcoran, Raymond E. Index to published geologic mapping in Oregon, 1898-1967. Portland, 1968. 20 p. (Oregon. Department of Geology and Mineral Industries. Miscellaneous paper 12.)

Oregon. Dept. of Geology and Mineral Industries. State of Oregon, map showing principal mineral deposits, 1:1, 000, 000. Portland, 1958.

Raisz, Erwin. Landform map of Oregon, ca. 1:1, 200, 000. Portland, Oregon State Dept. of Geology and Mineral Industries, 1955.

Walker, George W., and King, Philip B. Geologic map of Oregon, 1:2, 000, 000. Washington, 1969. (U.S. Geological Survey. Miscellaneous geologic investigations, Map I-595.)

Wells, Francis G., and Peck, Dallas L. Geologic map of Oregon west of the 121st meridian, 1:500, 000. Washington, 1967. (U.S. Geological Survey. Miscellaneous geologic investigations, Map I-325.)

PENNSYLVANIA

Lytle, William S., and Fairall, Virginia. Oil and gas fields of Pennsylvania, 1:316, 800. N.p., Pennsylvania Geological Survey, 1964.

Pennsylvania. Geological Survey. Atlas of Pennsylvania's mineral resources, Pt. 1- . Harrisburg, 1964- . Pt. 1, Limestones and dolomites; Pt. 3, Metal mines and occurrences.

Pennsylvania. Topographic and Geologic Survey. Geologic map of Pennsylvania, 1:250, 000. N.p., 1960. 2 sheets.

RHODE ISLAND

Quinn, A. W. Bedrock geologic map of Rhode Island, 1:125, 000. Washington, 1971. (Plate 1 of: U.S. Geological Survey. Bull. 1295.)

SOUTH CAROLINA

Cooke, C. Wythe. Cretaceous and Tertiary formations of South Carolina, 1:500, 000. N.p., 1936. (Plate 2 of: U.S. Geological Survey. Bull. 867, Geology of the coastal plain of South Carolina, by C. Wythe Cooke. Washington, G.P.O., 1936. 196 p.)

----------. Quaternary formations and terraces of South Carolina, 1:500,000. N.p., 1936. (Plate 1 of: U.S. Geological Survey. Bull. 867, Geology of the coastal plain of South Carolina, by C. Wythe Cook. Washington, G.P.O., 1936. 196 p.).

Overstreet, William C., and Bell, Henry, III. Geologic map of the crystalline rocks of South Carolina, 1:250,000. Washington, 1965. (U.S. Geological Survey. Miscellaneous geologic investigations, Map I-413.)

Smith, L. L. Economic mineral localities in South Carolina, ca. 1:780,000. N.p., 1961. (South Carolina. State development Board. Division of Geology. Map series, MS-5.)

SOUTH DAKOTA

Agnew, Allen F. Oil and gas tests in South Dakota, 1:1,000,000. N.p., 1959. (South Dakota, Geological Survey. Oil and gas investigations, Map 3.)

Darton, N. H. Geologic map of South Dakota, 1:500,000. 1951. Washington, U.S. Geological Survey, 1967.

Steece, Fred V. Preliminary map of the Precambrian surface of South Dakota, ca. 1:1,100,000. N.p., 1961. (South Dakota. Geological Survey. Mineral resources investigations, Map no. 2.)

Tipton, Merlin J. Bibliography of reports containing maps on South Dakota geology, published before January 1, 1959. Vermillion, 1966. 71 p. (South Dakota. Geological Survey. Circular no. 33.)

TENNESSEE

Hardeman, William D., and Miller, Robert A. Mineral resources and mineral industries of Tennessee, 1:500,000. N.p., State of Tennessee, Dept. of Conservation and Commerce, Division of Geology, 1959.

Hershey, Robert E., and Maher, Stuart W. Quarry map of Tennessee, 1:500,000. N.p., State of Tennessee, Dept. of Conservation and Commerce, Division of Geology, 1962.

Miller, Robert A., and others. Geologic map of Tennessee, 1:250,000. N.p., State of Tennessee, Dept. of Conservation, Division of Geology, 1966. 4 sheets.

Tennessee. Division of Geology. Status of detailed geologic mapping in Tennessee. Nashville, 1970.

TEXAS

Arbingast, Stanley Alan, and others. Atlas of Texas. Austin, Bureau of Business Research, University of Texas, 1967. 126 p. Includes generalized geologic, soil, vegetation, water, and mineral maps, 1:5,000,000 and 1:7,500,000.

Brown, Thomas E. Index to areal geologic maps in Texas, 1891-1961. Austin, University of Texas, Bureau of Economic Geology, 1963. Includes index diagram, 1:1,200,000.

Darton, N. H., and others. Geologic map of Texas, 1:500,000. Washington, U.S. Geological Survey, 1937. 4 sheets.

Flawn, P. T. Basement rocks of Texas and southeast New Mexico, 1:1,250,000. N.p., 1956? (Plate 1 of: Texas. University. Publ. no. 5605.)

Newton, L. W. Texas, oil, minerals, and manufactures, 1:774,080. Chicago, Denoyer-Geppert, 1952. (Texas series, T13.)

Oetking, Philip F., and Feray, Dan E. Geological highway map of Texas, ca. 1:1,800,000. Dallas, Dallas Geological Society, 1963.

Sellards, Elias Howard, and Hendricks, L. Structural map of Texas, 1:500,000. 3rd ed., 1946. Austin, University of Texas, Bureau of Economic Geology, 1951. 4 sheets.

Texas. Bureau of Economic Geology. Geologic atlas of Texas, 1:250,000. N.p., 1964- . 14 out of 44 sheets completed thus far; some have texts.

Vlissides, Sophie D. Map of Texas showing oil and gas fields, pipelines, and areas of exposed basement rocks,

1:1, 000, 000. Washington, 1964. (U.S. Geological Survey. Oil and gas investigations map OM-214.)

UTAH

Hintze, Lehi F., and Boardman, Leona. Geologic map index of Utah, 1:1, 000, 000. N.p., Utah Geological and Mineralogical Survey? 1964. 2 sheets.

Mardirosian, Charles A. Mining districts and mineral deposits of Utah, 1:750, 000. Salt Lake City, Utah, The Author, 1966.

Ridd, Merrill K. Landforms of Utah in proportional relief, 1:650, 000. N.p., 1963. (Association of American Geographers, Annals, vol. 53, no. 4, Dec. 1963, Map supplement no. 3.)

Utah. University. College of Mines and Mineral Industries. Geologic map of... Utah, 1:250, 000. N.p., 1963-1964. 4 sheets.

VERMONT

Doll, Charles G. and others. Centennial geologic map of Vermont, 1:250, 000. N.p., Vermont Geological Survey, 1961.

Pearre, Nancy C., and Calkins, J. A. Mineral deposits and occurrences in Vermont, exclusive of clay, sand and gravel, and peat, 1:500, 000. Washington, 1957. (U.S. Geological Survey. Mineral investigations resource map MR 5.)

Stewart, David P., and MacClintock, Paul. Surficial geology map of Vermont. N.p., Vermont Geological Survey, 1970.

VIRGINIA

Virginia. Division of Mineral Resources. Geologic map of Virginia, 1:500, 000. Charlottesville, 1963.

----------. Mineral industries and resources of Virginia, 1:500, 000. Charlottesville, 1963.

WASHINGTON

Huntting, Marshall T., and others. Geologic map of Washington, 1:500, 000. N.p., State of Washington, Division of Mines and Geology, 1961.

Weissenborn, A. E. Geologic map of Washington, 1:2, 000, 000. Washington, 1969. (U.S. Geological Survey. Miscellaneous geologic investigations map I-583.)

WEST VIRGINIA

Cardwell, Dudley H., and others. Geologic map of West Virginia, 1:250, 000. N.p., West Virginia Geological and Economic Survey, 1968. 2 sheets.

----------. Oil and gas fields of West Virginia, 1:250, 000. N.p., West Virginia Geological Survey, 1970. (Accompanied by: Oil and gas fields in West Virginia, by Dudley H. Cardwell. Morgantown, 1970. 31 p.)

Haught, Oscar L. Structure contour map, datum Greenbrier limestone in West Virginia, 1:500, 000. N.p., West Virginia Geological and Economic Survey, 1968.

Lotz, Charles W. Probable original minable extent of the bituminous coal seams in West Virginia, 1:1, 500, 000. N.p., West Virginia Geological and Economic Survey, 1970.

West Virginia. Geological and Economic Survey. Mineral resources and mineral industries of West Virginia, 1:500, 000. Morgantown, W. Va., 1958.

WISCONSIN

Bean, E. F. Geologic map of Wisconsin, 1:1, 000, 000. 1949. N.p., Wisconsin Geological and Natural History Survey, 1959.

WYOMING

Love, J. D., and others. Geologic map of Wyoming, 1:500, 000. 1955. Washington, U.S. Geological Survey, 1967.

Osterwald, F. W., and Dean, B. G. Preliminary tectonic map of Wyoming, east of the overthrust belt, showing the distribution of uranium deposits, 1:500,000. Washington, 1958. (U.S. Geological Survey. Mineral investigations field studies map MF 127.)

U.S. Geological Survey. Conservation Division. Branch of Mineral Classification. Reported occurrences of selected minerals in Wyoming, 1:500,000. Washington, 1964. (U.S. Geological Survey. Mineral investigations resource map, MR-42.)

Wyoming. Geological Survey. Mines and mineral map of Wyoming, 1:500,000. Laramie, 1970.

MIDDLE AMERICA

Instituto Centroamericano de Investigación y Tecnología Industrial (ICAITI). Mapa metálogenetico de América Central, 1:2,000,000. Guatemala, 1969. 2 sheets.

Pan American Union. Department of Economic Affairs. Natural Resources Unit. Annotated index of aerial photographic coverage and mapping of topography and natural resources. Washington, 1964- . 19 volumes indicate geologic map coverage for nearly all the Latin American (Middle and South American) countries.

U.S. Geological Survey. Geologic map of Central America, 1:1,000,000. Washington, N.d. (Plate 1 of U.S. Geological Survey. Bull. 1034.)

CANAL ZONE

Woodring, Wendell P. Geologic map of Canal Zone and adjoining parts of Panama..., 1:75,000. Washington, 1955. (U.S. Geological Survey. Miscellaneous geologic investigations, Map I-1; a similar map has been published as Plate 1 of U.S. Geological Survey. Professional paper 306.)

COSTA RICA

Dondoli, Cesar, and Dengo, Gabriel. Mapa geológico de Costa Rica, 1:700,000. N.p., Ministerio de Industria y Comercio, Dirección de Geológia, Minas y Petróleo, 1968.

CUBA

Academia de Ciencias de Cuba. Mapa carsológico de Cuba, 1:2,000,000. La Habana, 1968.

----------. Comisión Nacional. Mapa tectónico de Cuba; Tektonicheskaia karta Kuby, 1:1,250,000. Autores: YU.M. Pusharowski and others. Moskva, Geologicheskii

Institut, Akademiia Nauk SSSR; Glavnoe Upravlenie Geodezii i Kartografii Ministerstva Geologii SSSR, 1966.

Albear, J. F. Mapa hidrogeológico de Cuba, 1:1, 000, 000. N. p., Republica de Cuba, Instituto Cubano de Recursos Minerales, y Instituto Nacional de Recursos Hidraulicos, 1965.

Atlas nacional de Cuba. La Habana, Academia de Ciencias de Cuba, 1970. 132 p. Includes geologic, mineral deposit, ground water, geomorphology, karst, soil, and industrial mineral maps, 1:1, 500, 000; tectonic and coast type maps, 1:2, 500, 000. A Russian language edition of this atlas has been published.

Judoley, Constantion, and others. Mapa de yacimientos de Cuba, 1:500, 000. La Habana, Ministerio de Industrias, Instituto Cubano de Recursos Minerales, Departamento Cientifico de Geología, 1963. 4 sheets.

Nunez Jimenez, A., and others. Mapa geológico de Cuba, 1:1, 000, 000. N. p., Republica de Cuba, Ministerio de Industrias, Instituto Cubano de Recursos Minerales, 1962.

DOMINICAN REPUBLIC

Blesch, Robert R. Mapa geológico preliminar, con información del agua subterranea y datos mineralogicos República Dominicana, 1:250, 000. Washington, Organizacion de los Estados Americanos, 1967.

----------. Mapa geomorfológico República Dominicana, 1:500, 000. Washington, Organización de los Estados Americanos, 1967.

GUATEMALA

Bonis, Samuel, and others. Mapa geológico de la Republica de Guatemala, 1:500, 000. Guatemala, C. A., Instituto Geográfico Nacional, 1970. 4 sheets.

HAITI

Butterlin, Jacques A. Carte géologique de la République d'Haïti, 1:250, 000. Paris, SCIP, 1952. (Plate in Institut Français d'Haïti. Mémoires, no. 1, La géologie de la République d'Haïti et ses rapports avec celle des régions voisines, par Jacques Butterlin. N.p., 1954.

JAMAICA

Jamaica. Geological Survey. Jamaica, geology, 1:250,000. N.p., 1959. (Accompanied by Bulletin no. 4.)

MEXICO

Bonine, Michael E., and others. Atlas of Mexico. Austin, University of Texas, Bureau of Business Research, 1970. 138 p. Includes a geological map, 1:9,500,000.

Cserna, Zoltan de. Tectonic map of Mexico, 1:2,500,000. New York, Geological Society of America, 1961.

Mejorada, Santiago Hernandez Sanchez. Carta geológica de la República Mexicana, 1:2,000,000. N.p., Comité de la Carta Geológica de Mexico, 1968.

Raisz, Erwin. Landforms of Mexico, prepared for the Geography Branch of the Office of Naval Research, 1:3,000,000. Cambridge, Mass., The Author, 1959.

PUERTO RICO

Briggs, Reginald P. Provisional geologic map of Puerto Rico and adjacent islands, 1:240,000. Washington, 1964. (U.S. Geological Survey. Miscellaneous geologic investigations map, I-392.)

----------, and Akers, J. P. Hydrogeologic map of Puerto Rico and adjacent islands, 1:240,000. Washington, 1965. (U.S. Geological Survey. Hydrogeologic investigations atlas, HA-197.)

SOUTH AMERICA

Commission de la Carte Géologique du Monde. Carte géologique de l'Amérique du Sud... 1:5, 000, 000. Rio de Janeiro, 1964. 2 sheets. (Sold by Geological Society of America, Box 1719, Boulder, Colorado 80302, for $8.25.)

Organization of American States. Cuenca del Rio de la Plata; estudio para su planificación y desarrolo; Indice de mapas geológicos, 1:3, 000, 000. Washington, 1969. Three sheets; 9 other sheets cover topographic, soil, land use, vegetation, and other kinds of maps.

Pan American Union. Department of Economic Affairs. Natural Resources Unit. Annotated index of aerial photographic coverage and mapping of topography and natural resources. Washington, 1964- . 19 volumes indicate geologic map coverage for nearly all the Latin American (Middle and South American) countries.

ARGENTINA

Argentina. Dirección Nacional de Geología y Mineria. Mapa Geológico de la República Argentina, 1:2, 500, 000. Buenos Aires, 1964. 3 sheets.

Argentina. Dirección Nacional de Mineria. Mapa minero de la República Argentina, 1:4, 500, 000. N.p., 1957. With Anexo (text).

Argentina. Dirección Nacional de Geología y Mineria. División Geología Regional. Mapa geológico de la República Argentina, 1:5, 000, 000. N.p., 1964. With Explication.

----------. División Hydrogeológia. Mapa hidrogeológico ...de la República Argentina, 1:5, 000, 000. N.p., 1963. With Texto explicativo.

BOLIVIA

Ahlfeld, Federico. Mapa de los yacimientos minerales de Bolivia, 1:1, 500, 000. N.p., 1953. (Plate in: Ahlfeld, Federico. Los yacimientos minerales de Bolivia. N.p., Banco Minero de Bolivia y la Corporación Minera de Bolivia, 1954. 277 p.)

Bolivia. Servicio Geológico. Mapa geológico de la República de Bolivia, 1:2, 500, 000. N.p., 1968. With text.

BRAZIL

Brazil. Departamento Nacional da Producão Mineral. Divisão de Geologia y Mineralogia. Mapa geologico do Brasil, 1:5, 000, 000. N.p., 1960.

Instituto Brasileiro de Geografia. Divisão de Geografia. Mapa geomorfologico do Brasil, 1:5, 000, 000. N.p., 1968.

CHILE

Börgel O., Reynaldo. Mapa geomorfológico de Chile, 1:3, 300, 000. N.p., Universidad de Chile, Instituto de Geográfia, Sección Aplicada, n.d. 7 sheets.

Chile. Instituto de Investigaciónes Geológicas. Mapa geológico de Chile, 1:1, 000, 000. N.p., 1968. 7 sheets.

Ruiz F., Carlos. Mapa metálogenico de Chile, 1:1, 500, 000. Santiago, Instituto de Investigaciónes Geológicas, 1962.

COLOMBIA

Colombia. Instituto Geográfico Agustin Codazzi. Atlas de Colombia. Bogota, Litografia Arco, 1967. 203 p. Includes geology, structure, and earthquake maps.

Hubach, H., and others. Mapa geológico de Colombia, 1:1, 500, 000. (Ed. preliminar.) N.p., Ministerio de Minas y Petróleos, Servicio Geológico Nacional, 1962.

ECUADOR

Goossens, Pierre J. Mineral index map, Republic of Ecuador; Mapa índice mineralogico, República del Ecuador,

1:1, 000, 000. Quito, Servicio Nacional de Geología y Mineria, 1969. 2 sheets.

Granja Ballen, Julio. Mapa geológico de la Republica del Ecuador, 1:1, 000, 000. N. p., Servicio Nacional de Geología y Mineria, 1969. 2 sheets.

GUIANAS

Choubert, B. Esquisse géologique des Guyanes, ca. 1:2, 700, 000. N. p., 1969?

----------. Esquisse metallogenique des Guyanes, ca. 1:2, 700, 000. N. p., 1969?

FRENCH GUIANA

France. Service de la Carte Géologique. Carte géologique du Departement de la Guyane, 1:500, 000. N. p., 1960? 2 sheets.

GUYANA

British Guiana. Geological Survey. Provisional geological map of British Guiana, 1:1, 000, 000. Georgetown, 1962.

Guyana. Geological Survey. Mineral exploration map of Guyana, 1:1, 000, 000. Georgetown, 1966.

SURINAM

Fotogeologische Kaart van Suriname, 1:1, 000, 000. N. p., 1966. (Plate in: Nederlands Geologisch Mijnbouwkundig Genootschap, Verhandelingen, nr. 27; Proceedings of the 7th Guiana Geological Conference, November 1966, Paramaribo. N. p., 1969.)

O'Herne, L. Geomorfologische kaart van Suriname, 1:500, 000. N. p., Ministerie van Opbouw--Uitgave Geologisch Mijnbouwk. Dienst, 1966. 2 sheets.

PARAGUAY

Eckel, Edwin B. Geologic map and sections of Paraguay; mapa geológico y secciones del Paraguay, 1:1, 000, 000. Washington, 1959? (Plate 1 in: U. S. Geological Survey. Professional paper 327, Geology and mineral resources of Paraguay. Washington, 1959. 110 p.) Also issued separately.

PERU

Bellido B., Eleodoro, and others. Mapa geológico del Peru, 1:2,000,000. Lima, La Sociedad Geológica del Peru, 1956. 2 sheets.

Casas, Fernando de las. Peru, mapa metálogenetico, 1:1,000,000. Lima, Sociedad Nacional de Mineraria y Petróleo, 1969. 8 sheets.

Comité Peruano de la Carta Geológica del Mundo. Mapa geológico del Peru, 1:1,000,000. N.p., 1960. 8 sheets.

Peru. Servicio de Geología y Mineria. Mapa geológico generalizado del Peru, 1:4,000,000. N.p., 1969. (Plate in Boletín no. 22, Sinopsis de la geología del Peru, por Eleodoro Bellido B. Lima, 1969. 54 p.)

URUGUAY

Caorsi, Juan H. Mapa geológico de la República Oriental del Uruguay, 1:500,000. Montevideo, Instituto Geológico del Uruguay, División Geología Económica, 1957. 2 sheets.

VENEZUELA

Abreu, Ovidio. Mapa minero industrial y de materias primas de Venezuela, 1:2,000,000. Caracas, Camara Minera de Venezuela, 1967.

Congreso Geológico Venezolano, 4th, Caracas, 1969. Mapa geológico de la República de Venezuela, ca. 1:3,500,000. Comp.: Alirio Bellizzia G. N.p., República de Venezuela, Ministerio de Minas e Hidrocarburos, Dirección de Geología, 1969.

----------. Mapa tectónico de Venezuela, ca. 1:5,500,000. Comp.: Cecilia Martin B. N.p., República de Venezuela, Ministerio de Minas e Hidrocarburos, 1969.

----------. Recursos minerales de Venezuela, scale not given, por Alirio Bellizia G. N.p., Ministerio de Minas e Hidrocarburos, 1969.

Fiedler, G. Mapa de epicentros, superficie del volumen sísmico e isosistas, periodo 1530-1969. Caracas, República de Venezuela, Instituto Sismológico, 1969.

Laforest, Raul. Mapa geológico de la República de Venezuela, 1:1, 000, 000. Presentado ante el IV Congreso Mundial del Petróleo, Roma, 1955. N.p., República de Venezuela, Ministerio de Minas e Hidrocarburos, Dirección de Geología, 1958. 4 sheets.

Smith, Foster D. Mapa geológico-tectónico del norte de Venezuela, 1:1, 000, 000. Primer Congreso Venezolano de Petróleo, Marzo 1962. N.p., 1962. 2 sheets.

EUROPE

Beloussov, V. V. Seismo-tektonicheskaia karta Evropy... The seismotectonic map of Europe, 1:5, 000, 000. 1966, Moskva, Glavnoe Upravlenie Geodezii i Kartografii pri Sovete Ministrov SSSR, 1967. 2 sheets.

Carte internationale du Quaternaire de l'Europe; Internationale Quartär-Karte von Europe; International Quaternary Map of Europe, 1:2, 500, 000. Hannover, Bundesanstalt fur Bodenforschung, 1967- . 4 of 16 sheets issued; accompanied by texts.

Commission de la Carte Géologique du Monde. Sous-commission de la Carte Metallogenique. Carte internationale des gisements de fer de l'Europe, 1:2, 500, 000. N.p., Bundesanstalt für Bodenforschung, 1970- . 2 of 16 sheets issued.

Congres Geologique International. Commission de la Carte Géologique du Monde. Sous-Commission de la Carte Metallogenique. Comité de Redaction de la Carte Metallogenique de l'Europe. Carte metallogenique de l'Europe, 1:2, 500, 000. Coordonnateur general: Pierre Laffitte. Paris, Bureau de Recherches Géologiques et Minières, and UNESCO, 1968- . 2 of 9 sheets issued; accompanied by texts.

Europa geologi, 1:5, 000, 000. Stockholm, Generalstabens Litografiska Anstalt, 1956.

Haack, Hermann. Geologie von Mitteleuropa, 1:750, 000. Gotha, VEB Geographisch-Kartographische Anstalt, n.d.

International Geological Congress. Carte géologique internationale de l'Europe, 1:1, 500, 000. 2nd ed. Hannover, Bundesanstalt fur Bodenforschung, 19 - . Of the 2nd ed., 10 of 49 sheets have been issued.

International hydrogeological map of Europe; Carte hydrogéologique internationale de l'Europe; Internationale Hydrogeologische Karte von Europe, 1:1, 500, 000. Hannover, Bundesanstalt für Bodenforschung and UNESCO, 1970- . 1 of 35 sheets issued; accompanied by texts.

Lobeck, Armin Kohl. Geologic map of Europe, 1:5, 000, 000. N.Y., Columbia University, Geographical Press, 1942. Now distributed by C. S. Hammond Co., Maplewood, N. J.

Mezhdunarodnyi Geologicheskii Kongress. Komissiia po Geologicheskoi Karte Mira. Podkomissiia po Karte Uglenosnykh Otlozhenii Evropy. Mezhdunarodnaia karta uglenosnykh otlozhenii Evropy, 1:2, 500, 000. N.p., 1967. 16 sheets. Coal map.

Mitteleuropa, Geologische Karte, 1:4, 000, 000. Gotha, Justus Perthes, n.d. (Tafel 25 of Dorn, P. Geologie von Mitteleuropa. 2. Aufl. Stuttgart, E. Schweizerbart, 1960. 488 p.; also Sydow-Wagners Method, Schul-Atlas, nr. 20.)

Organisation for Economic Co-operation and Development. The search for and exploitation of crude oil and natural gas in the O. E. C. D. European area, Oil and gas map, 1:3, 750, 000, Carte géopetroliere. Paris, 1962.

Rein, Richard. Geologische Karte von Europe, 1:3, 000, 000. Gotha, Justus Perthes, n.d. (Haack, Herman. Physikalischer Wandatlas, Abt. II: Grund und Boden.)

Rühle, E. Europa: Geologia-stratygrafia, 1:3, 000, 000. Warszawa, Panstwowe Przedsieborstwo Wydawnictw Kartograficznych, 1960.

ALBANIA

Albania. Ministere d'Industrie et des Mines. Instut de Recherches Industrielles et Minières. Carte géologique de l'Albanie, 1:200, 000. Tirane, Imprime par Nish Mjete Mesimore e Sprotive "Hamid Shijaku," 1967. 3 sheets.

AUSTRIA

Akademie der Wissenschaften, Vienna. Atlas der Republik Österreich. Wien, Kartographische Anstalt Freytag-Berndt und Artaria, 1961- . v. Includes maps, some published separately and listed below.

Beck-Mannagetta, P. Geologische Übersichtskarte der Republik Österreich, mit tektonischer Gliederung, 1:1, 000, 000. Wien, Geologische Bundesanstalt, 1964.

Gattinger, T. Hydrogeologische Karte der Republik Österreich, 1:1, 000, 000. Wien, Geologische Bundesanstalt, 1969.

Lechner, K. Karte der Lagerstätten mineralischer Rohstoffe der Republik Österreich, 1:1, 000, 000. Wien, Geologische Bundesanstalt, 1964.

Österreichische Gesellschaft für Raumforschung und Raumplanung. Übersichtskarte der Mineral- und Heilquellen in Österreich, 1:500, 000. N.p., 1966. Accompanied by text.

Vetters, H. Geologische Karte der Republik Österreich und der Nachbargebiete, 1:500, 000. 2. Aufl. Wien, Geologische Bundesanstalt, 1968.

BELGIUM

Belgium. Comite National de Geographie. Atlas de Belgique...Atlas van België. N.p., Institut Geographique Militaire, 1948- . Contains maps, 1:500, 000, on: geomorphology, hydrology, geology, lithology, structure and seismology; mineral maps at other scales.

BRITISH ISLES

Bickmore, D. P., and others. The atlas of Britain and Northern Ireland. Oxford, Clarendon Press, 1963. 200 p. Includes geology, structure, lithology, surficial deposit, glacial, ground water, oil, gas, iron, and quarry maps, 1:2, 000, 000.

Dunning, F. W. Tectonic map of Great Britain and Northern Ireland, 1:1, 584, 000. Chessington, Surrey, publ. by the Director General of the Ordnance Survey...for the Institute of Geological Sciences, 1966.

Great Britain. Geological Survey. Geological map of Great Britain, 1:625, 000. 2nd ed., 1957. Chessington, Surrey, Ordnance Survey, 1964. 2 sheets.

----------. Geological map of the British Islands, 1:1, 584, 000. 5th ed. Chessington, Surrey, Ordnance Survey Office, 1969.

IRELAND

O'Brien, M. V. Geological map of Ireland, 1:750, 000. 3rd ed. Dublin, Ordnance Survey, 1962.

BULGARIA

Kamenoff, Boian G. Geolozhka karta na Bulgariia. . . Carte géologique de la Bulgarie, 1:500, 000. N. p., Direction pour les Recherches Géologiques et Minières de la Bulgarie, 1947.

CYPRUS

Bear, L. M. Geological map of Cyprus, 1:250, 000. N. p., Geological Survey Department, Cyprus, 1963.

Tullström, N. H. O., Hydrogeological map of Cyprus, 1:250, 000. Prep. in cooperation with the United Nations Survey of Ground Water and Mineral Resources, a project of the United Nations Development Programme (Special Fund). N. p., Geological Survey Department, Government of Cyprus, 1970.

CZECHOSLOVAKIA

Czechoslovak Republik. Ustredni Sprava Geodezie a Kartografie. Atlas Ceskoslovenska Socialisticke Republiky. Vydala Ceskoslovenska Akademie Ved. Redaktor atlasu Antonin Götz. Praha, 1966. 257 p. Includes maps, 1:1, 000, 000, on: Geology, tectonics, Quaternary, mineral deposits, geomorphology, metals, hydrogeology; these maps have been published separately, by the Ustredni Ustav Geologicky, Praha, in both English and Czech editions, as sheets of Regional geology of Czechoslovania, geological atlas 1:1, 000, 000.

Fusan, O. Geological map of Czechoslovakia, 1:500, 000. Praha, Ustredni Ustav Geologicky, 1967. 2 sheets.

DENMARK

Andersen, S. A. Det Danske landskabs historie, 1:500, 000.

København, Geodaetisk Institut, 1944. (Tavle 3 of: Det Danske landskabs historie; Danmarks geologi i almenfattelig fremstilling, by S. A. Andersen. I. Bind, undergrunden. København, Populaer-Videnskabeligt Forlag ved Andersen & Gotterup, 1944. 480 p.) Geology.

Denmark. Geodaetisk Institut. Landskabskort over Danmark, 1:250, 000. København, 1950. 4 sheets.

Denmark. Geologiske Undersøgelse. Det dansk Istidslandskab; Glaciomorphological map of Denmark. København, 1948. (Accompanied by: Det danske Istidslandskabs Terraenformer og deres Opstaaen... af V. Milthers. København, C. A. Reitzels Forlag, 1948. Danmarks Geologiske Undersogelse, 3. Raekke, Nr. 28.) 4 sheets.

FINLAND

Geografiska Sällskapet i Finland. Suomen kartasto; Atlas of Finland; Atlas över Finland, 1960. 4th ed. Helsinki, Kustannusosakeyhtio Otava, 1961. 39 pl. Includes geological and glacial maps, 1:2, 000, 000.

Suomen Maantieteellinen Seura. Suomen maaperäkartta, 1:2, 000, 000. Helsinki, Topografikunnan Paino, 1969.

FRANCE

Albinet, M. Carte de vulnerabilité a la pollution des nappes d'eau souterraine de France, 1:1, 000, 000. Orleans, Bureau de Recherches Géologiques et Minières, 1970. 2 sheets.

Aubouin, Jean. Manuel de travaux pratiques de cartographie. Paris, Dunod, 1970. 332 p. This and the following title are treatises on the making of geological maps, using sheets of the French map series, 1:50, 000 to 1:1, 000, 000, as examples.

Bonté, Antoine. Introduction a la lecture des cartes géologiques. 4th ed. Paris, Masson, 1969. 277 p.

Comite National de Géographie. Atlas de France. Paris, Editions Géographiques de France, 1945. 82 p. Sheets of second edition are appearing. Includes geological, mineral, and surficial maps, 1:2, 500, 000; geomorphological map, 1:1, 000, 000.

France. Bureau de Recherches Géologiques et Minières. Atlas des eaux souterraines de la France. Paris, Editions B.R.G.M., 1970. n.p.

----------. Carte des nappes d'eau souterraine de la France, 1:1,000,000. N.p., 1964?

France. Direction des Mines. Bureau de Documentation Minière. Carte des ressources minerales de la France, 1:1,000,000. Paris, Compagnie Française d'Éditions, 1963.

France. Service de la Carte Géologique. Carte géologique de la France, 1:1,000,000. 5ème ed. Orleans, Bureau de Recherches Géologiques et Minières, 1968. 2 sheets.

Goguel, Jean. Carte tectonique de la France, 1:2,500,000. Paris, Service de la Carte Géologique de la France, 1941? (Accompanies Goguel, Jean. Carte tectonique de la France. Paris, Imp. Nat., 1941. 15 p. Memoires pour servir d l'explication de la Carte géologique detailée de la France.)

Horon, O. Carte des gisements de fer de la France, 1:1,000,000. Paris, Bureau de Recherches Géologiques et Minières, 1962. 2 sheets.

GERMANY

Baden-Württemberg. Geologische Landesamt. Geologische Übersichtskarte von Baden-Württemberg, 1:200,000. N. p., 1956-1962. 4 sheets.

Bentz, A. Geologische Übersichtskarte von Nordwestdeutschland, 1:300,000. Hannover, Amt für Bodenforschung, 1951. 4 sheets.

Franke, Dietrich, and Söllig, Armin. 20 Jahre geologische Kartierung in der Deutschen Demokratischen Republik; ein Überblick über die geologischen Kartenwerke der DDR. (Geologie, Zeitschrift fur das Gesamtgebiet der geologischen Wissenschaften, Jg. 18, Heft 8, Okt. 1969, pp. 891-910.)

Geologische Karte von Bayern, 1:500,000. 2. Aufl. München, Bayerische Geologische Landesamt, 1964. With Erläuterung.

Grahmann, Rudolf. Grundwasserhoffigkeit; Groundwater possibilities; Nappes phreatiques eventuelles... N.p., Statistisches Bundesamt, Institut für Landeskunde, Institut für Raumforschung, 1966?. (Die Bundesrepublik Deutschland in Karten... 2311.)

Kölbel, H. Geologische Karte der Deutschen Demokratischen Republik, Karte ohne känozoische Bildungen, 1:500, 000. Berlin, Zentrale Geologische Institut der Stattlichen Geologischen Kommission der DDR, 1962. 2 sheets.

Rösing, Franz. Geologische Übersichtskarte von Hessen, 1:300, 000. Wiesbaden, Hessische Landesamt für Bodenforschung, 1960.

Schamp, Heinz. Die geologischen Übersichtskarten Deutschlands. (Geographisches Taschenbuch, 1960/61, pp. 181-191.) History of small and medium scale mapping.

Schumacher, K. H. Geologische Karte der Deutschen Demokratische Republik, 1:500, 000. N.p., Geologische Kommission, Forschungsinstitut für die Erkundung von Erdöl und Erdgas, 1964. 2 sheets.

Thews, Joe Dietrich. Übersichtskarte der Grundwasserbeschaffenheit in Hessen, 1:300, 000. Wiesbaden, Hessischen Landesamt für Bodenforschung, 1966.

Weber, Hans. Geologische Karte von Thuringen, 1:300, 000. Berlin, VEB Deutscher Verlag der Wissenschaften, 1955.

Weidenbach, F., and Vollrath, A. Geologische Übersichtskarte von Südwestdeutschland, 1:600, 000. N.p., Geologische Landesamt in Baden-Württemberg, 1954.

GREECE

Aronis, G. Ydrogeologikos chartes Ellados; Hydrogeological map of Greece, 1:1, 000, 000. Athenai, Ekdoseis "Delta" Spuros Demetrakos kai Sia, n.d.

Genikos chartes geologias kai oruktou ploutou tes Ellados, 1:600, 000. Athenai, S. Kontos & D. Philaktos, 1959? 2 sheets.

Metallogenetikos chartes tes Ellados; Carte metallogenique de la Grece, 1:1, 000, 000. Athenai, Institouton Geologias kai Ereunon Hypedaphous, 1965.

Renz, Carl. Geologikos chartes tes Ellados... Geologic map of Greece, 1:500, 000. Athens, Institute for Geology and Subsurface Research, 1954. 2 sheets.

HUNGARY

Dank, Viktor, and Fülop, Joszef. Magyarorszag Paleozoos es Mezozoos kepzodmenyeinek fedetlen földtani terkepe; Geological map of the Paleozoic and Mesozoic basement of Hungary. N.p., Magyar Allami Földtani Intezet, 1967.

Hungary. Foldtani Intezet. Magyarorszag földtani terkepe, 1:300, 000. Budapest, 1956. 4 sheets. Geology and mineral deposits shown.

Kartografiai Vallalat, Budapest. National atlas of Hungary. Prepared with the cooperation of the Geographical Committee of the Hungarian Academy of Sciences. Editor in chief: Sandor Rado. Budapest, Cartographical Establishment "Cartographia," 1967. 112 p. Includes geology map, 1:1, 000, 000, and tectonic, mineral resource, and geomorphology maps at 1:1, 500, 000.

Pecsi, L. Geomorphological map of Hungary, 1:1, 000, 000. Budapest, Cartographia, 1969. (Plate in: Hungarian Academy of Sciences. Geographical Research Institute. Studies in geography in Hungary, 6. Geomorphological regions of Hungary, by Marton Pecsi. Budapest, Akademiai Kiado, 1970. 45 p.)

Schmidt, Eligius Robert. Magyarorszag vizfoldtani atlasza. Budapest, M. All. Foldtani Intezet, 1962? 73 p. "Hydrogeological atlas of Hungary."

Vadasz, E. Esquisse de la grande tectonique de la Hongrie, 1:900, 000. N.p., 1959. (Plate in: Hungary. Földtani Intezet. Evkönyve, vol. 49, f. 2, 1960.)

ICELAND

Kjartansson, Gudmundur. Jardfraekikort af Islandi... Geological map of Iceland, 1:250, 000. Reykjavik, Museum of Natural History, Dept. of Geology and Geography, 1960- . 5 of 9 sheets have been published.

ITALY

Dainelli, Giotto. Atlante fisico economico d'Italia. Milano, Consociazione Turistica Italiana, 1940. 82 p. Includes maps on geology, lithology, volcanism, and geomorphology, 1:2, 500, 000.

Italy. Servizio Geologico. Carta geologica d'Italia, 1:1, 000, 000. Roma, 1961. 2 sheets.

NETHERLANDS

Atlas van Nederland. 's-Gravenhage, Staatsdrukkerij- en Uitgeverijbedrijf, Govt. Print. and Pub. Off., 1963- . Sheets appearing; includes map of mineral resources, 1:600, 000.

Kleine geologische overzichtskaart van Nederland, 1:600, 000. N.p., Geologische Stichting, 1947. With text.

Netherlands. Geologische Dienst. Geological map of the Netherlands (simplified), 1:500, 000. N.p., 1951.

Utrecht. Universite. Institut de Geographie. The Netherlands, 1:500, 000. Carte geomorphologique provisoire des Pays-Bas; Provisional geomorphological map of the Netherlands; Proeve van een geomorfologische kaart van Nederland. Utrecht, 1951.

NORWAY

Dons, Johannes A. Kart-katalog Norges berggrunn. Catalogue of maps. Bed-rock geology of Norway. Oslo, 1956. 146 p. (Norges Geologiske Undersøkelse. Skrifter, Nr. 193.)

Holtedahl, Olaf, and Anderson, B. G. Glacial map of Norway, 1:2, 000, 000. Oslo, 1960. (Accompanied by: Norway. Geologiske Undersokelse. Skrifter, Nr. 208, Geology of Norway, editor: Olaf Holtedahl. Olso, 1960. 540 p.)

----------, and Dons, J. A. Geologisk kart over Norge, berggrunnskart; geological map of Norway (bedrock),

1:1, 000, 000. Oslo, 1960. (Accompanied by: Norway. Geologiske Undersøkelse. Skrifter, Nr. 208, Geology of Norway, editor: Olaf Holtedahl. Oslo, 1960. 540 p.)

POLAND

Osika, Roman. Mapa mineralogeniczna Polski; Mineralogenicheskaia karta Pol'shchi; Mineralogenic map of Poland, 1:1, 000, 000. N.p., Wydawnictwa Geologiczne, 1969.

----------. Mineralogenic atlas of Poland, 1:2, 000, 000. Warsaw, 1970. 7 sheets and text describe coal, oil and gas, metals, nonmetals, and ground water.

Poland. Instytut Geologiczny. Atlas geologiczny Polski. N.p., 1954-58. Includes geology, drift, paleogeology, and tectonic maps, 1:1, 000, 000.

----------. Atlas géologique de Pologne. Warszawa, Wydawnictwa Geologiczne, 1956. Includes geological and drift maps, 1:2, 000, 000; sheets of the above two atlases have also been issued separately.

Znosko, Jerzy. Geological atlas of Poland. Warsaw, Geological Institute, 1968. 10 paleogeologic maps, 1:2, 000, 000.

PORTUGAL

Portugal. Estação Agronomica Nacional. Carta litológica de Portugal, 1:1, 000, 000. Lisbon, 1967.

Portugal. Serviços Geológicos. Carta das nascentes minerais de Portugal, 1:1, 000, 000. Lisbon, 1970.

----------. Carta geológica de Portugal, 1:1, 000, 000. N. p., 1968.

----------. Carta geológica do Quaternario de Portugal, 1:1, 000, 000. Lisbon, 1969.

----------. Carta hidrogeólogica de Portugal, 1:1, 000, 000. Lisbon, 1970.

----------. Carta Mineira de Portugal, 1:500, 000. N.p., 1960.

ROMANIA

Academia Republicii Populare Romîne. Institutul de Geologie si Geografie. Monografia geografica a Republicii Populare Romîne, I, Geografia fizica: Anexe. N.p., 1960. 27 p. Includes geologic and geomorphic maps, 1:1, 500, 000; tectonic, mineral, and drift maps, 1:3, 000, 000.

Liteanu, E. Harta Cuaternarului, 1:1, 000, 000. N.p., Republica Populara Romina, Comitetul Geologic, 1964.

Rumania. Institutul Geologic. Comitetul de Statal Geologeie. Atlas geologic, 1:1, 000, 000. Foaia 1- . N.p., 1966- . Published sheets include geology, ground water, and mineral resource maps, issued separately.

SPAIN

Aranguren Sabas, Felix, and Suarez Feito, José. Mapa geológico de la Peninsula Ibérica, Baleares y Canarias, 1:1, 000, 000. 5.A edicion. Madrid, Instituto Geológico y Minero de España, 1966. 2 sheets.

Lopez de Azcona, Juan Manuel, y Suarez Feito, José. Mapa geológico de España y Portugal penínsulares, Baleares y Canarias, 1:1, 250, 000. 2.A edicion. N.p., Instituto Geológico y Minero de España, con la colaboración de los Servicios Geológicos de Portugal, 1965.

Spain. Instituto Geográfico y Catastral. Mapa estratigráfico, 1:2, 000, 000. N.p., 1965. (Atlas nacional de España, Lamina 29.)

Spain. Instituto Geológico y Minero. Mapa minero de España, excluidos hierro y carbon. 2.A edición. Madrid, 1964.

----------. Mapa sismoestructural de la Península Ibérica, Baleares y Canarias, 1:1, 000, 000. Madrid, 1966.

SWEDEN

Svenska Sällskapet för Antropologi och Geografi. Atlas över Sverige. Stockholm, Generalstabens Litografiska Anstalts Förlag, 1953- . Sheets issued cover tectonics, drift and glaciation, mineral resources, geology, 1:2, 000, 000.

Sweden. Generalstabens Litografiska Anstalts Förlag. Sveriges lösa jordarter, samt östersjöns utvacklinghistoria, ca. 1:2, 000, 000. Stockholm, 1951.

Sweden. Geologiska Undersökning. Karta över landisens avsmältning och högsta kustlinjen i Sverige; the deglaciation and the highest shore-line in Sweden, 1:1, 000, 000. Stockholm, 1961. 3 sheets.

----------. Karta över Sveriges berggrund; Pre-Quaternary rocks of Sweden, 1:1, 000, 000. Stockholm, 1958. (Sveriges Geologiska Undersökning, Ser. Ba no. 16.) 3 sheets; with text.

----------. Karta över Sveriges jordarter; Quaternary deposits of Sweden, 1:1, 000, 000. Stockholm, 1958. (Sveriges Geologiska Undersökning, Ser. Ba no. 17.) 3 sheets; with text.

SWITZERLAND

Hotz, Walter. Die geologischen und tektonischen Karten der Schweiz... Bern, 1931. 101 p. (Schweizerische Geologische Kommission. Beiträge zur geologischen Karten der Schweiz, n.f., 62 Lieferung.) Bibliographic study.

Schweizerische Geologische Kommission. Carte géologique générale de la Suisse; Geologische Generalkarte der Schweiz; Carta geológica generale della Svizzera, 1:200, 000. Bern, 1942-1964. 8 sheets. With texts.

----------. Geotechnische Karte der Schweiz; Carte géotechnique de la Suisse; Carta geotecnica della Svizzera, 1:200, 000. Bern, Kummerly & Frey, 1963-1967. 4 sheets.

TURKEY

Ankara. Maden Tetkik ve Arama Enstitüsü Yayinlarindan. Türkiye jeoloji haritasi; geological map of Turkey, 1:500, 000. Ankara, 1961- . 18 of 21 sheets have been issued; with texts.

----------. Türkiye maden zuhurlari; distribution of mineral occurrences in Turkey, 1:2, 500, 000. Haziran, 1960.

Izburak, Resat. Türkiye jeomorfografik harita, 1:2, 000, 000. N.p., Harita Cenel Müdürlügü Matbaasi Döner Sermayesi, 1968.

Ketin, Ihsan. Tectonic map of Turkey, 1:2, 500, 000. N.p., 1966? (Plate in: Ankara. Maden Tetkik ve Arama Enstitüsü, Bull. of the Mineral Research and Exploration Institute of Turkey, foreign edition, no. 66, April 1966.)

YUGOSLAVIA

Cissarz, Arnold. Karta mineralnih lezista FNR Jugoslavije, i njihov genetski polozaj; Karte der Minerallagerstätten Jugoslaviens, und deren genetische Stellung, 1:750, 000. Beograd, Izdanje Zavoda za Geoloska Istanzivanja NRS, 1956. (Plate in: Yugoslavia. Geoloski Institut. Rasprave Zavoda za Geolosko i Geofizicko Istrazivanje N.R. Srbije, Sveska 6, Lagerstätten und Lagerstättenbildung in Jugoslavien...von Arnold Cissarz. Beograd, 1956. 152 p.) 2 sheets.

Mikincic, Vjekoslav. Geoloska karta FNR Jugoslavije i susednih zemalja; Carte géologique de la R. F. P. de Yougoslavie et des pays limitrophes, 1:500, 000. Beograd, Naucna Knijiga, 1953. 4 sheets.

THE SOVIET UNION

Only a few of the numerous general and regional geoscience maps of Russia are listed below.

Akademiia Nauk SSSR. Institut Geografii. Geomorfologicheskaia karta SSSR, 1:4,000,000. Moskva, Glavnoe Upravlenie Geodezii i Kartografii, 1960. 6 sheets.

Akademiia Nauk SSSR. Sovet po Seismologie. Atlas zemletriasenii v SSSR; rezul'taty nabliudenii seti seismicheskikh stantsii SSSR 1911-1957gg. Moskva, Izd-vo Akademii Nauk SSSR, 1962. 336 p. Earthquake atlas.

Atlas litologo-paleogeograficheskikh kart SSSR. Glavnyi redaktor: A. P. Vinogradov. Moskva, Glavnoe Upravlenie Geodezii i Kartografii Ministerstva Geologii SSSR, 1967-69. 4 vols. Paleogeologic maps, 1:7,500,000.

Atlas SSSR. Moskva. Glav. Upr. Geodezii i Kartografii pri Sovete Ministrov SSSR, 1969. 199 p. Includes geology, structure, mineral, and geomorphology maps, 1:17,000,000; geophysical and permafrost maps, 1:25,000,000.

Bespalov, V. F. Geologicheskaia karta Kazakhskoi Sovetskoi Sotsialisticheskoi Respubliki... 1965. Moskva, Vsesoiuznyi Aerogeologicheskii Trest Ministerstva Geologii SSSR, 1967. 6 sheets.

Davydova, Mariia Ivanovna. Fizicheskaia geografiia SSSR. Izd. 2. Moskva, Izd. "Prosveshchenie," 1966. 846 p. Atlas includes geologic, tectonic, and geomorphologic maps, 1:17,000,000.

Nalivkin, D. V. Geologicheskaia karta SSSR, 1:2,500,000... 1965. Moskva, Vsesoiuznyi Aerogeologicheskii Trest Ministerstva Geologii SSSR, 1968. 16 sheets.

Russia. Gosudarstvennyi Geologicheskii Institut. Vsesoiuznyi Nauchno-Issledovatel'skii Institut Gidrogeologii i Inzhenernoi Geologii. Karta podzemnogo stoka SSSR v protsentakh ot obshchego rechnogo stoka i keoffitsientov podzemnogo stoka v protsentakh ot osadkov. Map of the USSR ground water discharge... 1964. Redaktor: B. I. Kudelin. Moskva, Glavnoe Upravlenie Geodezii i Kartografii Ministerstva Geologii SSSR, 1965. 4 sheets.

Russia. Gosudarstvennyi Geologicheskii Komitet SSSR. Vsesoiuznyi Nauchno-Issledovatel'skii Institut Gidrogeologii i Inzhenernoi Geologii. Karta modulei ekspluatatsionnykh resursov presnykh i solonovatykh podzemnykh vod SSSR, 1:5, 000, 000... 1964. Glav. redaktor: N. N. Bindeman. Moskva, Glavnoe Upravlenie Geodezii i Kartografii Gosudarstvennogo Geologicheskogo Komiteta SSSR, 1965. 4 sheets.

Russia. Ministerstvo Geologii. Geologicheskaia karta severo-zapadnoi chasti Tikhookskogo podvizhnogo poisa; Geological map of the northwestern part of the Pacific mobile belt...(within the USSR)... 1:1, 500, 000. Editor in chief: L. I. Krasny. 1964. Moskva, Glavnoe Upravlenie Geodezii i Kartografii, 1966. 12 sheets.

----------. Karta metamorficheskikh fatsii SSSR, 1:7, 500, 000. Moskva, Glavnoe Upravlenie Geodezii i Kartografii, 1966. 2 sheets.

----------. Karta perspektiv neftegazonosnosti SSSR... 1:5, 000, 000... 1967. Redaktor: V. IA. Avrov. Moskva, Glavnoe Upravlenie Geodezii i Kartografii pri Sovete Ministrov SSSR, 1969. 5 sheets.

----------. Mestorozhdeniia poleznykh iskoraemykh SSSR, uchebnaia karta, 1:5, 000, 000. Moskva, Glavnoe Upravlenie Geodezii i Kartografii, 1966. 4 sheets.

Russia. Ministerstvo Geologii. Vsesoiuznyi Issledovatel'skii Institut Gidrogeologii i Inzhenernoi Geologii. Karta podzemnogo stoka SSSR (zona intensivnogo vodoobmena). Map of the USSR ground water discharge (zone of intensive water-exchange), 1:5, 000, 000... 1964. Glav. redaktor: B. I. Kudelin. Moskva, Glavnoe Upravlenie Geodezii i Kartografii Ministerstva Geologii SSSR, 1965. 4 sheets.

Russia. Ministerstvo Geologii. Vsesoiuznyi Naucho-Issledovatel'skii Geologicheskii Institut. Geologicheskaia karta Russkoi platformy i ee obramleniia, 1:1, 500, 000.

Glav. redaktor: D. V. Nalivkin... 1965. Moskva, Vsesoiuznyi Aerogeologicheskii Trest Ministerstva Geologii SSSR, 1970. 16 sheets.

----------. Geologicheskaia karta SSSR, 1:5, 000, 000... 1966. Moskva, Vsesoiuznyi Aerogeologicheskii Trest Ministerstva Geologii SSSR, 1970. 5 sheets.

----------. Geologicheskaia karta SSSR, 1:7, 500, 000... 1966. Redaktor: D. V. Nalivkin. Moskva, Vsesoiuznyi Aerogeologicheskii Trest Ministerstva Geologii SSSR, 1967. 2 sheets.

----------. Geomorfologicheskaia karta SSSR, 1:7, 500, 000... 1966. Redaktor: G. S. Ganeshchin. Moskva, Vsesoiuznyi Aerogeologicheskii Trest Ministerstva Geologii SSSR, 1967. 2 sheets.

----------. Gidrogeologicheskaia karta SSSR, 1:7, 500, 000... 1966. Redaktor: I. K. Zaitsev. Moskva, Vsesoiuznyi Aerogeologicheskii Trest Ministerstva Geologii SSSR, 1967. 2 sheets.

----------. Gidrokhimicheskaia karta SSSR mineral'nye vody, 1:5, 000, 000. Redaktor: I. K. Zaitsev. 1964. Moskva, Glavnoe Upravlenie Geodezii i Kartografii, Ministerstvo Geologii SSSR, 1966. 6 sheets.

----------. Karta Chetvertichnykh otlozhenii SSSR, 1:5, 000, 000... 1966. Glav. redaktor G. S. Ganeshchin. Moskva, Vsesoiuznyi Aerogeologicheskii Trest Ministerstva Geologii SSSR, 1969. 4 sheets. Quaternary.

----------. Karta Chetvertichnykh otlozhenii SSSR, 1:7, 500, 000... 1966. Redaktor G. S. Ganeshchin. Moskva, Vsesoiuznyi Aerogeologicheskii Trest Ministerstva Geologii SSSR, 1968. 2 sheets. Quaternary.

----------. Karta osnovnykh metallogenicheskikh zon territorii SSSR, 1:7, 500, 000. Redaktor: K. V. Il'in. 1967. Moskva, Vsesoiuznyi Aerogeologicheskii Trest Ministerstva Geologii SSSR, 1968. 2 sheets.

----------. Tektonicheskaia karta SSSR, 1:2, 500, 000... 1964. Glav. redaktor: T. N. Spizharskii. Moskva, Glavnoe Upravlenie Geodezii i Kartografii Ministerstva Geologii SSSR, 1966. 19 sheets.

----------. Tektonicheskaia karta SSSR, 1:7, 500, 000... 1966. Redaktor: T. N. Spizharskii. Moskva, Vsesoiuznyi Aerogeologicheskii Trest Ministerstva Geologii SSSR, 1968. 4 sheets.

Russia. Ministerstvo Vysshego i Srednego Spetsial'nogo Obrazovaniia. Geomorfologicheskaia karta Evropeiskoi chasti SSSR i Kavkaza, 1:2, 500, 000. Karta sostavlena na Kafedre Geomorfologii, Moskovskogo Gosudarstvennogo Universiteta im. M. V. Lomonosova. Avtori: M. V. Karandeeva... 1966. Moskva, Vsesoiuznyi Aerogeologicheskii Trest Ministerstva Geologii SSSR, 1970. 4 sheets.

Spizharski, T. N. Geologicheskaia karta Sibirskoi Platformy, 1:1, 500, 000. N.p., Ministerstvo Geologii i Okhrany Nedr SSSR, Vsesoiuznyi Nauchno-Issledovatel'skii Geologicheskii Institut, 1959. 6 sheets.

Strelkov, S. A. Geomorfologicheskaia karta Sovetskoi Arktiki, 1:2, 500, 000... 1959. N.p., Ministerstvo Geologii i Okhrany Nedr. Nauchno-Issledovatel'skii Institut Geologii Arktiki, 1961. 4 sheets.

----------. Karta Chetvertichnykh otlozhenii Sovetskoi Arktiki, 1:2, 500, 000. N.p., Ministerstvo Geologii i Okhrany Nedr SSSR, Nauchno-Issledovatel'skii Institut Geologii Arktiki, 1958.

Vardaniants, L. A. Geologicheskaia karta kristallicheskogo fundamenta, Russkoi Platformii, 1:2, 500, 000... 1964. Moskva, Ministerstvo Geologii SSSR, Vsesoiuznii Naucho-Issledovatel'skii Geologicheskii Institut, Zapadn'ii Geofizicheskii Trest, 1966. 4 sheets.

Zaitsev, I. K. Gidrogeologicheskaia karta SSSR, 1:2, 500, 000. N.p., Ministerstvo Geologii i Okhrany Nedr SSSR, 1959. 19 sheets.

ASIA

Markovskii, A. P. Geologicheskaia karta srednei Azii i prilegaiushchikh territorii, 1:1, 500, 000... 1964. Moskva, Glavnoe Upravlenie Geodezii i Kartografii Ministerstvo Geologii SSSR, 1966. 4 sheets.

United Nations. Economic Commission for Asia and the Far East. Geologic map of Asia and the Far East, 1:5, 000, 000. Bangkok, 1959. 6 sheets. With text.

----------. Mineral distribution map of Asia and the Far East, 1:5, 000, 000. Bangkok, 1963. 4 sheets. With text.

----------. Oil and natural gas map of Asia and the Far East, 1:5, 000, 000. N.p., 1962. 4 sheets. With text.

AFGHANISTAN

Afghanistan. Geological Survey Department. Geological map of Afghanistan, 1:2, 500, 000. N.p., Afghan Cartographic Institute, 1969.

ARABIAN PENINSULA

Raisz, Erwin. Landforms of Arabia, prepared for the Environmental Protection Section of the Office of the Quartermaster General, ca. 1:3, 500, 000. Cambridge, Mass., Institute of Geographical Exploration, Harvard University, 1951.

U.S. Geological Survey, and the Arabian American Oil Co. Geologic map of the Arabian peninsula, 1:2, 000, 000. Washington, 1963. (Miscellaneous geologic investigations map, I-270 A.)

CHINA

China. Academy of Sciences. Geological Scientific Research Institute. Tektonicheskaia karta Kitaia, sostavlena Geologicheskim Nauchno-Issledovatel'skim Institutom Akademii Nauk Kitaia, 1:4, 000, 000. N.p., Kartograficheskoi Fabrike Gos. Ministerstva Geologii i Okhrany Nedr SSSR, 1962. 4 sheets.

Meng, Chao-yi, and Li, William Y. L. Geologic map of the Republic of China, 1:4, 000, 000. N.p., Chinese Petroleum Corp., 1970.

INDIA

Eremenko, N. A., and others. Tectonic map of India, 1:2, 000, 000. N.p., Oil and Natural Gas Commission, India, 1968. 4 sheets. With text.

Hoy, Harry E. The landforms of India and Pakistan, 1:3, 800, 000...for the Office of the Quartermaster General. Norman, Okla., University of Oklahoma Research Institute, 1955.

India. Geological Survey. Geology and minerals of India (localities of principal minerals are shown), 1:6, 082, 560. Calcutta, 1961.

Ray, D. K. Metallogenic-minerogenetic map of India, 1:2, 000, 000. N.p., Geological Survey of India, 1963. 4 sheets. With text.

Roy, B. C., and others. Geological map of India, 1:2, 000, 000. (6th ed.--1962.) N.p., Geological Survey of India, 1963. 4 sheets. With text.

----------. Geological map of India, 1:5, 000, 000. 1962. N.p., Geological Survey of India, 1963.

Subramanyam, Sarvashri V., and others. Geohydrological map of India, 1:2, 000, 000. N.p., Geological Survey of India, 1969. 4 sheets.

INDOCHINA

Fromaget, Jacques, and Saurin, E. Carte géologique Viet Nam-Cambodge-Laos, 1:2, 000, 000. 2nd ed. N.p., Service Géologique de l'Indochine, 1952. 2 sheets.

INDONESIAN ARCHIPELAGO

BORNEO

Borneo, British Territories. Geological Survey Department. Sketch map showing the geology of Borneo, 1:2, 000, 000. N.p., Survey Department, Federation of Malaya, 1957.

Borneo. Geological Survey Department. Geological map of the British Territories in Borneo, 1:1, 000, 000. 2nd ed. N.p., Survey Department, Federation of Malaya, 1959.

Collenette, P. Mineral resources of Sarawak, Brunei, and North Borneo. 2nd ed. N.p., Geological Survey Department, British Territories in Borneo, 1962.

INDONESIA

Indonesia. Direktorat Geologi. Geologic map of Indonesia. 1:2, 000, 000. Washington, 1965. (U.S. Geological Survey. Miscellaneous geologic investigations map I-414.) 2 sheets.

JAVA

Indonesia. Geological Survey. Geological map of Djawa and Madura, 1:500, 000. Bandung, 1963. 2 sheets.

NEW GUINEA

Australia. Bureau of Mineral Resources, Geology, and Geophysics. Geological map of the Territory of Papua and New Guinea, 1:2, 534, 400. Canberra, 1963?

IRAN

Abdalian, S. Carte seismo-tectonique de l'Iran, 1:4, 000, 000. N.p., Institut Géophysique de l'Université de Tehran, n. d. (Plate in: Tehran. University. Institute of Geophysics. Publ. no. 16, Seismo-tectonique de l'Iran, par Sedrak Abdalian. Tehran, Institut Géophysique de l'Université de Tehran, n.d.)

Bariand, P. Preliminary metallogenic map of Iran, 1:2, 500, 000. Tehran, Geological Survey of Iran, 1965? (Plate in: Iran. Geological Survey. Report no. 7, Preliminary metallogenic map of Iran, by P. Bariand.

N.p., 1965. 48 p.)

Iran Oil Co. Geological Staff. Geological map of Iran, 1:2,500,000. N.p., National Iranian Oil Co., 1959? With text.

IRAQ

Iraq. Ministry of Development. Mineral Survey Project. Geological map of Iraq, 1:1,000,000. N.p., 1960.

ISRAEL

Israel. Survey of Israel. Atlas of Israel. Amsterdam, Elsevier, 1970. v.p. Includes geology, geomorphology, structure, and ground water maps, 1:500,000; mineral and lithology maps, 1:1,000,000.

Picard, I. L. Geological map, Israel, 1:500,000. N.p., Survey of Israel, 1964. Based on Israel Atlas map.

JAPAN

Isomi, Hiroshi. Tectonic map of Japan, 1:2,000,000. N.p., Geological Survey of Japan, 1968. 3 sheets.

Isshiki, Naoki. Volcanoes of Japan, 1:2,000,000. N.p., Geological Survey of Japan, 1968.

Japan. Geological Survey. Coal fields of Japan, 1:2,000,000. N.p., 1957.

----------. Distribution map of oil and gas fields in Japan, 1:2,000,000. N.p., 1959.

----------. Geological map of Japan, 1:2,000,000. N.p., 1964.

----------. Index to the geological maps of Japan. N.p., 1963-64. 4 vols.

----------. Mineral provinces of Japan, 1:2,000,000. N.p., 1957-60. 4 sheets.

Murashita, Toshio. Hydrogeological map of Japan, 1:2,000,000. N.p., Geological Survey of Japan, 1964.

JORDAN

Deutsche Geologische Mission in Jordanien. Geologische Karte von Jordanien; Geological map of Jordan... 1:250, 000. Geologische Gesamtbearbeitung: F. Bender. Hannover, Bundesanstalt für Bodenforschung, 1968. 5 sheets.

KOREA

Korea. Geological Survey. Geologic map of Korea, 1:1, 000, 000. N.p., 1956. 2 sheets.

KUWAIT

Austria. Geological Survey. ...Synoptic geologic map of the State of Kuwait, 1:250, 000. N.p., Kuwait Ministry of Commerce and Industry, 1966? With text.

LEBANON

Dubertret, Louis. Carte géologique du Liban, 1:200, 000. N.p., République Libanaise, Ministère des Travaux Publics, 1955.

United Nations. Carte hydrogéologique du Liban, 1:200, 000. N.p., Nations Unies, 1967.

MALAYSIA

Bean, J. H. Malaya, mineral distribution map, 1:500, 000. N.p., Geological Survey, Federation of Malaysia (Malaya), 1966. 2 sheets.

Malaya. Geological Survey. Geological map of Malaya, 1:500, 000. 6th ed. (Diamond Jubilee) 1963. Ipoh, 1965. 2 sheets.

Malaysia, Singapura & Brunei, 1:1, 140, 480. Kuala Lumpur, Di-Terbitkan Oleh Dewan Bahasa Dan Pustaka, 1966. 2 sheets cover Malay Peninsula and northern Borneo.

PAKISTAN

Bakr, M. Abu, and Jackson, Roy O. Geological map of Pakistan, 1:2, 000, 000. Rawalpindi, Geological Survey of Pakistan, 1964.

PHILIPPINES

Philippine Islands. Bureau of Mines. Mineral distribution map of the Philippines, 1:2,500,000. Manila, 1964. 6 sheets.

----------. Geological Survey Division. Geological map of the Philippines, 1:1,000,000. Manila, 1964. 8 sheets.

SYRIA

Dubertret, Louis. Carte géologique, Liban, Syrie, et bordure des pays voisins, 1:1,000,000. Paris, Institute Géographique National, 1962. (Accompanied by: Museum d'Histoire Naturelle, Paris, Notes et Mémoires sur le Moyen-Orient, T. 8; also in Lexique Stratigraphique International, vol. 3, F. 10 c'.)

TAIWAN

Li-Sho, Chang. Geologic map of Taiwan, 1:300,000. N.p., Geological Survey of Taiwan, 1953. 2 sheets.

THAILAND

Siam. Department of Mineral Resources. Geological map of Thailand, 1:1,000,000, comp. by Jumchet C. Javanaphet. N.p., Royal Thai Survey Department, 1970. 2 sheets.

AFRICA

Ady, Peter H. Oxford regional economic atlas: Africa. Oxford, Clarendon Press, 1965. 164 p. Includes geology, structure, and mineral maps.

Association des Services Géologiques Africains. Carte géologique de l'Afrique; Geological map of Africa, 1:5, 000, 000. Paris, ASGA-UNESCO, 1963. 9 sheets, with a text.

Choubert, G., and others. Carte tectonique de l'Afrique; Tectonic map of Africa, 1:15, 000, 000. N.p., UNESCO, 1968.

----------. Carte tectonique internationale de l'Afrique; International tectonic map of Africa, 1:5, 000, 000. Comp. under the aegis of the International Geological Congress, Commission for the World Geological Map, Sub-Commission for the World Tectonique (sic) Map... N.p., Association of African Geological Surveys, and UNESCO, 1968. 9 sheets.

Feys, R. Carte des dépôts houillers en Afrique; Coal map of Africa, 1:10, 000, 000. Paris, Economic Commission for Africa and Association des Services Géologiques Africains, 1965. 2 sheets.

France. Bureau de Recherches Géologiques et Minières. Afrique occidentale: Carte géologique... N.p., 1956-   . 7 sheets.

----------. Carte géologique du Sahara, 1:500, 000; Massif du Hoggar. N.p., 1961. 12 sheets.

Gerard, Georges. Carte géologique de l'Afrique Equatoriale Française, 1:2, 000, 000. N.p., Direction des Mines et de la Géologie de l'A. E. F., 1956. 4 sheets, with text.

Haughton, S. H. Geological map of southern Africa, 1:5, 000, 000. N.p., 1968. (Plate in: Geological history of southern Africa, by S. H. Haughton. N.p., Geological Society of South Africa, 1969. 535 p.)

International Geological Congress, 19th, Algiers, 1952. Carte géologique du nord-ouest de l'Afrique, 1:2, 000, 000. N.p., 1952-1962. 4 of 6 sheets issued.

Lombard, Jean. Afrique; Africa; Geologic background to serve as a basis for various synthesis or thematic maps, 1:10, 000, 000. Paris, United Nations Educational, Scientific, and Cultural Organization, and Association of African Geological Surveys, 1967. Also available at scale 1:20, 000, 000.

----------. Carte des gisements de fer en Afrique; Iron ore deposits in Africa, 1:10, 000, 000. Paris, Association des Services Géologiques Africains, 1970. 2 sheets.

----------. Carte minérale de l'Afrique; Mineral map of Africa, 1:10, 000, 000. Paris, Association of African Geological Surveys-UNESCO, 1968. With text.

Mestraud, J.-L. Carte géologique de la République Centrafricaine, 1:1, 500, 000. N.p., Bureau de Recherches Géologiques et Minières, 1964.

Mohr, P. A. Geological map of Horn of Africa, 1:2, 000, 000. London, Philip & Tacey, Ltd., 1963. (Accompanied by: The geology of Ethiopia, by Paul A. Mohr. N.p., University College of Addis Ababa Press, 1962? 268 p.)

Raisz, Erwin. Landform map of North Africa, 1:15, 000, 000. Base map prepared for Environmental Protection Branch, Office of the Quartermaster General. N.p., 1952. 2 sheets.

Russia. Vsesoiuznyi Nauchno-Issledovatel'skii Geologicheskii Institut. Geologicheskaia karta Afriki; Geological map of Africa, 1:10, 000, 000. Redaktor: S. A. Muzylev; sostavil L. A. Sniatkov... 1966. Moskva, Vsesoiuznyi Aerogeologicheskii Trest Ministerstva Geologii SSSR, 1969. 2 sheets. With text.

----------. Karta poleznykh iskopaemykh Afriki; Map of mineral resources of Africa, 1:10, 000, 000. Redaktor: S. A. Musylev; sostavil A. Z. Aleiner... 1966. Moskva, Vsesoiuznyi Aerogeologicheskii Trest Ministerstva

Geologii SSSR, 1969. 2 sheets. With text.

Saggerson, E. P. East Africa, geology, 1:4,000,000. N.p., Directorate of Overseas Surveys, 1961. (Accompanied by: Handbook of natural resources of East Africa. Nairobi, East African Literature Bureau, 1961?)

ALGERIA

Carte des gîtes mineraux de l'Algerie, 1:500,000. N.p., Service Géologique de l'Algerie, 1965- . 1 of 6 sheets published.

Cornet, A. Carte géologique de l'Algerie, 1:500,000. 2ème ed. N.p., Service de la Carte Géologique, 1951-1952. 7 sheets issued, covering northern half of Algeria.

ANGOLA

Angola. Direcção Provincial dos Serviços de Geologia e Minas. Ocorrencias minerais, 1:2,000,000. Luanda, 1966. With text.

Portugal. Ministério do Ultramar. Junta das Missões Geográficas e de Investigações do Ultramar. Esboço geológico de Angola, 1:2,000,000. Lisbon, Instituto Geográfico e Cadastral, 1954. With text.

CAMEROUN

Gazel, J. Carte géologique du Cameroun, 1:1,000,000. N. p., Direction des Mines et de la Géologie du Cameroun, 1956. (Accompanies its Bulletin no. 2, Carte géologique du Cameroun...Notice explicative... Paris, 1956. 62 p.) 2 sheets.

CHAD

Wolff, J.-P. Carte géologique de la République du Tchad, 1:1,500,000. N.p., Bureau de Recherches Géologiques et Minières, 1964. 2 sheets.

CONGO

Cahen, L. Esquisse tectonique du Congo Belge et du Ruanda-Urundi, 1:3,000,000. Bruxelles, Royaume de Belgique, Ministère des Colonies, Commission de Géologie, 1952.

----------, and Lepersonne, J. Carte géologique du Congo Belge et du Ruanda-Urundi, 1:2, 000, 000. Bruxelles, Royaume de Belgique, Ministère des Colonies, Commission de Géologie, 1952. 4 sheets, with text.

République Democratique du Congo. Institut Géographique. Carte des Mines et des industries, 1:3, 000, 000. N.p., 1969.

Snel, M. J. Carte hydrogéologique du Congo Belge et du Ruanda Urundi, 1:5, 000, 000. N.p., 1957? (Plate in: Congo Belge. Service Géologique. Bull. no. 7, fasc. 2, Contribution a l'étude hydrogéologique du Congo Belge, par M. J. Snel. N.p., 1957. 31 p.)

DAHOMEY

France. Bureau de Recherches Géologiques et Minières. Carte géologique de la République du Dahomey, 1:1, 000, 000. N.p., 1960. With text. 2 sheets.

GABON

Hudeley, H. Carte géologique de la République Gabonaise, 1:1, 000, 000. Orleans, Bureau de Recherches Géologiques et Minières, 1970? (Accompanied by Mémoire no. 72, Carte géologique de la République Gabonaise... Notice explicative. Paris, 1970. 191 p.)

GHANA

Ghana. Geological Survey. Ghana...geological, 1:2, 000, 000. Accra, 1969.

IVORY COAST

Bararre, E., and Tagini, B. Carte géologique de la Côté d'Ivoire, 1:1, 000, 000. Abidjan, Côté d'Ivoire, Direction des Mines et de la Géologie, 1965.

KENYA

Kenya. Mines and Geological Department. Kenya, geological map, 1:3, 000, 000. 2nd ed. Nairobi, Survey of Kenya, 1962.

Kenya Colony. Survey. Atlas of Kenya. Nairobi, 1959. 44 p. Includes geology and mineral maps, 1:3, 000, 000.

LESOTHO

Stockley, G. M. Geological map of Basutoland, 1:380, 160. N.p., 1947. (Accompanied by: Report on the geology of Basutoland, by G. M. Stockley. Maseru, Basutoland Government, 1947. 114 p.)

LIBYA

Conant, Louis C., and Goudarzi, Gus H. Geologic map of the Kingdom of Libya, 1:2, 000, 000. Washington, 1964. (U.S. Geological Survey. Miscellaneous geologic investigations map I-350-A; reprinted in 1969 as Plate 2 of its Professional paper 660.)

MALAGASY REPUBLIC

Besairie, Henri. Carte tectonique de Madagascar, 1:3, 000, 000. Tananarive, Service Géologique de Madagascar, 1961.

----------. Madagascar, carte géologique, 1:1, 000, 000. 1964. Tananarive, Service Géologique de Madagascar, 1965. 3 sheets.

----------. Madagascar, carte minière, 1:2, 500, 000. 1960. Tananarive, Service Géologique de Madagascar, 1961.

MALAWI

Bloomfield, K. Geological map of Malawi, 1:1, 000, 000. N.p., Geological Survey, 1966.

MOROCCO

Comité de Géographie du Maroc. Atlas du Maroc. Rabat, Institut Scientifique Cherifien, 1955- . Sheets appearing. Includes metallogenic and mineral maps, 1:2, 000, 000.

Emberger, André. Carte des minéralisations plombo-zincifères du Maroc; gîtes et indices renfermant du plomb ou du zinc en élément majeur ou mineur, 1:2, 000, 000. N.p., 1969. (Morocco. Service Géologique. Notes et mémoires, no. 215.)

Morocco. Direction des Mines et de la Géologie. Division de la Géologie. Service d'Etudes des Gîtes Minéraux. Carte des gites minéraux du Maroc, 1:500,000. N.p., 1968- . One of 6 sheets has appeared.

MOZAMBIQUE

Dias, Manuel Bettencourt. Provincia de Mocambique... Esboço tectónico, 1:3,000,000. N.p., 1956. (Plate in: Mozambique. Serviços de Indústria e Geologia. Boletim no. 18. Esboço tectónico de Moçambique, 1:3,000,000... Lourenço Marques, 1956. 34 p.)

Freitas, A. J. de. República Portugueza, Provincia de Moçambique, Esboço geológico, 1:2,000,000. N.p., Reparticao Tecnica de Industria e Geologia, 1956. (Plate in: Mozambique. Servicos de Indústria e Geologia. Boletim no. 23, Noticia explicativa do Esboço geológico de Mocambique... por A. J. de Freitas. Lourenço Marques, 1957. 82 p.)

NIGER

Greigert, J., and Pougnet, R. République du Niger, carte géologique, 1:2,000,000. N.p., Bureau de Recherches Géologiques et Minières, 1967? With text.

NIGERIA

Nigeria. Geological Survey. Geological map of Nigeria, 1:2,000,000. N.p., Director of Federal Surveys, 1965.

Nigeria. Federal Surveys. Nigeria, geology and mineral deposits, 1:3,000,000. N.p., 1967.

PORTUGUESE GUINEA

Portugal. Junta de Investigações do Ultramar. Carta geológica da Guine, 1:500,000. N.p., 1968?

RHODESIA

Rhodesia and Nyasaland. Dept. of Trigonometrical and Topographical Survey. Provisional geological map of the Federation of Rhodesia and Nyasaland, 1:2,500,000.

Causeway, 1961. (Federal atlas map no. 4.)

----------. Provisional mineral map of Rhodesia and Nyasaland, 1:2,500,000. Causeway, 1960. (Federal atlas map no. 5.)

Southern Rhodesia. Geological Survey. Provisional geological map of Southern Rhodesia, 1:1,000,000. Rev. ed. Salisbury, 1962. (Accompanied by: Southern Rhodesia. Geological Survey. Bull. no. 50, An outline of the geology of Southern Rhodesia, by W. H. Swift. Salisbury, 1961. 73 p.)

----------. Provisional mineral map of Southern Rhodesia, 1:2,000,000. 6th ed. Salisbury, 1963.

SENEGAL

France. Bureau de Recherches Géologiques et Minières. Carte géologique de la République du Sénégal et de la Gambie, 1:500,000. N.p., 1962. (Plate in: Senegal. Service des Mines et de la Géologie. Carte géologique du Sénégal... N.p., B.R.G.M., 1962. 36 p.) 4 sheets.

Permingeat, F. Carte des gîtes mineraux de la République du Sénégal, 1:500,000. N.p., Bureau de Recherches Géologiques et Minières, 1966. 4 sheets.

SIERRA LEONE

Sierra Leone. Geological Survey Department. Geological, Sierra Leone, 1:1,000,000. N.p., Directorate of Overseas Surveys, 1960.

SOMALILAND

Azzaroli, A., and Merla, G. Carta geologica della Penisola Somala, 1:4,000,000. N.p., Agip Mineraria, Consiglio Nazionale delle Ricerche, 1960.

SOUTH AFRICA

Coertze, F. J., and others. Geologiese kaart van die Republiek van Suid-Afrika en die Koninkryke van Lesotho en Swaziland; Geological map of the Republic of South Africa and the Kingdoms of Lesotho and Swaziland,

1:1, 000, 000. Pretoria, Geological Survey, 1970. 4 sheets; symbols indicate mineral deposits.

SOUTH WEST AFRICA

Union of South Africa. Geological Survey. Geological map of South West Africa, showing mineral occurrences and gravity contours (Bouguer anomalies), 1:1, 000, 000. Pretoria, 1964. 4 sheets.

SPANISH GUINEA

Pacheco, Francisco Hernandez. Mapa geológico esquematico de la Guinea Continental Española, 1:500, 000. N.p., Dirección General de Marruecos y Colonias e Instituto de Estudios Africanos, 1954.

SPANISH MOROCCO

Comision de Estudios Geológicos de Marruecos. Bosquejo geológico de la zona del Protectorado Español, 1:400, 000. N.p., 1952.

SPANISH SAHARA

Medina, Manuel Alia. Bosquejo geológico del Sahara Español, 1:2, 000, 000. N.p., Dirección General de Maruecos y Colonias, Instituto de Estudios Africanos, Servicio Geológico del Africa Occidental Española, 1952.

SUDAN

Sudan. Survey Department. Geological map, Sudan, 1:4, 000, 000. Khartoum, 1967.

SWAZILAND

Hunter, D. R. Geological map of Swaziland, 1:250, 000. Mbabane, Geological Survey and Mines Department, 1966.

TANZANIA

Tanganyika. Department of Lands and Survey. Atlas of Tanganyika, East Africa. 3rd ed. Dar es Salaam, 1956. 29 p. Includes geology and mineral maps, 1:3, 000, 000.

Tanganyika. Geological Survey. Geological map of Tanganyika, 1:2, 000, 000. Dodoma, 1959. With text.

TUNISIA

Castany, G. Carte géologique de la Tunisie, 1:500, 000. N. p., Service des Mines, de l'Industrie et de l'Energie, 1951- . Two of three sheets published, with text.

Nicolini, P. Carte des gîtes mineraux de la Tunisie. 1:500, 000. N.p., Service Géologique, 1966- . Two of three sheets published, with text.

UGANDA

Uganda. Lands and Surveys Department. Atlas of Uganda. 2nd ed. Entebbe, 1967. Includes geomorphology, geology, and mining maps, 1:1, 500, 000.

UNITED ARAB REPUBLIC

Egypt. Department of Survey and Mines. Egypt, geological, 1:2, 000, 000. Giza, 1937.

ZAMBIA

Northern Rhodesia. Geological Survey. Geological map of Northern Rhodesia, 1:1, 000, 000. Lusaka, 1961. (Accompanied by: Northern Rhodesia. Geological Survey. Bull. no. 3, vol. 1, The geology and mineral resources of Northern Rhodesia, by W. H. Reeve. Lusaka, 1963. 213 p.) 4 sheets.

Zambia. Survey Department. The Republic of Zambia: Minerals, 1:2, 500, 000. Lusaka, 1968. (Atlas sheet no. 5.)

OCEANIA

Australia. Bureau of Mineral Resources, Geology, and Geophysics. Geological map of the world: Australia and Oceania, 1:5,000,000. Canberra, 1967. In 13 sheets, 10 so far published.

AUSTRALIA

Australia. Bureau of Mineral Resources, Geology, and Geophysics. Geological map of the world: Australia and Oceania, 1:5,000,000. Canberra, 1965. Sheets 6, 7, 11, and 12 cover Australia, New Guinea, Tasmania.

Australia. Department of National Development. Resources Information and Development Branch. Geology, 1:6,000,000. 2nd ed. Canberra, 1966. (Atlas of Australian resources, 2nd series.)

Australia. Water Resources Council. Australia, underground water, 1:5,000,000. Canberra, Department of National Development, Water, Power, and Geographic Branch, 1965. With text.

----------. Relief and landform map of Australia, with symbols for rock types of general hydrologic significance, 1:5,000,000, by E. Löffler. Canberra, 1969. With text.

Geological Society of Australia. Tectonic Map Committee. Tectonic map of Australia, 1:2,534,400. Canberra, Bureau of Mineral Resources, Geology, and Geophysics, Department of National Development, 1960. 4 sheets. Text.

Lobeck, Arnim Kohl. Physiographic diagram of Australia, 1:7,500,000. N. Y., Geographic Press, Columbia University, 1951. 2 sheets, text on verso.

Plumb, T. W. Mineral deposits, 1:6,000,000. 2nd ed. Canberra, Geographic Section, Department of National

Development, 1969. (Atlas of Australia, 2nd series.)

NEW SOUTH WALES

Lloyd, J. C. Mineral map of New South Wales, 1:3, 000, 000. N.p., Department of Mines, N.S.W., 1962.

New South Wales. Geological Survey. Geological map of New South Wales, 1:3, 000, 000. Sydney, Department of Mines, 1968?

----------. Division of Regional Geology. Geological map of New South Wales, 1:1, 000, 000. N.p., 1962. 2 sheets.

QUEENSLAND

Queensland. Geological Survey. Geological map of Queensland, 1:2, 534, 400. N.p., 1953.

----------. Prospector's map of Queensland, ca. 1:5, 200, 000. Brisbane, 1966?

----------. Queensland, showing distribution of principal mineral occurrences, ca. 1:5, 200, 000. Brisbane, 1966.

SOUTH AUSTRALIA

Sprigg, R. C., and Dickinson, S. B. Geological map, South Australia, 1:2, 027, 520. Adelaide, Geological Survey of South Australia, 1953.

----------. Structural geological map, South Australia, 1:2, 027, 520. Adelaide, Geological Survey of South Australia, 1953.

TASMANIA

Geological Society of Australia. Tasmania Division. Tectonic Map Committee. Structural map of Tasmania, 1:500, 000. N.p., Tasmania Department of Mines, 1960.

Tasmania. Geological Survey. Geological map of Tasmania, 1:506, 880. Hobart, 1961.

VICTORIA

Victoria. Geological Survey. Mineral deposits, Victoria, 1:1, 000, 000. N.p., 1970.

----------. Victoria, geological map, 1:1, 000, 000. Melbourne, 1963.

WESTERN AUSTRALIA

Horwitz, R. C. Geological map of Western Australia, 1:2, 500, 000. N.p., Geological Survey of Western Australia, 1966.

Western Australia. Department of Mines. Regional divisions, reported mineral occurrences, Western Australia, 1:2, 534, 400. Perth, 1968.

NEW ZEALAND

New Zealand. Geological Survey. Geological map of New Zealand, 1:2, 000, 000. Wellington, Department of Scientific and Industrial Research, 1958.

ANTARCTICA

Craddock, Campbell, and others. Geologic maps of Antarctica. N. Y., 1970. v.p. (American Geographical Society, Folio 12.) Contains geological, tectonic, and geomorphological maps, 1:10, 000, 000.

Russia. Ministerstvo Geologii SSSR. Nauchno-Issledovatel'skii Institut Geologii Arktiki. Tektonicheskaia karta poliarnykh oblastei Zemli; Tectonic map of polar regions of the earth, 1:10, 000, 000. Glavnyi redaktor: B. Kh. Egiazarov. Moskva, Vsesoiuznyi Aerogeologicheskii Trest Ministerstva Geologii SSSR, 1971. 4 sheets.

Sovetskaia Antarkticheskaia Ekspeditsiia. Atlas Antarktiki, vyp. 1- . Moskva, Glavnoe Upravlenie Geodezii i Kartografii, 1966- . Vol. 1 contains geological, bottom deposit, glacial, and tectonic maps.

SUBJECT INDEX

GEOGRAPHIC INDEX